PROCEEDINGS OF SYMPOSIA IN PURE MATHEMATICS
VOLUME XVI

GLOBAL ANALYSIS

AMERICAN MATHEMATICAL SOCIETY
Providence, Rhode Island
1970

Proceedings of the Symposium in Pure Mathematics
of the American Mathematical Society

Held at the University of California
Berkeley, California
July 1–26, 1968

Prepared by the American Mathematical Society
under National Science Foundation Grant GP–8410

SHIING-SHEN CHERN
STEPHEN SMALE
Editors

Standard Book Number 8218–1416–0
Library of Congress Catalog Number 70–95271

AMS 1968 Primary Subject Classification 5750
Printed in the United States of America

CONTENTS

PREFACE

The papers in these Proceedings grew out of lectures given at the fifteenth Summer Mathematical Institute of the American Mathematical Society, whose topic was global analysis. The Institute was held at the University of California at Berkeley from July 1 to July 26, 1968, and was partially financed by the National Science Foundation.

Notes of lectures were distributed at the time of the conference and some of the papers here are just as in those notes. These volumes, however, can be distinguished from the notes in the sense that in general the papers here are not just expositions of material that has or will appear elsewhere; most of the articles could just as well have appeared in Journals.

The unity given by the subject matter makes it desirable to collect them here. It is hoped that the volumes will provide an important start to the scientist who wishes to learn what is going on in that part of mathematics called global analysis.

The organizing committee for the institute consisted of: F. Browder, S.-S. Chern, L. Hörmander, I. Singer, and S. Smale, with the co-editors serving as co–chairmen.

Seminar organizers were: F. Browder, E. Calabi, H. Goldschmidt, R. Hermann, C. Morrey, R. Palais, C. Pugh, I. Singer, and D. Spencer.

Finally the editors would like to thank the many people who made the institute and volumes possible. Of especially direct help to ourselves were Celeste Andrade, Ann Harrington, Gordon and Jacqueline Walker.

S.-S. Chern

December 1968

S. Smale

EXISTENCE THEOREMS FOR NONLINEAR PARTIAL DIFFERENTIAL EQUATIONS

FELIX E. BROWDER

PART I

Introduction. It is my purpose in the present discussion to describe the application of some recently developed general techniques in nonlinear functional analysis to the study of the existence theory for solutions of various classes of nonlinear partial differential equations, specifically nonlinear elliptic boundary value problems, nonlinear elliptic eigenvalue problems, and nonlinear equations of evolution of parabolic and wave equation type. Corresponding to these three classes of problems, we shall consider three related abstract theories: the theory of nonlinear operators of monotone type from a reflexive Banach space X to its conjugate space X^*, the application of the Lusternik-Schnirelman category in infinite dimensional spaces, and the theory of semigroups of nonlinear nonexpansive operators in Hilbert and Banach spaces.

Our partial differential equations will be defined for the sake of simplicity on an open subset G of an Euclidean n-dimensional space R^n rather than upon a manifold and we shall speak for convenience of notation of a single equation rather than a system of equations. We shall make no use, however, of the maximum principle for second order scalar elliptic or parabolic equations or of the Di Giorgi-Nash estimates or their consequences so that there is no conceptual loss of generality in the restriction.

The points of the open set G will be denoted by $x = (x_1, \ldots, x_n)$ and the elementary differential operators by $D^\alpha = \prod_{j=1}^n (\partial/\partial x_j)^{\alpha_j}$ for an ordered n-tuple $\alpha = (\alpha_1, \ldots, \alpha_n)$ of nonnegative integers, with the order of the operator D^α being written as $|\alpha| = \sum_{j=1}^n \alpha_j$. To write nonlinear partial differential operators in a convenient form, we introduce the vector space R^{s_m} whose elements are $\xi_m = \{\xi_\alpha | \, |\alpha| \leq m\}$, and divide each such ξ into two parts $\xi_m = (\zeta, \eta)$ where $\eta = \{\eta_\beta | \, |\beta| \leq m - 1\} \in R^{s_{m-1}}$ is the lower order part of ξ_m, and $\zeta = \{\zeta_\alpha | \, |\alpha| = m\}$ is the part of ξ corresponding to the mth derivatives. In this notation, a general (nonlinear) partial differential operator of order m is defined by a mapping $F: G \times R^{s_m} \to R^1$ and defines an operator on functions u on G which assigns to each such u, another function v on G with

$$v(x) = F(x, \xi_m(u)(x)), \quad x \in G,$$

where $\xi_m(u)(x) = \{D^\alpha u(x) | \, |\alpha| \leq m\}$.

The systematic study of boundary value problems for nonlinear partial differential equations goes back as far at least as the work of Sergei Bernstein in the first decade of the twentieth century who made a determined and difficult attack upon second-order nonlinear elliptic equations in the plane using perturbation results based upon iteration methods or power series expansions together with a continuation argument. He considered a family of nonlinear elliptic equations of the form

$$F_\lambda(x, \xi_2(u)(x)) = 0 \tag{1}$$

in our present notation on $G \times R^2$ with Dirichlet boundary conditions prescribed on the boundary of the smoothly bounded set G, where F_0 might be the Laplace operator and F_1 is the genuinely nonlinear elliptic equation whose solution is desired. Assuming that F does not depend on the Oth order jet of u, Bernstein found solutions for (1) for λ close to any given λ_0 for which a solution was given. If all the solutions obtained in this fashion could be shown to lie in a determined compact subset of an appropriate function space of possible solutions, then the interval $|\lambda - \lambda_0| < \varepsilon$ of permissible perturbations could be chosen independent of λ_0 and the continuation process would pass from $\lambda = 0$ to $\lambda = 1$ in a finite number of steps. The determination or proof of the existence of the desired compact subset became known under the canonical terminology as the problem of finding an *a priori bound for solutions.*

Within the framework of this kind of method, it was often necessary to treat more general situations where the uniqueness of continuation in the parameter was not automatically assured but in which at certain parameter values, *bifurcation* took place and several branches of solutions appeared. In the context of partial differential equations and integral equations, this was treated first by Liapounov and Erhard Schmidt.

In the work of Schauder in the late 1920's in a process culminating in the celebrated paper of Leray and Schauder in 1934, Bernstein's analytical method was replaced by a more general and more powerful set of topological tools. Leray and Schauder considered the same equation (1) under Dirichlet boundary conditions but without assumptions yielding uniqueness of solutions for general λ. They showed that this equation was equivalent to a family of nonlinear functional equations in a suitable Banach space X of the form $(I - C_\lambda)(u) = 0$, where C_λ is a compact mapping of the closure of an open set G_λ of X into X (compact denoting that C_λ is continuous and that its image is a relatively compact subset of X). Using the topological degree defined by L. E. J. Brouwer for self-mappings of Euclidean spaces, a concept of degree of mapping was defined in Banach spaces for mappings of the form $I - C$ with C a compact map of cl(G) into X provided that $(I - C)$ has no zeroes on the boundary of G. This degree remains invariant under continuous deformation of data provided that it remains defined. Under suitable hypotheses of uniqueness at $\lambda = 0$, Leray and Schauder showed that the degree over 0 of $I - C_0$ was nonnull, and if the deformation defined by the equation (1) was permissible, the degree of $(I - C_1)$ would be different from zero and the problem for $\lambda = 1$ would have a solution. The problem of showing the permissibility of the deformation leads again to the problem of proving an *a priori bound*, this time in the appropriate Banach space which for the completely nonlinear problem is the space $C^{2,h}(G)$ of functions with second derivatives satisfying a Hölder condition and for the quasilinear case, $C^{1,h}(G)$. This *a priori* bound has to be established for all solutions of the family of nonlinear equations (1) and presents a basic difficulty to the application of this topological method in any concrete boundary value problem. In practice, it makes it necessary to rest the proof of the existence of solutions upon extremely difficult problems which amount to proving strong regularity theorems for these solutions. Except for quasilinear second-order equations in the plane, this remains in general an unconsummated and, in some cases, an impossible program.

An independent method for treating nonlinear partial differential equations of certain types is the direct method of the calculus of variations which goes back to Riemann and was revitalized in the twentieth century by Hilbert and Tonelli. In its classical form, this was restricted to those partial differential equations which were Euler-Lagrange equations of multiple integral problems. If we consider a general mth order multiple integral of the type

$$\phi(u) = \int_G F(x, \xi_m(u)(x))dx$$

over a space of functions V on the domain G, then u is (at least formally) a critical point of ϕ in V whenever u satisfies an appropriate boundary value condition on the boundary of G together with the partial differential equation

$$A(u) = \sum_{|\alpha| \leq m} (-1)^{|\alpha|} D^\alpha F_{\xi_\alpha}(x, \xi_m(u)) = 0 \tag{2}$$

where F_{ξ_α} is the partial derivative of the function F with respect to ξ_α.

The direct method of the calculus of variations attempts to obtain solutions of the variational boundary value problem for the Euler-Lagrange equation (2) by trying to construct critical points and in particular maxima or minima of the functional ϕ. A natural outgrowth of this line of ideas is the attempt to use the topological theory of variational problems (like the Morse theory or the Lusternik-Schnirelman theory) to find solutions of appropriate kinds of nonlinear partial differential equations. We shall return very explicitly to this consideration in our discussion of nonlinear elliptic eigenvalue problems.

It has become clear in recent years that for treating variational problems for multiple integrals, the appropriate space of functions u to use is a subspace V of an appropriately chosen Sobolev space $W^{m,p}(G)$, where $W^{m,p}(G)$ is the Banach space of all L^p functions u on G (with respect to Lebesgue n-measure) whose distribution derivatives up to order m are also L^p functions. $W^{m,p}(G)$ becomes an uniformly convex (and hence reflexive) Banach space with respect to the norm

$$||u||_{m,p}^p = \sum_{|\alpha| \leq m} ||D^\alpha u||_{L^p}^p$$

for $1 < p < +\infty$. The subspace V with which we are concerned must include the testing functions $C_c^\infty(G)$ with compact supports in G and must be closed in $W^{m,p}(G)$. Hence it must include the smallest such space $W_0^{m,p}(G)$, the closure of the testing functions in $W^{m,p}(G)$, i.e. $W_0^{m,p}(G) \subset V \subset W^{m,p}(G)$.

An important property of the functional ϕ that appears in all arguments to prove the existence of minima for ϕ in infinite dimensional spaces is its convexity or its almost-convexity. Let us consider a functional ϕ from the Banach space V to the reals which we assume to be once differentiable in the sense of Fréchet on V, and let

$$\phi'_x(y) = \lim_{t \to 0} t^{-1}(\phi(x + ty) - \phi(x)).$$

For each x in V, ϕ'_x is a bounded linear functional of y in V and hence an element of V^*, the space of bounded linear functionals on the Banach space V. Thus ϕ' defines a mapping $x \to \phi'_x$ of V into V^* and the zeroes of this mapping are the critical points of ϕ. We may ask the following question: What property of ϕ' corresponds to the property of convexity of ϕ? The answer (first explicitly stated by the Russian mathematician Kachurovski in 1960) is the following: ϕ is convex if and only if

$$(3) \qquad (\phi'_x - \phi'_u, x - u) \geq 0$$

for all x and u in V (where we let (w, u) denote the pairing between a functional w which is an element of V^* and its argument u which is an element of V). We can formulate this result in terms of the following definition (which seems to have first been given explicitly by the Russian mathematician Vainberg in 1960):

DEFINITION. *Let T be a mapping of a subset $D(T)$ of the Banach space V into V^*. Then T is said to be monotone if for each pair of elements u and v of $D(T)$,*

$$(4) \qquad (T(u) - T(v), u - v) \geq 0.$$

Our preceding assertion is then the following:

PROPOSITION 1. *$T = \phi'$ is monotone if and only if ϕ is convex.*

PROOF OF PROPOSITION 1. Suppose first that ϕ is convex. Then for any pair of elements x and y of V and for any λ with $0 < \lambda < 1$, we have

$$\phi(y + \lambda(x - y)) - \phi(y) \leq \lambda[\phi(x) - \phi(y)].$$

Dividing by λ and letting $\lambda \to 0+$, we obtain $(\phi'_y, x - y) \leq \phi(x) - \phi(y)$. Similarly, $(\phi'_x, y - x) \leq \phi(y) - \phi(x)$. Adding the two inequalities, we obtain $(\phi'_y - \phi'_x, x - y) \leq 0$, which is exactly the monotonicity property of $T = \phi'$.

Conversely, suppose that $T = \phi'$ is monotone. Let x and y be elements of V, and form the function

$$p(\lambda) = \phi(\lambda x + (1 - \lambda)y) - \lambda\phi(x) - (1 - \lambda)\phi(y), \quad 0 \leq \lambda \leq 1.$$

The function p is differentiable and vanishes at $\lambda = 0$ and $\lambda = 1$. To show that ϕ is convex, it suffices to show that $p(\lambda) \leq 0$ in $[0, 1]$. Suppose not. Then $p(\lambda)$ must have a maximum at some point ξ in $(0, 1)$ and $p'(\xi) = 0$. Let λ be a point in

[0, 1] with $\xi < \lambda$. Then

$$\begin{aligned} p'(\lambda) - p'(\xi) &= (T(\lambda x + (1-\lambda)y) - T(\xi x + (1-\xi)y), x - y) \\ &= (\lambda - \xi)^{-1}(T(\lambda x + (1-\lambda)y) - T(\xi x + (1-\xi)y, [\lambda x + (1-\lambda)y] \\ &\qquad -[\xi x + (1-\xi)y]) \\ &\geq 0. \end{aligned}$$

Hence p does not decrease for $\lambda > \xi$. It follows that $p(\xi) \leq p(1) = 0$, and $p(\lambda) \leq 0$ for all λ in [0, 1]. Hence ϕ is convex. Q.E.D.

One can consider the problem of treating the direct method of the calculus of variations by considering the properties of ϕ' rather than those of ϕ and translating problems concerning critical points into problems of the zeroes of mappings of V into V^*. More generally, however, since our primary interest in ϕ in the present context is to obtain solutions of its Euler-Lagrange equations which means to obtain zeroes of ϕ', we can abandon the consideration of ϕ altogether and consider more general mappings T of V into V^* which have the basic property of derivatives of convex functionals, i.e. study the properties of monotone mappings T of V into V^*.

There is another set of considerations here which serves as a useful guideline to the development of this line of ideas, and that is the analogy with the situation for linear equations. In the study of strongly elliptic linear partial differential equations and their boundary value problems of variational type, or what is sometimes called the Hilbert space approach (or one of the Hilbert space approaches) to elliptic boundary value problems, one began in the early 1950's with the theory of self-adjoint linear elliptic operators L, i.e. linear elliptic operators which are Euler-Lagrange operators of quadratic multiple integral problems treated by the direct method of the calculus of variations, and extended the theory to general strongly elliptic linear operators L by writing each such L in generalized divergence form

$$Lu = \sum_{|\alpha|, |\beta| \leq m} (-1)^{|\alpha|} D^\alpha(a_{\alpha\beta}(x) D^\beta u).$$

Corresponding to each such representation of L (and every linear elliptic operator L has at least one such representation if its coefficients are of class C^m at least), one defines the related *Dirichlet form*

$$a(u, v) = \sum_{|\alpha|, |\beta| \leq m} (a_{\alpha\beta}(x) D^\beta u, D^\alpha v),$$

which for testing functions v and smooth u and $a_{\alpha\beta}$ is just the result of integrating by parts in the expression for (Lu, v). Hence it extends the meaning of the Dirichlet problem for the equation $Lu = f$ if we replace it by the conditions:

(5) $$a(u, v) = (f, v), (v \in W_0^{m,2}(G)), \qquad u \in W_0^{m,2}(G).$$

A solution u of the equations (5) is termed a variational solution of the original boundary value problem. If we assume the coefficients $a_{\alpha\beta}$ bounded, then $a(u, v)$

is a bounded bilinear form on $W_0^{m,2}(G)$ and hence can be written in the form

$$a(u, v) = (Tu, v), (Tu \in V^*), \quad \text{for all } v \text{ in } V = W_0^{m,2}(G).$$

T is then a continuous linear mapping of V into V^*. The equation $Tu = f$ is equivalent to the conditions (5) above for the variational solution u. If we postulate for simplicity, the strong form of the Gårding inequality

$$a(u, u) \geq c\|u\|_{m,2}^2, \qquad u \in W_0^{m,2}(G), \quad (c > 0),$$

then T satisfies the condition $(Tu, u) \geq c\|u\|^2$ for all u in $V = W_0^{m,2}(G)$. It then follows that the equation $Tu = f$ has a solution, and exactly one for each f in V^*. Indeed, $c\|u\|^2 \leq \|u\| \cdot \|Tu\|$ implies that $c\|u\| \leq \|Tu\|$, $u \in V$, which shows that T is injective and has closed range in V^*. Hence the range of T is precisely the annihilator in V^* of the null space of the adjoint T^* of T, where T^* maps $V^{**} = V$ into V^*. Since $(Tu, u) = (T^*u, u)$, $(u \in V)$, T^* satisfies the same condition as T and hence has a trivial null space. Hence $R(T) = V^*$, and all equations of the form $T(u) = f$ have exactly one solution u for each f in V^*.

It is our purpose, then, to extend the variational theory of multiple integral problems to nonvariational problems in such a way as to obtain as a special case the results of the existence theory for strongly elliptic linear problems. The class of nonlinear elliptic operators for which such an extension makes sense is the class of operators A which can be written in generalized divergence form

$$A(u) = \sum_{|\alpha| \leq m} (-1)^{|\alpha|} D^\alpha A_\alpha(x, \xi_m(u)(x)) \tag{6}$$

where for each α, A_α is a mapping of $G \times R^{s_m}$ into R^1. As opposed to the linear case in which every linear differential operator can be written in divergence form if its coefficients are sufficiently smooth, the condition that a given operator A can be put in the form of equation (6) is rather restrictive and implies in particular that the operator A is quasilinear and has a relatively special character. On the other hand, since this class does include all equations obtained from variational problems and all linear equations, a relatively sharp existence theory for such equations without use of difficult regularity theorems for solutions has a considerable systematic value in the context of a theory of nonlinear equations where existence theorems are few and hard to obtain.

Equations of the form (6) were first considered by Visik in 1962–1963 and treated using compactness arguments together with *a priori* estimates for higher derivatives. Starting in 1963, the writer considered these equations in the context of the then just-initiated theory of monotone operators from V to V^*.

1. **Nonlinear elliptic boundary value problems and operators of monotone type.** We shall carry through our discussion under simple concrete hypotheses on the functions $A_\alpha(x, \xi)$ defining our nonlinear elliptic operators in divergence form, beginning with the simplest cases and reaching the more general cases by a step-by-step process of successive complications of hypothesis.

For the initial part of our discussion, we shall assume the following three conditions:

ASSUMPTIONS (A). (1) *Each* $A_\alpha(x, \xi_m)$ *is measurable in* x *with respect to Lebesgue* n*-measure on* G, *for fixed* ξ_m *in* R^{s_m}, *and continuous in* ξ_m *on* R^{s_m} *for almost all* x *in* G. *There exists a real number* p *with* $1 < p < \infty$ *such that for a given function* g *in* $L^{p'}(G)$, $p' = p/(p-1)$ *and a constant* $c > 0$, *we have*

$$|A_\alpha(x, \xi_m)| \leq c|\xi_m|^{p-1} + g(x). \tag{7}$$

(2) *For each* x *in* G *and each pair* ξ_m, ξ'_m *in* R^{s_m},

$$\sum_{|\alpha| \leq m} [A_\alpha(x, \xi_m) - A_\alpha(x, \xi'_m)][\xi_\alpha - \xi'_\alpha] \geq 0. \tag{8}$$

(3) *There exists a positive constant* $c_0 > 0$ *and a function* h *in* $L^{p'}(G)$ *such that for all* ξ_m *in* R^{s_m}, *and all* x *in* G,

$$\sum_{|\alpha| \leq m} A_\alpha(x, \xi_m)\xi_\alpha \geq c_0|\xi_m|^p - h(x). \tag{9}$$

Conditions (2) and (3) replace the strong ellipticity condition applied in the theory of linear elliptic problems. We note two facts about these conditions. First, they bear on the dependence of the differential operators A_α on all the derivatives of order less than or equal to m, not just on the highest order derivatives. Second, the characteristic forms involved in conditions (2) and (3) depend upon the components of the jet ξ_m and not upon products ξ_1^α of the components of the 1-jet as in the linear case. The first phenomenon will disappear in the later results of our discussion which will assume conditions only upon the dependence of $A_\alpha(x, \xi)$ on the highest order components ζ and only for $|\alpha| = m$. The second phenomenon, however, seems to be ineradicable since the Fourier transform arguments, by which the consideration of m-jets in the linear case is reduced to that of products of components of the 1-jet, are not applicable in the nonlinear case.

Let us note in addition that we make no assumption in (A) on the properties of the open set G, neither upon its boundedness nor upon the smoothness of its boundary. As one negative feature of the later results which free the hypotheses from the lower order derivatives, we shall have to impose conditions upon G being bounded and smoothly bounded which suffice for the validity of the full form of the Sobolev Imbedding Theorem.

Let us now state our problem precisely in the following form: We define the generalized Dirichlet form

$$a(u, v) = \sum_{|\alpha| \leq m} (A_\alpha(\cdot, \xi_m(u)), D^\alpha v) \tag{10}$$

for all u and v in $W^{m,p}(G)$, where $(\,,)$ denotes the usual L^2 inner product $(f, g) = \int_G fg$ for real-valued functions. By condition (1) of the Assumptions (A), $A_\alpha(x, \xi(u)(x))$ is a measurable function of x on G for each u in $W^{m,p}(G)$ and lies in $L^{p'}(G)$. Hence, it follows from Hölder's inequality that $a(u, v)$ is well defined for any pair u, v in $W^{m,p}(G)$ and satisfies an inequality of the form

$$|a(u, v)| \leq s(\|u\|_{m,p})\|v\|_{m,p}. \tag{11}$$

In particular, for each fixed u in V, $a(u, v)$ is a continuous linear functional of v in V, for any closed subspace V of $W^{m,p}(G)$. This linear functional, we denote by $T(u)$, so that $T(u)$ is an element of V^* and T is a mapping of V into V^* determined by the generalized Dirichlet form $a(u, v)$ and the subspace V of $W^{m,p}(G)$.

VARIATIONAL BOUNDARY VALUE PROBLEM. *Let V be a closed subspace of $W^{m,p}(G)$ with $W_0^{m,p}(G) \subset V$, and let $a(u, v)$ be a generalized Dirichlet form as defined in equation* (10) *above. Let f be a given element of V^*, the conjugate space of V. We ask for an element u of V which satisfies the condition*

$$a(u, v) = (f, v) \tag{12}$$

for all v in V.

The equation (12) together with the restriction that u lies in V, as in the linear case, has the force not only of requiring that u should satisfy the partial differential equation $A(u) = f$ (at least in a generalized sense) but also of imposing boundary conditions upon u. These boundary conditions arise from two sources: The restriction that u lies in V may impose a boundary condition if V is significantly smaller than the total space $W^{m,p}(G)$. (In the case where $V = W_0^{m,p}(G)$, the Dirichlet problem, this is the only boundary condition imposed by the problem and is the null Dirichlet boundary condition in the variational sense.) Second, the equation (12) imposes boundary conditions, the so-called natural boundary conditions, as soon as V is significantly larger than the minimal subspace $W_0^{m,p}(G)$. These arise from the fact that $f = A(u)$, and hence that $a(u, v) = (A(u), v)$, an effective integration by parts which imposes a vanishing condition upon certain expressions in the boundary terms. In the simple case of $m = 1$, where $A(u) = -\sum_{j=1}^n D_jA_j(x, \xi_1(u))$, the natural boundary conditions become the nonlinear analogue of the Neumann boundary condition for the nonlinear operator A in its divergence form, namely on boundary (G).

$$\sum_{j=1}^{n} A_j(x, \xi_1(u)) \cos(N, x_j) = 0,$$

(N = exterior normal at the boundary.)

Thus, one observes that in the Variational Boundary Value Problem as defined above, except in the Dirichlet problem, the boundary conditions as well as the partial differential equation involved are also nonlinear.

THEOREM 1. *Let $A(u)$ be a quasilinear operator of order $2m$ in generalized divergence form satisfying the Assumptions* (A) *on an open subset G of R^n, V a closed subspace of $W^{m,p}(G)$ which contains the testing functions with compact support in G.*

Then for each f in V^*, the Variational Boundary Value Problem for $A(u) = f$ with respect to V in the sense defined above always has a solution. The family of solutions u for a given f is a closed bounded convex subset of V.

We reduce the proof of Theorem 1 to the proof of a general theorem concerning nonlinear monotone operators T acting from the reflexive Banach space V to its dual space V^*. The operator T is the one defined above, such that for each u in V, $a(u, v) = (T(u), v)$ for all v in V (where (w, v) is the pairing between V^* and V). The condition (1) of the Assumptions (A) implies that T is a continuous mapping of V into V^*. (This follows by standard arguments as in Chapter 1 of Krasnoselskiĭ's book, *Topological methods in the theory of nonlinear integral equations.*) The inequality (11) above implies that T maps bounded sets of V into bounded sets of V^*. It is more essential, however, to see what the conditions (2) and (3) of Assumptions (A) imply for the form $a(u, v)$ and the mapping T.

Let u and v be elements of V. Then

$$a(u, u - v) - a(v, u - v)$$
$$= \int \sum_{|\sigma| \leq m} [A_\sigma(x, \xi_m(u)(x)) - A_\sigma(x, \xi_m(v)(x))][\xi_\sigma(u)(x) - \xi_\sigma(v)(x)]dx$$
$$\geq 0$$

since the integrand is nonnegative by condition (2). Putting this in terms of T, we find that

$$(T(u) - T(v), u - v) = a(u, u - v) - a(v, u - v) \geq 0, \qquad (u, v \in V),$$

i.e. T is a monotone mapping of V into V^*.

Let u be any element of V. Then by condition (3)

$$\begin{aligned} a(u, u) &= \int \sum_{|\alpha| \leq m} A_\alpha(x, \xi_m(u)(x))\xi_\alpha(u)(x)dx \\ &\geq c_0 \int |\xi_m(u)(x)|^p dx - ||h||_{L^{p'}}||u||_{m,p} \\ &\geq c_0||u||^p_{m,p} - c_1||u||_{m,p}. \end{aligned}$$

Since $p > 1$, it follows that

$$\frac{(T(u), u)}{||u||_{m,p}} = \frac{a(u, u)}{||u||_{m,p}} \geq c_0||u||^{p-1}_{m,p} - c_1 \to +\infty, \qquad (||u||_{m,p} \to +\infty).$$

We say that *the mapping T is coercive* if

$$(Tu, u)/||u|| \to +\infty \quad \text{as } ||u|| \to +\infty,$$

and we have verified from the above argument that condition (3) implies the coercivity of T. (Indeed, condition (3), or mild variants thereof which allow lower order derivatives in the negative term, is the only simple concrete analytical condition which seems to imply coercivity of T.)

Hence, Theorem 1 will follow from the following abstract theorem.

THEOREM 2. *Let V be a reflexive Banach space, T a continuous monotone mapping of V into V^* which is coercive. Then the range of T is all of V^*.*

Theorem 2 was first proved independently by the writer and George Minty in 1963 and is the prototype of the more delicate and more general theorems on monotone operators in Banach spaces. The condition that T is continuous is unnecessarily strong, since it suffices to assume that T is hemicontinuous, i.e. continuous from each line segment in V to the weak topology of V^*.

PROOF OF THEOREM 2. For the sake of greater intuitiveness and because of the separability of the space V in our application, we give the proof only for the case of V separable.

Since V is separable, we can construct an increasing sequence $\{V_n\}$ of finite dimensional subspaces of V whose union is dense in V. For each n, let ψ_n be the injection mapping of V_n into V and ψ_n^*, its dual, the corresponding projection mapping of V^* onto V_n^*. We construct mappings T_n of V_n into V_n^* for each n by setting $T_n(u) = \psi_n^* T \psi_n(u)$ $(u \in V_n)$. Then each T_n is a continuous mapping of V_n into V_n^* (since it is the composition of continuous mappings). The sequence of mappings $\{T_n\}$ is called the sequence of Galerkin approximation mappings for T corresponding to the sequence $\{V_n\}$.

The principal property of the mappings T_n which is used in the proof of our theorem is the following: For each u and v in V_n,

$$(T_n u - T_n v, u - v) = (\psi_n^*(Tu - Tv), u - v) = (Tu - Tv, u - v) \geq 0$$

and $(T_n u, u) = (Tu, u)$. Since T is assumed to be coercive, there exists a function $c(r)$ of the real argument $r \geq 0$ such that $c(r) \to +\infty$ as $r \to +\infty$ and for each u in V, $(Tu, u) \geq c(\|u\|)\|u\|$. Hence

$$(T_n u, u) \geq c(\|u\|)\|u\| \quad (u \in V_n)$$

with the function $c(r)$ independent of n. We now apply the following lemma.

LEMMA 1. *Let F be a finite dimensional Banach space, T a continuous coercive mapping of F into F^* with*

$$(Tu, u) \geq c(\|u\|)\|u\| \quad (u \in F)$$

for a given function $c(r)$ with $c(r) \to +\infty$ as $r \to +\infty$.

Then T maps F onto F^*. For each w in F^*, there exists u in F with $Tu = w$ and for all such u $\|u\| \leq k(\|w\|)$ and the function k depends only upon the function c.

PROOF OF LEMMA 1. Suppose that $T(u) = w$ for a given w. Then

$$c(\|u\|)\|u\| \leq (Tu, u) = (w, u) \leq \|w\| \cdot \|u\|,$$

so that (except for the trivial case $u = 0$), $c(\|u\|) \leq \|w\|$. If we set

$$k(s) = \{\sup r |\, c(r) \leq s\},$$

then, since $c(r) \to +\infty$ as $r \to +\infty$, $k(s) < +\infty$ for each s, and $\|u\| \leq k(\|w\|)$.

To prove that $R(T) = F^*$, it suffices to prove that $0 \in R(T)$. Indeed if w is a given element of F^* and if we set, $T_w(u) = Tu - w$, then

$$(T_w u, u) = (Tu, u) - (w, u) \geq c(\|u\|)\|u\| - \|w\| \cdot \|u\|$$

and T_w is also coercive. Since $T_w(u) = 0$ if and only if $T(u) = w$, it suffices to identify T_w with T and prove that each coercive mapping T from F to F^* has a zero. Since the conditions of the lemma are invariant if one replaces F by an equivalent Banach space and since every finite dimensional space is equivalent to a finite dimensional Hilbert space, we may assume without loss of generality that F is a Hilbert space and that $F^* = F$. Let $S = I - T$ mapping F into F. Since

$$(Su, u) = \|u\|^2 - (Tu, u),$$

we may choose $R > 0$ so large that for $\|u\| = R$, $(Su, u) < \|u\|^2$. We consider the closed ball B_R of radius R about the origin in F and define a continuous self-mapping of B_R by setting $f(u) = r(S(u))$, where r is the radial retraction of F on B_R given by

$$\begin{aligned} r(v) &= v, \quad v \in B_R, \\ &= \frac{Rv}{\|v\|}, \quad v \text{ outside of } B_R. \end{aligned}$$

By the Brouwer fixed point theorem, f has a fixed point u in B_R. For u in the interior of B_R, $u = f(u)$ implies that $f(u) = S(u)$ lies in B_R since every point of F outside B_R is mapped by r on the boundary of B_R. If u lies on the boundary of B_R, then $f(u) = \lambda Su$, $\lambda \leq 1$, and hence

$$(u, u) = (f(u), u) = \lambda(Su, u) < \|u\|^2$$

which contradicts $\|u\| = R$. Hence $Su = u$, and $Tu = 0$. Q.E.D.

PROOF OF THEOREM 2 CONTINUED. By Lemma 1, there exists a constant $M > 0$ and for each integer n, an element u_n of V_n such that $T_n u_n = 0$, where $\|u_n\| \leq M$ for all n.

Since V is reflexive, each bounded closed convex subset of V is sequentially weakly compact. Hence we may pass to a subsequence of the sequence $\{u_n\}$ and assume without loss of generality that u_n converges weakly to u_0 in V as $n \to \infty$. We wish to show that $Tu_0 = 0$.

For v in V_m for a given m and for any $n \geq m$, we have

$$\begin{aligned} 0 \leq (Tv - Tu_n, v - u_n) &= (Tv, v - u_n) - (T_n u_n, v - u_n) \\ &= (Tv, v - u_n) - (0, v - u_n) \\ &= (Tv, v - u_n). \end{aligned}$$

Since $(Tv, v - u_n)$ converges as $n \to \infty$ to $(Tv, v - u_0)$ as $v - u_n$ converges weakly to $v - u_0$, it follows that for each such v $(Tv, v - u_0) \geq 0$. The union of the spaces V_m is dense in V. Hence, for any element v of V, we may find a sequence $\{v_j\}$ of elements of $\bigcup_m V_m$ such that v_j converges to v. By the continuity of T, $Tv_j \to Tv$. Hence $(Tv_j, v_j - u_0) \to (Tv, v - u_0)$ and therefore

$$(Tv, v - u_0) \geq 0 \quad v \in V.$$

LEMMA 2. *If T is a continuous monotone mapping of V into V^*, then for given elements u_0 of V and w of V^*, $Tu_0 = w$ if and only if*

$$(13) \qquad (Tv - w, v - u_0) \geq 0 \quad (v \in V).$$

PROOF OF LEMMA 2. If T is monotone, then the inequality (13) follows from $Tu_0 = w$. Suppose on the other hand that the inequality (13) holds. Let z be any element of V and for $t > 0$, let $v_t = u_0 + tz$. Replacing v in (13) by v_t, we obtain $t(Tv_t - w, z) \geq 0$, or, cancelling $t > 0$, $(Tv_t - w, z) \geq 0$. Letting $t \to 0+$, $Tv_t \to Tu_0$. Hence

$$(Tu_0 - w, z) \geq 0 \quad (z \in V)$$

so that $Tu_0 = w$. Q.E.D.

PROOF OF THEOREM 2 COMPLETED By Lemma 2, $Tu = 0$, and we have proved that $R(T) = V^*$. By Lemma 2, moreover, we know that $u_0 \in T^{-1}(w)$ if and only if

$$(13) \qquad (Tv - w, v - u_0) \geq 0, \quad (v \in V).$$

For each fixed v of V, the inequality (13) constrains u_0 to lie in a translate of a half-space which is convex. Hence the inequality (13) is satisfied exactly for all v in V in the intersection of these half-spaces which is a closed convex set. Hence for each w in V^*, $T^{-1}(w)$ is a nonempty, closed, bounded, convex subset of V. Q.E.D.

As we remarked above, the conclusion of Theorem 1 about the existence of a solution to the Variational Boundary Problem for $A(u) = f$ on V for any f in V^* under Assumptions (A) now follows from the abstract existence theorem for the range of the monotone operator T in V^*.

When we wish to pass from Assumptions (A) involving the dependence properties of the A_α on lower order derivatives and for $|\alpha| < m$, to other assumptions involving only the highest order terms, the context in which we treat the boundary value problems involved must be changed in at least two ways. First, we must make assumptions on the domain G which ensure that the lower order derivatives are qualitatively trivial with respect to the norm in $W^{m,p}(G)$, and second, we must modify the ellipticity or monotonicity assumption (2) of Assumptions (A) which represents a sort of degenerate ellipticity to another assumption which actually assures us that the highest order terms of A are elliptic.

For the first of these steps, we apply the well-known Sobolev Imbedding Theorem.

THEOREM 3. *Let G be a bounded open subset of R^n whose boundary is locally a $(n - 1)$-dimensional manifold with a well-defined tangent hyperplane at each point and a locally Lipschitzian normal direction. Then*

(a) *If $0 \leq k < j$, then $W^{j,p}(G) \subset W^{k,r}(G)$ for $r^{-1} \geq p^{-1} - n^{-1}(j - k)$, with a continuous imbedding mapping. The imbedding mapping is compact if $r^{-1} > p^{-1} - (j - k)n^{-1}$ and $r < \infty$.*

(b) *If* $0 \le k < j$ *and* $p^{-1} - n^{-1}(j-k) < 0$, *then* $W^{k,p}(G) \subset C^{k,h}(G)$ *for any* h *with* $0 < h < 1$ *such that* $p^{-1} - n^{-1}(j-k-h) < 0$. *The imbedding mapping is always compact in this case.*

We refer to an open set G for which the conclusions of Theorem 3 hold as being *smoothly bounded.* We refer for the proof of Theorem 3 to any of the many proofs in the literature of the Sobolev Imbedding Theorem.

We can now formulate our new assumptions which will replace the Assumptions (A).

ASSUMPTIONS (B). (1) *Each* $A_\alpha(x, \xi_m)$ *is measurable in* x *for fixed* ξ_m *in* R^{s_m} *and continuous in* ξ_m *on* R^{s_m} *for almost all* x *in* G. *Let* $b_{m,n,p}$ *be the greatest integer less than* $m - n/p$, *and let* ξ_b *denote the vector* $\{\xi_\alpha | \, |\alpha| \le b_{m,n,p}\}$, *from the vector space* R^{s_b}. *There exist continuous functions* c_α *and* c_1 *from* R^{s_b} *to* $L^{p_\alpha}(G)$ *and* R^1, *respectively, such that the following inequalities hold:*

$$|A_\alpha(x, \xi)| \le c_\alpha(\xi_b)(x) + c_1(\xi_b) \sum_{m - n/p \le |\beta| \le m} |\xi_\beta|^{p_{\alpha\beta}}, \tag{14}$$

with the exponents p_α *and* $p_{\alpha\beta}$ *satisfying the inequalities:*

$$\begin{aligned} p_\alpha &= p' \quad \text{for } |\alpha| = m; \\ p_\alpha &> s'_\alpha \quad \text{for } m - n/p \le |\alpha| < m',\ s_\alpha^{-1} = p^{-1} - n^{-1}(m - |\alpha|); \\ p_\alpha &= 1 \quad \text{for } |\alpha| < m - n/p \end{aligned} \tag{15}$$

and

$$\begin{aligned} p_{\alpha\beta} &\le p - 1 \quad \text{for } |\alpha| = |\beta| = m; \\ p_{\alpha\beta} &< s_\beta (s'_\alpha)^{-1} \quad \text{for } m - n/p \le |\alpha| \le m,\ |\beta| < m; \\ p_{\alpha\beta} &\le s_\beta \quad \text{for } |\alpha| < m - n/p. \end{aligned} \tag{16}$$

(2) *If* $\xi_m = (\zeta_m, \eta_{m-1})$ *is the division of* ξ_m *into its* mth *order components* ζ_m *and the corresponding* $(m-1)$*st order jet* η_{m-1}, *then for each* x *in* G *and each* η_{m-1} *in* $R^{s_{m-1}}$,

$$\sum_{|\alpha| = m} [A_\alpha(x, \zeta_m, \eta_{m-1}) - A_\alpha(x, \zeta'_m, \eta_{m-1})], [\zeta_\sigma - \zeta'_\sigma] > 0 \text{ for } \zeta_m \neq \zeta'_m. \tag{17}$$

(3) *There exist constants* $c_0 > 0$ *and* c_2 *such that for all* x *and* ξ_m

$$\sum_{|\alpha| \le m} A_\alpha(x, \xi)\xi_\alpha \ge c_0 |\xi|^p - c_2. \tag{18}$$

LEMMA 3. *Under the Assumptions* (B) *if* G *is bounded and smoothly bounded, the generalized Dirichlet form*

$$a(u, v) = \sum_{|\alpha| \le m} (A_\alpha(\cdot, \xi_m(u)), D^\alpha v)$$

is well defined for each pair of elements u *and* v *of* $W^{m,p}(G)$. *If* V *is a closed subspace of* $W^{m,p}(G)$ *with* $W_0^{m,p}(G) \subset V$, *then there exists a continuous mapping* T *of* V *into*

V^* such that for all u and v of V, $a(u, v) = (T(u), v)$. The mapping T is coercive, carries bounded subsets of V into bounded subsets of V^* and satisfies the following generalized monotonicity condition:

(S) *If $\{u_n\}$ is a sequence in V which converges weakly in V to an element u and if $(Tu_n - Tu, u_n - u) \to 0$, then u_n converges strongly to u in V.*

We give the detailed proof of Lemma 3 in the Appendix to §1 and shall assume the Lemma here. We note that the first part of the lemma asserting that the generalized Dirichlet form $a(u, v)$ is well defined makes it possible to state the Variational Boundary Value Problem for $A(u) = f$ with respect to the space V as we have done above under the conditions given in the Assumptions (B). We can therefore state the basic existence theorem in this case.

THEOREM 4. *Let G be a bounded, smoothly bounded domain in R^n, V a closed subspace of $W^{m,p}(G)$ which contains $W_0^{m,p}(G)$, $A(u)$ a quasilinear elliptic operator of order $2m$ on G in the generalized divergence form*

$$A(u) = \sum_{|\alpha| \leq m} (-1)^{|\alpha|} D^\alpha A_\alpha(x, \xi_m(u)),$$

where the functions A_α satisfy the Assumptions (B) *for the given exponent p. Then for each f in V^*, there exists at least one solution u in V of the variational boundary value problem for the equation $A(u) = f$ with respect to V, i.e. such that*

$$a(u, v) = (f, v) \quad (v \in V).$$

The proof of Theorem 4 follows from the following abstract theorem on non-linear operators in Banach spaces.

THEOREM 5. *Let V be a separable reflexive Banach space, T a continuous coercive mapping of V into V^* which maps bounded sets into bounded sets and which satisfies the following condition:*

(S) *If $\{u_n\}$ is a weakly convergent sequence in V with limit u and if*

$$(Tu_n - Tu, u_n - u) \to 0,$$

then u_n converges strongly to u.

Then the range of T is all of V^*.

PROOF OF THEOREM 5. We follow the general lines of the proof of Theorem 2 by constructing an increasing sequence $\{V_n\}$ of finite dimensional subspaces of V whose union is dense in V. For each n, we let ψ_n be the injection mappings of V_n into V, ψ_n^* the dual projection of V^* on V_n^*, and define the sequence of Galerkin approximants $\{T_n\}$ for T by setting $T_n = \psi_n^* T \psi_n : V_n \to V_n^*$. Since T is coercive and since $(T_n u, u) = (Tu, u)$ for all u in V_n, it follows that the mappings T_n are uniformly coercive, i.e. there exists a fixed function c on R^+ such that $c(r) \to +\infty$ as $r \to \infty$ such that for all n and each u in V_n, $(T_n u, u) \geq c(\|u\|)\|u\|$.

By Lemma 1, each T_n maps V_n onto V_n^*. Moreover, if u_n is any solution in V_n of the equation $T_n u_n = \psi_n^* w$ for a given element w of V^*, then there exists a constant M independent of n such that $||u_n|| \leq M$. Since the space V is reflexive and closed balls in V are weakly sequentially compact, we may pass to a subsequence of the sequence $\{u_n\}$ which is weakly convergent to an element u_0 of V and assume without loss of generality that u_n converges weakly to u_0.

It suffices to show that $Tu_0 = w$, since w was an arbitrarily chosen element of V^*. For this purpose, we note that $||Tu_n||$ is uniformly bounded for all n since $||u_n||$ is uniformly bounded and by hypothesis T maps bounded sets of V into bounded sets in V^*. For any v in a subspace V_m and for any $n \geq m$, we have

$$(Tu_n, v) = (T_n u_n, v) = (\psi_n^* w, v) = (w, v),$$

i.e. $(Tu_n, v) \to (w, v)$. Since the union of the spaces V_m is dense in V and $||Tu_n||$ is uniformly bounded, it follows that Tu_n converges weakly to w in V^*. Now consider

$$(Tu_n - Tu_0, u_n - u_0) = (Tu_n, u_n) - (Tu_n, u_0) - (Tu_0, u_n - u_0).$$

Since $u_n - u_0$ converges weakly to 0 as $n \to \infty$, we know that $(Tu_0, u_n - u_0) \to 0$. Since Tu_n converges weakly to w as $n \to \infty$, we know that $(Tu_n, u_0) \to (w, u_0)$. For the first term, we know that

$$(Tu_n, u_n) = (T_n u_n, u_n) = (\psi_n^* w, u_n) = (w, u_n) \to (w, u_0)$$

as $n \to \infty$, since u_n converges weakly to u_0. Hence $(Tu_n - Tu_0, u_0 - u_0) \to (w, u_0) - (w, u_0) - 0 = 0$. By property (S), it follows that u_n converges strongly to u_0. Since T is continuous, Tu_n converges strongly to Tu_0. Since Tu_n converges weakly to w, we conclude that $Tu_0 = w$. Q.E.D.

The proof of Theorem 4 now follows from Theorem 5 and Lemma 3.

One question that naturally arises when we compare the preceding results with the corresponding results for the linear theory is whether one can obtain some sort of analogue for the results that hold in the linear case without coerciveness (i.e. positivity) but with some added hypothesis of the sort that arises from a uniqueness condition.

Results of this sort can be obtained of two different sorts depending on whether one imposes Assumptions of type (A) or type (B). Let us reformulate the conditions of (A) without the coercivity condition:

ASSUMPTIONS A′. *The conditions* (1) *and* (2) *of Assumptions* (A) *without the coercivity condition* (3).

THEOREM 6. *Let G be an open set of R^n, $A(u)$ a quasilinear elliptic differential operator of order $2m$, ($m \geq 1$) in the generalized divergence form*

$$A(u) = \sum_{|\alpha| \leq m} (-1)^{|\alpha|} D^\alpha A_\alpha(\cdot, \xi_m(u))$$

where the functions A_α satisfy the conditions (1) *and* (2) *of Assumptions* (A′), *i.e. the conditions of Assumptions* (A) *without the coercivity condition* (3). *Let V be a closed subspace of $W^{m,p}(G)$ and suppose that for each w in V^*, there exists a constant M_w and a neighborhood N of w in V^* such that for any w' in N and any solution* *u of the variational boundary value problem for* $A(u) = w'$ *with respect to* V, $\|u\| \leq M_w$.

Then there exists a solution u of the variational boundary value problem for $A(u) = w$ *with respect to* V *for every* w *in* V^*.

We derive Theorem 6 from the following abstract theorem.

THEOREM 7. *Let* V *be an uniformly convex Banach space with a norm continuously once differentiable on the unit sphere. Let* T *be a continuous monotone mapping of* V *into* V^*. *Suppose that* T^{-1} *is locally bounded, i.e. each point* w_0 *of* V^* *has a neighborhood* N *in* V^* *such that* $T^{-1}(N)$ *is bounded in* V.

Then the range of T *is all of* V^*.

We note that Theorem 6 follows from Theorem 7 since V being a closed subspace of the product of L^p spaces for $p > 1$ is certainly uniformly convex and has a once continuously differentiable norm on the complement of 0.

PROOF OF THEOREM 7. We consider the convex function $f(u) = \frac{1}{2}\|u\|^2$ on V. Since the norm of V is continuously differentiable on the complement of the origin, $f(u)$ is continuously differentiable on all of V. Its derivative f', by Proposition 1 is a monotone mapping from V to V^* which we denote by J. For every y in V, we know by the proof of Proposition 1 that $f(x) \geq f(y) + (J(y), x - y)$. Setting $x = (1 \pm \varepsilon)y$ in this inequality, we obtain

$$\|y\|^2 + o(1) \leq (Jy, y) \leq \|y\|^2 + o(1),$$

or letting $\varepsilon \to 0$, $(Jy, y) = \|y\|^2$. Hence $\|Jy\| \geq \|y\|$. Obviously $J(cy) = cJ(y)$ for all real numbers c. For $\|y\| = 1$,

$$(Jy, x) - \|y\|^2 \leq \tfrac{1}{2}[\|x\|^2 - \|y\|^2],$$

i.e. $(Jy, x) \leq 1$, for all x with $\|x\| = 1$. Hence $\|Jy\| \leq 1$ for $\|y\| = 1$, so that $\|Jy\| \leq \|y\|$. Combining this with the opposite inequality, we see that J has the following two properties:

$$(Jy, y) = \|Jy\| \cdot \|y\|, \qquad \|Jy\| = \|y\|, \quad (y \in V). \tag{19}$$

A mapping J of V into V^* satisfying the two equations of (19) for every y in V is called a (normalized) *duality mapping* of V into V^*.

We now consider our original monotone mapping T and approximate it by a family of coercive monotone mappings $T_\varepsilon = T + \varepsilon J$, for $\varepsilon > 0$. Each T_ε is obviously monotone since both T and J are monotone. The coercivity of T_ε follows from the following calculation: for u in V

$$\begin{aligned}(T_\varepsilon u, u) &= (Tu, u) + \varepsilon\|u\|^2 \\ &= (Tu - T(0), u) + (T(0), u) + \varepsilon\|u\|^2 \\ &\geq \varepsilon\|u\|^2 - \|T(0)\| \cdot \|u\| \\ &= \|u\|[\varepsilon\|u\| - \|T(0)\|].\end{aligned}$$

Since T and J are both continuous, we may apply Theorem 2 and conclude that for each $\varepsilon > 0$, T_ε has all of V^* as its range.

Suppose now that we have a sequence $\{u_n\}$ in V for which $(Ju_n - Ju, u_n - u) \to 0$ for a given element u of V. We note that

$$\begin{aligned}(Ju_n - Ju, u_n - u) &= (Ju_n, u_n) + (Ju, u) - (Ju_n, u) - (Ju, u_n) \\ &= [||u_n||\cdot||Ju_n|| + ||u||\cdot||Ju|| - ||Ju_n||\,||u|| - ||Ju||\,||u_n||] \\ &\quad + [||u_n||\,||Ju|| - (u_n, Ju)] + [||u||\cdot||Ju_n|| - (u, Ju_n)].\end{aligned}$$

Each of the terms in square brackets is nonnegative, the first being simply $[||u|| - ||u_n||]^2$. If their sum converges to zero, each of the summands must converge to 0. Hence $||u_n|| \to ||u||$ and $(u_n, Ju) \to ||u||^2$. We show that u_n must converge strongly to u. Since V is uniformly convex and $||u_n|| \to ||u||$, it suffices to show that u_n converges weakly to u. To show the latter, since V is reflexive, it suffices to assume that u_n converges weakly to v and show that $v = u$. (This would show that every weakly convergent subsequence converges to u, and hence that the original sequence converges weakly to u.) However, we would then have $(v, Ju) = ||u||^2$ implying that $||v|| \geq ||u||$ and $||v|| \leq \lim ||u_n|| = ||u||$. Hence $(v, Ju) = ||v||^2 = ||Ju||^2$. The set of all such elements v is a convex subset of a sphere in V and hence must consist of a single point set at most. The point u lies in this set, so that $v = u$ and u_n converges strongly to u.

We assert that for each $\varepsilon > 0$, T_ε is an injective mapping and T_ε^{-1} is continuous from V^* to V in the strong topologies. To show this, it suffices to assume that $T_\varepsilon u_n - T_\varepsilon u \to 0$ for a sequence $\{u_n\}$ and to show that u_n must converge to u. Since T_ε is coercive, $T_\varepsilon^{-1}(B)$ is bounded for any bounded set B in V^*. In particular, $\{u_n\}$ is bounded. Hence $(T_\varepsilon u_n - T_\varepsilon u, u_n - u) \to 0$. However,

$$\begin{aligned}(T_\varepsilon u_n - T_\varepsilon u, u_n - u) &= (Tu_n - Tu, u_n - u) + \varepsilon(Ju_n - Ju, u_n - u) \\ &\geq \varepsilon(Ju_n - Ju, u_n - u).\end{aligned}$$

Hence $(Ju_n - Ju, u_n - u)$, being nonnegative for every n, must converge to 0, and u_n converges strongly to u by the result of the preceding paragraph. Hence T_ε^{-1} is well defined and continuous for each $\varepsilon > 0$.

To show that $R(T)$ is all of V^*, it suffices to show that every point of the boundary of $R(T)$ must be an interior point of $R(T)$ since then $R(T)$, being a nonempty open and closed subset of the connected space V^*, must be all of V^*. Suppose then that w_0 lies in the closure of $R(T)$. By the hypothesis of the local boundedness of T^{-1}, there exists a ball B_0 about w_0 and a bounded set B in V such that $T^{-1}(B_0) \subset B$. We choose a larger ball B_1 about 0 in V which contains the closure of B in its interior. Choose $d > 0$ so small that the ball of radius $4d$ about w_0 is contained in B_0. Since w_0 lies in the closure of $R(T)$, we may find a point of $R(T)$ in $B_d(w_0)$, i.e. a point $w_1 = T(x_1)$ where by the choice of d, x_1 lies in B. Since J maps bounded sets into bounded sets, we may choose ε so small that $||\varepsilon J(x)|| < d$ for all x in B_1. Then $T_\varepsilon(x_1)$ lies in the ball of radius $2d$ about

w_0. Let w be any other point in $B_{2d}(w_0)$ and let C_ε be the line segment in this ball joining w with $T_\varepsilon(x_1)$. If $C'_\varepsilon = T_\varepsilon^{-1}(C_\varepsilon)$, then C'_ε is a continuous curve for each $\varepsilon > 0$ which begins at x_1 and ends at $T_\varepsilon^{-1}(w)$. We assert that for ε sufficiently small (with the measure of smallness independent of w), C'_ε must lie entirely within the set B_1. Indeed, otherwise this curve would have to pass out of B_1 at some point and would intersect the boundary of B_1. For such an intersection point x, $T(x) = T_\varepsilon(x) - \varepsilon J(x)$ would differ by at most d from a point of $B_{2d}(w_0)$ and hence would lie in $B_{3d}(w_0)$ which is contained in B_0. Hence x would lie in $T^{-1}(B_0) \subset B$, which is entirely contained in the interior of B_1. This contradicts the fact that x lies on the boundary of B_1.

Thus for each w in $B_{2d}(w_0)$, $\varepsilon < \varepsilon_0$, $T_\varepsilon^{-1}(w)$ lies in the ball B_1. Choose a sequence $\varepsilon_j \to 0$, and let $x_j = T_{\varepsilon_j}^{-1}(w)$. Then $T(x_j) = T_{\varepsilon_j}(x_j) - \varepsilon_j J(x_j)$ converges to w as $j \to \infty$. Since V is reflexive, we may pass to a subsequence and assume that x_j converges weakly to some x in V. For any v in V, we know that $(Tv - Tx_j, v - x_j) \geq 0$. Since $Tv - Tx_j$ converges strongly to $Tv - w$ while $v - x_j$ converges weakly to $v - x$, it follows that $(Tv - Tx_j, v - x_j) \to (Tv - w, v - x) \geq 0$. By Lemma 2, it follows that $Tx = w$, i.e. $B_{2d}(w_0) \subset R(T)$. Hence the proof of Theorem 7 is complete. Q.E.D.

A result for operators satisfying Assumptions of the type (B) demands a weakening of the Assumptions (B) that eliminates the coercivity condition (3) but keeps the basic property of the operator T that follows from condition (3), namely the strong property (S). Such a set of assumptions is the following:

ASSUMPTIONS (B′). *These are the conditions* (1) *and* (2) *of the Assumptions* (B) *with condition* (3) *replaced by the new condition*

(3′) *There exist two continuous functions c_0 and c from R^{s_b} to R^1 with $c_0(\eta_b) > 0$ for each η_b, such that for all x in G, all ζ_m and all η_{m-1} in $R^{s_{m-1}}$, we have*

$$\sum_{|\alpha|=m} A_\alpha(x, \zeta_m, \eta_{m-1})\zeta_\alpha \geq c_0(\eta_b)|\zeta_m|^p - c(\eta_b) \sum_{m-n/p \leq |\beta| \leq m-1} |\eta_\beta|^{p_{\alpha\beta}}.$$

Under the Assumptions (B′), we have the following extension of Lemma 3, which we write in an even stronger form:

LEMMA 3′. *Let G be a bounded, smoothly bounded open set in R^n and $A(u)$ a quasilinear elliptic differential operator in generalized divergence form satisfying the Assumptions* (B′). *Let $a(u, v)$ be the generalized Dirichlet form for the operator $A(u)$, V a closed subspace of $W^{m,p}(G)$, T the mapping of V into V^* given by $(T(u), v) = a(u, v)$, $(u, v \in V)$. Then T satisfies the following stronger form of condition* (S):

(S+) *If u_n is a weakly convergent sequence in V with limit u and if*

$$\overline{\lim}\, (Tu_n - Tu, u_n - u) \leq 0,$$

then u_n converges strongly to u in V.

The proof of Lemma 3′ is given in the Appendix to §1.

THEOREM 8. *Let G be a bounded, smoothly bounded open set in R^n, $A(u)$ a quasilinear elliptic operator of order $2m$ in the generalized divergence form*

$$A(u) = \sum_{|\alpha| \leq m} (-1)^{|\alpha|} D^\alpha A_\alpha(\cdot, \xi_m(u)),$$

with the functions A_α satisfying the Assumptions (B′). *Suppose that the following additional conditions hold:*

(a) *A is odd, i.e. $A_\alpha(x, -\xi_m) = -A_\alpha(x, \xi_m)$ for all x, ξ_m.*

(b) *A is homogeneous of some positive order, i.e. for all x, ξ_m and all $c > 0$,*

$$A_\alpha(x, c\xi_m) = c^a A_\alpha(x, \xi_m)$$

for a fixed $a > 0$.

(c) *There exists no nonzero solution of the variational boundary value problem for $A(u) = 0$ with respect to a given closed subspace V of $W^{m,p}(G)$.*

Then for every w in V^*, there exists at least one solution u of the variational boundary value problem for $A(u) = w$ with respect to the space V.

Theorem 8 follows from Lemmas (3) and (3′) and the following abstract theorem, Theorem 9. Both Theorems 8 and 9 are modifications of a recent result of Pohojhayev (J. Functional Anal. Appl., 1967).

THEOREM 9. *Let V be a separable reflexive Banach space, T a continuous mapping of V into V^* which maps bounded sets of V onto bounded sets of V^* and satisfies condition* (S). *Suppose that T is odd (i.e. $T(-u) = -T(u)$) and homogeneous of some positive degree a (i.e. $T(cu) = c^a T(u)$ for all u in V and all $c > 0$). Suppose further that 0 is the only solution of $T(u) = 0$.*

Then the range of T is all of V^*.

Since the homogeneity of degree a of all the A_α and their oddness leads to the same property for T, it follows that Theorem 9 and Lemma 3′ together imply Theorem 8.

PROOF OF THEOREM 9. As in the proofs of Theorems 2 and 5, we form the Galerkin approximants $\{T_n\}$ for T with respect to an increasing sequence of finite dimensional subspaces $\{V_n\}$ of V with dense union in V. Since T is odd and $T_n = \psi_n^* T \psi_n$ is the composition of T with linear mappings, each approximant T_n is an odd mapping of V_n into V_n^*.

Consider T_n as a mapping of the unit ball $B_1(0, V_n)$ into V_n^*. If S_n is the boundary of this ball and if $T_n(S_n)$ does not contain 0, it follows from the Antipodal Point Theorem of Borsuk-Ulam-Lusternik-Schnirelman that the degree of T_n on $B_1(0, V_n)$ over 0 is an odd integer. In particular, it follows that if $d_n = \operatorname{dist}(0, T_n(S_n))$, then the ball of radius d_n about 0 in V_n^* lies in the image of T_n on $B_1(0, V_n)$.

LEMMA 4. *Let T be a continuous mapping of the reflexive space V into V^* which maps bounded sets into bounded sets and which satisfies condition* (S). *Suppose that $T(x) \neq 0$ for all x with $\|x\| = 1$. Then for all n sufficiently large, there exists $d_0 > 0$ such that $d_n \geq d_0$, where $d_n = \operatorname{dist}(0, T_n(S_n))$ as above.*

PROOF OF LEMMA 4. Suppose the contrary. Then there would exist a sequence of integers $n_j \to +\infty$, a sequence of points $x_j \in V_{n_j}$ with $\|x_j\| = 1$ such that

$||T_{n_j}(x_j)|| \to 0$. Since V is reflexive, we may assume without loss of generality that x_j converges weakly to some element x of V as $j \to \infty$. For each v in some V_m and $n_j \geq m$, we have

$$(Tx_j, v) = (T_{n_j}x_j, v) \to 0, \quad (j \to \infty).$$

Hence, since $||Tx_j||$ are uniformly bounded, Tx_j converges weakly to 0 in V^*. Thus $(Tx_j - Tx, x_j - x) = (Tx_j, x_j) - (Tx_j, x) + (Tx, x - x_j)$, where $(Tx_j, x_j) = (T_{n_j}x_j, x_j) \to 0$ since $T_{n_j}x_j$ goes to zero in norm and $||x_j||$ is uniformly bounded. Therefore

$$(Tx_j - Tx, x_j - x) \to 0, \quad (j \to \infty).$$

Applying condition (S), we see that x_j converges strongly to x. Hence $||x|| = 1$, and Tx_j converges strongly to Tx since T is continuous. Since Tx_j converges weakly to 0, it follows that $Tx = 0$, contradicting the hypothesis. Q.E.D.

Proof of Theorem 9 Completed. By Lemma 4, there exists n_0 and a positive constant d_0 such that for $||w|| < d_0$ and all $n \geq n_0$, there exists a solution u_n with $||u_n|| \leq 1$ of the equation $T_n u_n = \psi_n^* w$, $u_n \in V_n$. For v in some V_m, $(Tu_n, v) = (w, v)$ for $n \geq m$. Since $||Tu_n||$ is uniformly bounded, it follows that Tu_n converges weakly to w as $n \to \infty$. Hence if we choose a weakly convergent subsequence and identify it with our original sequence, we may assume that u_n converges weakly to u, Tu_n converges weakly to w, and $(Tu_n, u_n) = (w, u_n)$ converges to (w, u). Hence $(Tu_n - Tu, u_n - u) \to (w, u) - (w, u) = 0$. Applying condition (S), it follows that u_n converges strongly to u, Tu_n converges to Tu, hence $Tu = w$. Thus the range of T contains the ball of radius d_0 about 0.

On the other hand, the homogeneity of T implies that $R(T)$ is a cone with vertex at 0. Hence $R(T) = V^*$. Q.E.D.

An important extension of the existence theory of solutions of variational boundary value problems for our quasilinear elliptic equations is that given by the theory of nonlinear variational inequalities. In intuitive terms, this is analogous in the case of variational problems for a functional ϕ to the passage from the problem of finding critical points of ϕ on the whole space V to the more general problem of finding extrema or critical points in an appropriate sense for ϕ restricted to a convex subset C of V.

Let us consider this latter problem in closer detail. Suppose that u_0 in C affords a minimum to ϕ with respect to all points of C. Then for any u in C and any λ with $0 \leq \lambda \leq 1$, we have

$$\phi(u_0) \leq \phi(\lambda u + (1 - \lambda)u_0) = \phi(u_0 + \lambda(u - u_0)).$$

If we assume that ϕ has a derivative ϕ', then for λ small

$$\phi(u_0 + \lambda(u - u_0)) = \phi(u_0) + \lambda(\phi'_{u_0}, u - u_0) + o(\lambda).$$

In these terms, the inequality at the minimum point becomes

$$(\phi'_{u_0}, u - u_0) + o(1) \geq 0, \quad (\lambda \to 0+),$$

or letting $\lambda \to 0$, $0 \leq (\phi'_{u_0}, u - u_0)$, $\quad u \in C$.

DEFINITION. *Let T be a mapping of V into V^*, C a closed convex subset of V, w an element of V^*. Then u_0 in C is said to be a solution of the variational inequality given by the triple (T, w, C) if u_0 satisfies the condition*

$$(Tu_0 - w, u - u_0) \geq 0 \tag{20}$$

for all u in C.

Similarly, in the more concrete context of partial differential equations, one can extend the definition of a variational boundary value problem for a quasilinear elliptic operator in divergence form to the following more general definition of a nonlinear elliptic variational inequality:

DEFINITION. *Let G be an open subset of R^n, $A(u)$ a quasilinear elliptic partial differential operator of order $2m$, $(m > 1)$, on G in the generalized divergence form*

$$A(u) = \sum_{|\alpha| \leq m} (-1)^{|\alpha|} D^\alpha A_\alpha(\cdot, \xi_m(u)),$$

with the corresponding generalized Dirichlet form

$$a(u, v) = \sum_{|\alpha| \leq m} (A_\alpha(\cdot, \xi_m(u)), D^\alpha v).$$

Let V be a closed subspace of a Sololev space $W^{m,p}(G)$ on which the form $a(u, v)$ is well defined, C a closed convex subset of V, w an element of V^*.

Then the nonlinear variational inequality for the equation $A(u) = w$ on C with respect to V asks for an element u_0 of C which satisfies the following inequality:

$$a(u_0, v - u_0) \geq (w, v - u_0), \quad v \in C. \tag{21}$$

Using these definitions for both the abstract and concrete cases, we can state the generalizations of our previous existence theorems for solutions of elliptic boundary value problems and monotone operator equations to the more general context of elliptic variational inequalities and variational inequalities for operators of monotone type.

THEOREM 10. *Let G be an open subset of R^n, $A(u)$ a quasilinear elliptic operator of order $2m$ on G in the divergence form*

$$A(u) = \sum_{|\alpha| \leq m} (-1)^{|\alpha|} D^\alpha A_\alpha(\cdot, \xi_m(u))$$

with the functions A_α satisfying the Assumptions (A) *on G for a given exponent p with $1 < p < \infty$. Let V be a closed subspace of $W^{m,p}(G)$, C a closed bounded convex subset of V.*

Then for each w in V^*, there exists at least one solution u_0 in C of the nonlinear variational inequality $a(u_0, v - u_0) \geq (w, v - u_0)$, $v \in C$.

The set of such solutions is a nonempty closed bounded convex subset of C.

THEOREM 11. *Let V be a reflexive Banach space, C a closed bounded convex subset of V with $0 \in C$, T a continuous monotone coercive mapping of C into V^*. Then for each w in V^*, there exists at least one solution u_0 in C of the nonlinear variational inequality*

$$(Tu_0, v - u_0) \geq (w, v - u_0), \quad v \in C,$$

and the set of such solutions u_0 for a given w is a nonempty closed bounded convex subset of C.

THEOREM 12. *Let G be a bounded, smoothly bounded open subset of R^n, $A(u)$ a quasilinear elliptic operator of order $2m$ on G of the divergence form*

$$A(u) = \sum_{|\alpha| \le m} (-1)^{|\alpha|} D^\alpha A_\alpha(\cdot, \xi_m(u))$$

with the functions A_α satisfying the Assumptions (B) *on G for a given exponent p with $1 < p < \infty$. Let V be a closed subspace of $W^{m,p}(G)$, C a closed bounded convex subset of V.*

Then for each w in V^*, there exists at least one solution u_0 in C of the nonlinear elliptic variational inequality $a(u_0, v - u_0) \ge (w, v - u_0)$, $(v \in C)$.

THEOREM 13. *Let V be a reflexive Banach space, C a closed bounded convex subset of V with $0 \in C$, T a continuous coercive mapping of C into V^* which maps bounded sets into bounded sets. Suppose that T satisfies the following condition:*

(S+) *If $\{u_n\}$ is a weakly convergent sequence in C with limit u and if $\overline{\lim}\,(Tu_n - Tu, u_n - u) \le 0$, then u_n converges strongly to u.*

Then for each w in V^*, there exists at least one solution u_0 in C of the variational inequality

$$(Tu_0 - w, v - u_0) \ge 0, \quad (v \in C).$$

We have verified in the preceding discussion that the concrete hypotheses of Theorem 10 imply the abstract hypotheses of Theorem 11 and (using Lemma 3′ of the Appendix) that the concrete hypotheses of Theorem 12 imply the abstract hypotheses of Theorem 13 for the particular Banach spaces V and operators T from V to V^* generated by the quasilinear operator $A(u)$ and its Dirichlet form. Hence it suffices to give the proofs of Theorems 11 and 13.

Both proofs employ the following auxiliary result.

LEMMA 5. *Let C_0 be a compact convex subset in a Banach space X, S a continuous mapping of C_0 into X^*. Then there exists an element u_0 of C_0 such that for all v of C_0, $(Su_0, v - u_0) \ge 0$.*

PROOF OF LEMMA 5. Suppose the conclusion of the lemma were not true. Then for each u in C_0, there would exit at least one element v_u of C_0 such that $(Su, v_u - u) < 0$.

For each v in C_0, form the subset of C_0 given by $N_v = \{u | u \in C_0, (Su, v - u) < 0\}$. Since $(Su, v - u)$ is a continuous function of u on C_0, N_v is an open subset of C_0 for each v. By the preceding remark, it follows from our negative hypothesis that the family $\{N_v : v \in C_0\}$ covers the compact set C_0. Hence we can find a finite subcovering $\{N_{v_1}, \dots, N_{v_s}\}$ and a partition of unity $\{\beta_1, \dots, \beta_s\}$ subordinated to this finite subcovering, where $0 \le \beta_j(x) \le 1$ for all j and x, β_j has its support in N_{v_j}, and $\sum_j \beta_j(x) = 1$ on C_0. Using this partition of unity, we define a mapping f of c_0 into itself by setting

$$f(x) = \sum_{j=1}^{s} \beta_j(x) v_j.$$

We note that f is continuous and that for all x in C_0,

$$(Sx, f(x) - x) = \sum_{j=1}^{s} \beta_j(x)(Sx, v_j - x) < 0$$

since on the support of β_j, $(Sx, v_j - x) < 0$, and each point x of C_0 is contained in the support of at least one β_j.

On the other hand, f is a continuous selfmapping of C_0 and indeed of the convex span of the finite set $\{v_1, \ldots, v_s\}$. Hence by the Brouwer fixed point theorem, it has a fixed point x_0 in $C_0, f(x_0) = x_0$. For this point, we have $0 = (Sx_0, f(x_0) - x_0) < 0$, which is a contradiction following from the negative hypothesis. Hence the conclusion of Lemma 5 is true. Q.E.D.

Proof of Theorem 11. Let $\{v_j\}$ be an infinite sequence of points of C whose union is dense in C with $v_1 = 0$, and for each n, let C_n be the span of $\{v_1, \ldots, v_n\}$. Then $\{C_n\}$ is an increasing sequence of compact convex subsets of C, and we can apply Lemma 5 to each C_n. If we do this, we obtain the result that the set

$$F_n = \{u | u \in C_n, (Tu, v - u) \geq 0, \text{ for all } v \text{ in } C_n\}$$

is nonempty for each n.

By hypothesis, T is monotone. Hence $(Tv, v - u) \geq (Tu, v - u) \geq 0$ for any u in F_n and any v in C_n. Moreover, T is coercive by hypothesis so that there exists a function $c(r)$ for $r \geq 0$ with $c(r) \to +\infty$ as $r \to +\infty$ such that for all u in C, $(Tu, u) \geq c(||u||)||u||$. If u lies in F_n, $(Tu, u) \leq (Tu, v_1) = 0$, and hence $c(||u||) \leq 0$, $||u|| \leq M, u \in F_n$, for a suitable constant M independent of the index n.

Choosing a point u_n from each F_n and using the weak sequential compactness of closed balls in a reflexive Banach space, we may assume by thinning out the sequence that u_n converges weakly to u_0 in V. Since C is closed and convex, it is weakly closed. Therefore, u_0 is an element of C.

For any v in $\bigcup_n C_n$, and n sufficiently large $(Tv, v - u_n) \geq 0$. Taking the limit as $n \to \infty$, we obtain

$$(Tv, v - u_0) \geq 0, \quad v \in \bigcup_n C_n.$$

By construction, each point v of C is the strong limit of a sequence $\{v^k\}$ from the union of the C_n. Since T is continuous, Tv^k converges strongly to Tv. Hence $(Tv^k, v^k - u_0)$ converges to $(Tv, v - u_0)$, so that $(Tv, v - u_0) \geq 0$ for all v in C.

Let v be any element of C, and for $0 < \lambda < 1$, let $v_\lambda = (1 - \lambda)u_0 + \lambda v$. Since u_0 and v lie in C and C is convex, v_λ lies in C and hence

$$0 \leq (Tv_\lambda, v_\lambda - u_0) = (Tv_\lambda, \lambda(v - u_0))$$

$$= \lambda(Tv_\lambda, v - u_0).$$

Cancelling $\lambda > 0$, we see that

$$(Tv_\lambda, v - u_0) \geq 0, \quad (0 < \lambda < 1).$$

Letting $\lambda \to 0+$, we note that Tv_λ converges to Tu_0 by the continuity of T. Hence $(Tu_0, v - u_0) \geq 0$, $v \in C$.

This shows that the variational inequality with $w = 0$ has a solution in C and that the set of solutions coincides with the set of all u in C such that

$$(Tv, v - u) \geq 0, \quad (v \in C).$$

The latter is a closed convex subset of C (being the intersection of C and a family of flats). It is bounded by the coerciveness of T as above.

The same conclusion follows for a general w by replacing T by the new monotone coercive mapping T_w where $T_w u = Tu - w$, $u \in C$. Q.E.D.

PROOF OF THEOREM 13. Again, we form the increasing sequence of compact convex subsets $\{C_n\}$ of C, and verify by Lemma 5 that the subset

$$F_n = \{u | u \in C_n, (Tu, v - u) \geq 0, v \in C_n\}$$

is a nonempty subset of C_n. If we choose u_n in F_n for each n, it follows from the coerciveness of T as in the proof of Theorem 11 that there exists a constant $M > 0$, independent of n, such that $||u_n|| \leq M$. Passing to a convenient subsequence, we may assume without loss of generality that u_n converges weakly to an element u_0 of V, which must lie in C since the closed convex set C is weakly closed.

For any v in $\bigcup_m C_m$, $(Tu_n, u_n) \leq (Tu_n, v)$ for all n sufficiently large by the definition of the set F_n. Let v be an arbitrary element of C, and consider a sequence v^k from $\bigcup_m C_m$ converging strongly to v. For each k, we have $\overline{\lim}\,(Tu_n, u_n - v^k) \leq 0$. For each n,

$$(Tu_n, u_n - v) = (Tu_n, u_n - v^k) + (Tu_n, v^k - v).$$

Since T maps bounded sets in C into bounded sets in V^*, there exists a constant $M_1 > 0$ such that $||Tu_n|| \leq M_1$ for all n. If we are given $\varepsilon > 0$, we may choose k so large that $||v - v^k||M_1 < \varepsilon$. Holding k thus chosen fixed and considering $n \to \infty$, we obtain

$$\overline{\lim}\,(Tu_n, u_n - v) \leq \overline{\lim}\,(Tu_n, u_n - v^k) + \sup_n |(Tu_n, v^k - v)|$$

where

$$\overline{\lim}\,(Tu_n, u_n - v^k) \leq 0, \qquad \sup_n |(Tu_n, v^k - v)| \leq M_1||v_k - v|| < \varepsilon.$$

Hence

$$\overline{\lim}\,(Tu_n, u_n - v) \leq \varepsilon, \quad \varepsilon > 0,$$

i.e.

$$\lim\,(Tu_n, u_n - v) \leq 0, \quad (v \in C).$$

We now let $v = u_0$, and consider $(Tu_n - Tu_0, u_n - u_0) = (Tu_n, u_n - u_0) + (Tu_0, u_0\ \overline{\lim} - u_n)$. Since u_n converges weakly to u_0, $(Tu_0, u_0 - u_n) \to 0$ as $n \to \infty$. Hence $\lim\,(Tu_n - Tu_0, u_n - u_0) \leq 0$. Applying the hypothesized condition (S+), it follows that u_n converges strongly to u_0. Since T is continuous, Tu_n converges strongly to Tu_0. Therefore,

$$(Tu_n, u_n - v) \to (Tu_0, u_0 - v), \quad v \in C,$$

and we see that $(Tu_0, v - u_0) \geq 0$, $v \in C$, i.e. u_0 is the solution of the variational inequality for T on C with $w = 0$. The corresponding solution for general w is obtained by replacing T by a new monotone, coercive mapping T_w with $T_w u = Tu - w$. Q.E.D.

REMARKS. Existence theorems of the type of Theorems 11 and 13 were first obtained independently by Hartman, Stampacchia (Acta, 1966) and the writer (Bull. Amer. Math. Soc., 1965, 780–785) and were extended to various classes of partially monotone maps (see, for example, the writer's Montreal notes). The widest generalization for everywhere defined mappings is to the class of pseudo-monotone mappings by H. Brézis (Annales de Fourier, to appear), which includes Theorem 13 as a special case. The writer has studied the theory of variational inequalities from a more general viewpoint as a special case of the theory of maximal monotone mappings from the reflexive Banach space V to V^* (see Proc. Nat. Acad. Sci. 1967, and the writer's recent papers in Math. Annalen 1967–68).

Hartman and Stampacchia used the theory of variational inequalities together with estimates for second-order operators of the Di Giorgi type to obtain strong existence theorems for second-order strongly elliptic boundary value problems. The theory of variational inequalities can also be used in higher order equations to substantially extend the range of permissible growth conditions on the lower order terms in the above existence theorems (see a forthcoming paper of the writer).

A more detailed discussion of aspects of this theory is given in the writer's forthcoming paper in the Proc. Amer. Math. Soc. Symposium on Nonlinear Functional Analysis, April 1968.

Appendix to § 1. Proof of Lemmas (3) **and** (3′). We now turn to the detailed concrete analytical argument necessary to establish the conclusions of Lemmas 3 and 3′.

The mapping D^α carries $W^{m,p}(G)$ continuously into $W^{m-|\alpha|,p}(G)$, and if G is bounded and smoothly bounded, it follows from the Sobolev Imbedding Theorem, (Theorem 3 above), that D^α carries $W^{m,p}(G)$ continuously into $L^{s_\alpha}(G)$ for $s_\alpha^{-1} = p^{-1} - n^{-1}(m - |\alpha|)$, $(s_\alpha < \infty)$, and D^α is a compact linear mapping into $L^{t_\alpha}(G)$ for any $t_\alpha < s_\alpha$. (All this, of course, when $|\alpha| \geq m - n/p$. In the other case, the mapping is compact into $C^0(\bar{G})$ and hence into $L^s(G)$ for any $s \leq \infty$.)

Let us consider the individual terms making up the expression for the generalized Dirichlet form

$$a(u, v) = \sum_{|\alpha| \leq m} (A_\alpha(\cdot, \xi_m(u)), D^\alpha v).$$

If $|\alpha| = m$, $v \to D^\alpha v$ maps $W^{m,p}(G)$ continuously into $L^p(G)$. For the inner product $(A_\alpha(\cdot, \xi_m(u)), D^\alpha v)$ to be well defined, the first term in the inner product $A_\alpha(\cdot, \xi_m(u))$

must lie in $L^{p'}(G)$, where p' is (as always) the conjugate exponent to p, ($p' = p/(p-1)$). If this were true, it would follow in addition by the Hölder inequality that

$$|(A_\sigma(\cdot, \xi_m(u)), D^\sigma v)| \leq \|A_\sigma(\cdot, \xi_m(u))\|_{L^{p'}} \|v\|_{m,p}.$$

By condition (1) of the Assumptions (B), $A_\alpha(\cdot, \xi_m(u))$ is a measurable function on G which satisfies the inequality:

$$|A_\alpha(x, \xi_m)| \leq c_\alpha(\xi_b)(x) + c_1(\xi_b) \sum_{m-n/p \leq |\beta| \leq m} |\xi_\beta|^{p_{\alpha\beta}},$$

where for each ξ_b (b being the greatest integer less than $m - n/p$), $c_\sigma(\xi_b)$ is an element of $L^{p'}(G)$ depending continuously in that space upon ξ_b, c_1 is a continuous function from R^{s_b} to R^+, and the exponents satisfy the inequalities

$$p_{\alpha\beta} \leq s_\beta(p-1)/p.$$

For u in $W^{m,p}(G)$, $\xi_b(u)(x)$ is a continuous function whose absolute value is bounded by a function of $\|u\|_{m,p}$. Hence $\|c_\alpha(\xi_b(u))\|_{L^{p'}} + |c_1(\xi_b(u))| \leq h(\|u\|_{m,p})$. Similarly, $\xi_\beta(u)$ lies in $L^{s_\beta}(G)$ and

$$|\xi_\beta(u)|^{p_{\alpha\beta}} \in L^{s_\beta/p_{\alpha\beta}}(G).$$

Under the given inequality on the exponents $p_{\alpha\beta}$, it follows that $(p_{\alpha\beta})^{-1}s_\beta \geq p'$ and $A_\alpha(\cdot, \xi_m(u))$ lies in $L^{p'}(G)$ for the bounded set G and satisfies an inequality of the form $\|A_\alpha(\cdot, \xi_m(u))\|_{L^{p'}(G)} \leq h(\|u\|_{m,p})$ for some continuous function h. Hence

$$|(A_\alpha(\cdot, \xi_m(u)), D^\alpha v)| \leq h(\|u\|_{m,p})\|v\|_{m,p}.$$

Similarly, if $|\alpha| < m$ with $|\alpha| \geq m - n/p$, $D^\alpha v$ lies in $L^{s_\alpha}(G)$ for any $s_\alpha < \infty$ with $s_\alpha^{-1} \geq p^{-1} - n^{-1}(m - |\alpha|)$ for any element v of $W^{m,p}(G)$. On the other hand, A_α for such α satisfies the inequality:

$$|A_\alpha(x, \xi_m)| \leq c_\alpha(\xi_b)(x) + c_1(\xi_b) \sum_{m-n/p \leq |\alpha| \leq m} |\xi_\beta|^{p_{\alpha\beta}}.$$

Here c_α is a continuous function from R^{s_b} to $L^{p_\alpha}(G)$, c_1 is a continuous function from R^{s_b} to R^+, and the exponents satisfy the inequalities

$$p_\alpha > (s_\alpha)', \quad p_{\alpha\beta} < s_\beta(s'_\alpha)^{-1}.$$

It follows that $c_\alpha(\xi_b(u))$ lies in $L^{s'_\alpha}$ and satisfies the inequality $\|c_\alpha(\xi_\beta(u))\|_{L^{s'_\alpha}} \leq h(\|u\|_{m,p})$. Similarly, $|c_1(\xi_b(u))| \leq h(\|u\|_{m,p})$. On the other hand,

$$|\xi_\beta(u)|^{p_{\alpha\beta}} \in L^{s_\beta/p_{\alpha\beta}}(G)$$

and since $s_\beta(p_{\alpha\beta})^{-1} \geq s'_\alpha$ by hypothesis, $A_\alpha(\cdot, \xi_m(u))$ lies in $L^{s'_\alpha}$ for each u in $W^{m,p}(G)$ and satisfies an inequality of the form $\|A_\alpha(\cdot, \xi_m(u))\|_{L^{s'_\alpha}} \leq h(\|u\|_{m,p})$.

Finally,

$$|(A_\alpha(\cdot, \xi_m(u)), D^\alpha v)| \leq \|A_\alpha(\cdot, \xi_m(u))\|_{L^{s'_\alpha}}\|v\|_{L^{s_\alpha}} \leq h(\|u\|_{m,p})\|v\|_{m,p}.$$

A similar and even simpler calculation verifies the same conclusion for $|\alpha| < m - n/p$.

Hence $a(u, v)$ is well defined under condition (1) of Assumptions (B) for all u and v of $W^{m,p}(G)$ and satisfies an inequality of the form $|a(u, v)| \leq h(||u||_{m,p})||v||_{m,p}$. If V is a closed subspace of $W^{m,p}(G)$, it follows that for fixed u in V, $a(u, v)$ is a bounded linear functional of v in V and defines an unique element $T(u)$ of V^* such that for all u and v of V, $a(u, v) = (T(u), v)$ and T satisfies the inequality $||T(u)||_{V^*} \leq h(||u||_{m,p})$, i.e. T maps bounded sets in V into bounded sets in V^*. The continuity of T under our present hypotheses follows from standard theorems on Niemytski operators on L^p spaces (see the previous reference to the book by Krasnoselskiĭ), and T is a well-defined continuous mapping of V into V^* which carries bounded sets into bounded sets. Under condition (3) of Assumptions (B), it follows easily as before that T is coercive.

It remains to show that under the condition (2) of Assumptions (B) and either condition (3) of Assumptions (B) or condition (3′) of Assumptions (B′), T possesses the property (S) or the stronger property (S+). Since (S+) is indeed stronger than S, it suffices to verify (S+).

Suppose then that $\{u_n\}$ is a sequence of elements of V which converges weakly to an element u of V and that

$$\overline{\lim}\,(Tu_n - Tu, u_n - u) \leq 0.$$

By the definition of T,

$$\begin{aligned}(Tu_n - Tu, u_n - u) &= a(u_n, u_n - u) - a(u, u_n - u)\\ &= \sum_{|\alpha| \leq m} (A_\alpha(\cdot, \xi_m(u_n)) - A_\alpha(\cdot, \xi_m(u)), D^\alpha u_n - D^\alpha u).\end{aligned}$$

Under our hypothesis on the exponents in the inequalities for A_α, for $|\alpha| < m$, there exists an exponent $t_\alpha < s_\alpha$ such that

$$||A_\alpha(\cdot, \xi_m(u))||_{L^{t'_\alpha}} \leq h(||u||_{m,p}), \quad (u \in W^{m,p}(G))$$

and the mapping $v \to D^\alpha v$ is a compact linear mapping of V into $L^{t_\alpha}(G)$. Since a compact linear mapping maps a weakly convergent sequence into a strongly convergent sequence, we know that for such α,

$$||D^\alpha u_n - D^\alpha u||_{L^{t_\alpha}(G)} \to 0, \quad (n \to \infty).$$

Hence

$$\sum_{|\alpha| < m} (A_\alpha(\cdot, \xi_m(u_n)) - A_\alpha(\cdot, \xi_m(u)), D^\alpha u_n - D^\alpha u) \to 0, \quad (n \to \infty),$$

and our original hypothesis implies that

$$\overline{\lim} \sum_{|\alpha| = m} (A_\alpha(\cdot, \xi_m(u_n)) - A_\alpha(\cdot, \xi_m(u)), D^\alpha u_n - D^\alpha u) \leq 0.$$

We shall now introduce explicitly the decomposition of the m-jet ξ_m in the form (ζ, η) where ζ denotes the components of order m, η the components of order less than m. (We omit the subscripts on ζ and η). We write $A_\alpha(\cdot, \xi_m)$ as $A_\alpha(\cdot, \zeta, \eta)$.

To prove that u_n converges strongly to u, it suffices to show that one can choose a strongly convergent infinite subsequence. (Indeed, if the original sequence did not converge strongly, one could obtain a first infinite subsequence, whose elements have distance from u bounded from below by a positive constant. Replacing our original sequence by this first subsequence, we obtain a new sequence satisfying the same hypothesis as the original one. But this new sequence has no infinite subsequences converging strongly since the limit of these further subsequences would have to coincide with their weak limit, namely u.) This fact enables us to simplify by passing to subsequences.

Since u_n converges weakly to u in $W^{m,p}(G)$ and for $|\beta| < m$ the map D^β of $W^{m,p}(G)$ into $L^{t_\beta}(G)$ for $t_\beta < s_\beta$ is compact, $D^\beta u_n$ converges strongly to $D^\beta u$ in $L^{t_\beta}(G)$ for $|\beta| < m$. Passing to an infinite subsequence, we may assume without loss of generality that for each β with $|\beta| < m$, $D^\beta u_n$ converges almost everywhere in G to $D^\beta u$, and indeed on the complement of a fixed subset N of G of measure zero. We shall consider only points x of G outside of N in the further argument.

Since the condition (3) of Assumptions (B) is stronger than the condition (3′) of the Assumptions (B′), we shall assume the latter for the rest of the argument, namely

$$\sum_{|\alpha|=m} A_\alpha(x, \zeta, \eta)\zeta_\alpha \geq c_0(\eta_b)|\zeta|^p - c(\eta_b) \sum_{m-n/p \leq |\beta| \leq m-1} |\eta_\beta|^{p_{\alpha\beta}}.$$

We note that

$$\sum_{|\alpha|=m} (A_\alpha(\cdot, \zeta(u), \eta(u_n)) - A_\alpha(\cdot, \zeta(u), \eta(u)), D^\alpha u_n - D^\alpha u) \to 0$$

since $\eta(u_n)$ converges strongly to $\eta(u)$ in the appropriate L^s spaces by the compactness of injections in the Sobolev theorem and hence $A_\alpha(\cdot, \zeta(u), \eta(u_n))$ converges strongly in $L^{p'}(G)$ to $A_\alpha(\cdot, \xi_m(u))$. Hence, our hypothesis reduces to the condition:

$$\overline{\lim} \sum_{|\alpha|=m} (A_\alpha(\cdot, \zeta(u_n), \eta(u_n)) - A_\alpha(\cdot, \zeta(u), \eta(u_n)), D^\alpha u_n - D^\alpha u) \leq 0,$$

which in turn implies that for any measurable subset H of G,

$$\overline{\lim} \int_H \sum_{|\alpha|=m} A_\alpha(x, \xi_m(u_n)(x))D^\alpha u_n(x)dx$$
$$\leq \overline{\lim} \int_H \{|A_\alpha(x, \xi_m(u_n)(x))D^\alpha u| + |A_\alpha(x, \zeta(u)(x), \eta(u_n)(x))| \cdot |D^\alpha u_n - D^\alpha u|\}dx$$

while by the inequality of condition (3′),

$$\int_H \sum_{|\sigma|=m} |D^\sigma u_n(x)|^p dx \leq c \int_H \sum_{|\sigma| \leq m} A_\sigma(x, \xi_m(u_n)(x))D^\sigma u_n(x)dx$$
$$+ c_1 \sum_{|\beta| \leq m-1} \int_H |D_\beta u_n(x)|^{p_{\alpha\beta}} dx.$$

Combining these two inequalities, we obtain an inequality for $\overline{\lim} \int_H \sum_{|\alpha|=m} |D^\alpha u(x)|^p dx$ in terms of the terms on the right of the two inequalities.

More precisely, given $\varepsilon > 0$, we can find an integer n_ε such that for $n \geq n_\varepsilon$,

$$\sum_{|\alpha|=m} \int_G [A_\alpha(x, \xi_m(u_n)(x)) - A_\alpha(x, \zeta(u)(x), \eta(u_n)(x))] [D^\alpha u_n(x) - D^\alpha u(x)]dx < \varepsilon.$$

For $n \geq n_\varepsilon$ and all measurable subsets H of G, we have

$$\begin{aligned}\sum_{|\alpha|=m} \int_H |D^\alpha u_n(x)|^p dx \leq c \sum_{|\alpha|=m} \Big\{ \int_H |A_\alpha(x, \xi_m(u_n)(x))| \cdot |D^\alpha u(x)| dx \\ + \int_H |A_\alpha(x, \zeta(u)(x), \eta(u_n)(x))| \cdot |D^\alpha u_n - D^\alpha u| dx \\ + c \int_H \sum_{|\beta| \leq m-1} |D^\beta u_n|^{p_{\alpha\beta}} dx \Big\} + \varepsilon.\end{aligned}$$

The integrals on the right of this last inequality have the following crucial property: They are absolutely continuous in H, uniformly in n, i.e. given $\varepsilon > 0$, there exists $\delta > 0$ such that if meas $(H) < \delta$, then the sum of all the integrals on the right is less than ε for all n. Hence for meas $(H) < \delta$ and $n \geq n_\varepsilon$,

$$\sum_{|\alpha|=m} \int_H |D^\alpha u_n(x)|^p dx < 2\varepsilon.$$

For each n, the last integral is absolutely continuous in H. Hence choosing δ even smaller, it follows that for meas $(H) < \delta$ the inequality holds for all n.

To show that u_n converges strongly to u in $W^{m,p}(G)$, it suffices to show that $D^\alpha u_n$ converges strongly in $L^p(G)$ for each α with $|\alpha| = m$. We know that $D^\alpha u_n$ converges weakly to $D^\alpha u$ in $L^p(G)$ and by the preceding argument that $\int_H |D^\alpha u_n(x)|^p dx$ are uniformly absolutely continuous in H. To show that $D^\alpha u_n$ converges strongly, it suffices by the Vitali Convergence Theorem to show that an infinite subsequence converges almost everywhere in G. We have already implicitly used the fact given by condition (2) of Assumptions (B) that

$$\sum_{|\alpha|=m} [A_\alpha(x, \zeta, \eta) - A_\alpha(x, \zeta', \eta)] \cdot [\zeta_\alpha - \zeta'_\alpha] > 0$$

if $\zeta \neq \zeta'$, in arguing that an integral involving an expression of the form above (with $\zeta = \zeta(u_n)$, $\eta = (u_n)$, $\zeta' = \zeta(u)$) over a subset H of G could not exceed the corresponding integral over G. We now use this condition in a sharper way, noting first that

$$\overline{\lim} \int_G [A_\alpha(x, \zeta(u_n)(x), \eta(u_n)(x)) - A_\alpha(x, \zeta(u)(x), \eta(u_n)(x))][D^\alpha u_n - D^\alpha u] dx \leq 0$$

implies that

$$\lim \int_G [A_\alpha(x, \zeta(u_n), \eta(u_n)) - A_\alpha(x, \zeta(u), \eta(u_n))] \cdot [D^\alpha u_n - D^\alpha u] dx = 0$$

since the integrand is nonnegative. This implies that the nonnegative integrands converge to 0 in $L^1(G)$. If we pass to an infinite subsequence, there exists a set N' of measure zero such that the integrands converge to 0 on $G - N'$, i.e. for $x \in G - (N \cup N')$,

$$\sum_\alpha [A_\alpha(x, \zeta(u_n)(x), \eta(u_n)(x)) - A_\alpha(x, \zeta(u)(x), \eta(u_n))][D^\alpha u_n(x) - D^\alpha u(x)] \to 0$$

while $D^\beta u_n(x) \to D^\beta u(x)$ for each β with $|\beta| < m$. If we combine this with the inequality of condition (3′), we see that

$$|\zeta(u_n)(x)|^p \le c \sum_{|\alpha|=m} \{A_\alpha(x, \zeta(u_n)(x), \eta(u_n)(x)D^\alpha u_n(x) + c_1 \sum_{|\beta|\le m} |D^\beta u_n(x)|^{p_{\alpha\beta}}\}$$

while

$$\sum_{|\alpha|=m} A_\alpha(x, \zeta(u_n)(x), \eta(u_n)(x))D^\alpha u_n(x) \le \sum_{|\alpha|=m} [A_\alpha(x, \zeta(u_n)(x), \eta(u_n)(x))D^\alpha u(x)$$
$$+ A_\alpha(x, \zeta(u)(x), \eta(u_n)(x)[D^\alpha u_n(x) - D^\alpha u(x)]] + o(1)$$
$$\le c|\zeta(u_n)(x)|^{p-1} + c_2(x) \sum_{|\beta|<m} |D^\beta u_n(x)|^{p_{\alpha\beta}}(1 + |\zeta(u_n)(x)|).$$

The two inequalities combined imply that for each x in the complement of $N \cup N'$, we have $|\zeta(u_n)(x)|^p \le c|\zeta(u_n)(x)|^{p-1} + c'|\zeta(u_n)(x)|$ and hence that for each such, $|\zeta(u_n)(x)|$ is bounded over all n. Fixing such a value of x and passing to a subsequence for which $\zeta(u_n)(x)$ converges to some value ζ for the given point x, we see from the limit relation above (since $\eta(u_n)(x) \to \eta(u)(x)$) that

$$\sum_{|\alpha|=m} [A_\alpha(x, \zeta, \eta(u)(x)) - A_\alpha(x, \zeta(u)(x), \eta(u(x)))](\zeta_\alpha - D^\alpha u(x)) = 0.$$

Using the full force of condition (2) of Assumptions (B), this implies that $\zeta = \zeta(u)(x)$ for the given x. Thus any convergent subsequence of the bounded sequence of vectors $\zeta(u_n)(x)$ converges to $\zeta(u)(x)$. Therefore $\zeta(u_n)(x)$ converges for the original sequence to $\zeta(u)(x)$, and the almost everywhere convergence is proved.

The almost everywhere convergence together with the uniform absolute continuity implies the strong convergence of $D^\alpha u_n$ to $D^\alpha u$ in $L^p(G)$ and the proof of the property (S+) of the mapping T is thereby complete.

PART II

2. Nonlinear elliptic eigenvalue problems and the Lusternik-Schnirelman category. We now turn our attention from the variational boundary value problems for quasilinear elliptic operators A of order $2m$ to the closely related but somewhat different class of nonlinear elliptic eigenvalue problems of variational type. Here, instead of looking for a solution of the elliptic equation $A(u) = f$ under null variational boundary conditions, we look for solutions u of the eigenvalue problem $A(u) = \lambda B(u)$ under null variational boundary conditions for some real parameter λ. We assume that A is an elliptic operator of order $2m$ in divergence form

$$A(u) = \sum_{|\alpha|\le m} (-1)^{|\alpha|} D^\alpha A_\alpha(\cdot, \xi_m(u)),$$

and that B is an operator of lower order in divergence form

$$B(u) = \sum_{|\beta|\le m-1} (-1)^{|\beta|} D^\beta B_\beta(\cdot, \xi_{m-1}(u)).$$

We can be guided by two intuitive considerations in the formulation of this theory. The first is the example of the linear case, where A is a linear elliptic operator of order $2m$ and B, for simplicity, is the identity operator. Here we pass into the domain of the generalized Sturm-Liouville problem for linear elliptic differential operators. We know from the wide experience that has been accumulated in this domain that such a theory can only be constructed on a reasonably general basis (so far as construction of actual eigenfunctions satisfying the original boundary conditions is concerned) if the domain G is bounded and smoothly bounded (in the sense of Theorem 3 of §1) and if the linear operator A is selfadjoint. Selfadjointness for linear operators is equivalent to the condition that such an operator is the Euler-Lagrange operator of a quadratic multiple integral operator.

Since, in addition, the closely related eigenfunction theory of compact selfadjoint operators is based very firmly upon the variational formulation of such an eigenvalue problem, it seems both reasonable and prudent to restrict our attention in the nonlinear case to selfadjoint problems, i.e. to the case when A and B are the Euler-Lagrange operators of multiple integral problems.

We assume given two functionals f and h in the form of multiple integral operators over a bounded domain G in R^n, where $f(u) = \int_G F(\cdot, \xi_m(u))dx$, and $h(u) = \int_G H(\cdot, \xi_{m-1}(u))dx$. We consider these functionals to be defined on a closed subspace V of $W^{m,p}(G)$ with V containing the testing functions with support in G. If we assume that the functions F and H defining the functionals f and h are once continuously differentiable in ξ_m, we can define the Euler-Lagrange operators of f and h by considering their differentials f' and h', i.e. formally for u and v in V,

$$(f'(u), v) = \lim_{t\to 0} t^{-1}(f(u + tv) - f(u)) = \sum_{|\alpha|\le m} (F_\alpha(\cdot, \xi_m(u)), D^\alpha v)$$

and

$$(h'(u), v) = \lim_{t\to 0} t^{-1}(h(u + tv) - h(u)) = \sum_{|\beta|\le m-1} (H_\beta(\cdot, \xi_{m-1}(u), D^\beta v)$$

where

$$F_\alpha(x, \xi) = \partial F(x, \xi)/\partial \xi_\alpha, \qquad H_\beta(x, \eta) = \partial H(x, \eta)/\partial \eta_\beta.$$

We see that $(f'(u), v)$ is the Dirichlet form in the sense of §1 for a quasilinear partial differential operator of order $2m$, namely

$$A(u) = \sum_{|\alpha|\le m} (-1)^{|\alpha|} D^\alpha F^\alpha(\cdot, \xi_m(u)), \tag{1}$$

while $(h'(u), v)$ is the corresponding Dirichlet form for an operator B in divergence of order $\le 2m - 2$, namely

$$B(u) = \sum_{|\beta|\le m-1} (-1)^{|\beta|} D^\beta H_\beta(\cdot, \xi_{m-1}(u)). \tag{2}$$

DEFINITION. *Let G be an open subset of R^n, A and B two quasilinear partial differential operators on B in the divergence forms* (1) *and* (2) *above, V a closed subspace of a Sobolev space $W^{m,p}(G)$. Then by a solution of the eigenvalue problem for the pair of operators (A, B) with respect to the variational boundary conditions defined by V, we mean an element u of V such that for some real number λ,*

$$a(u, v) = \lambda b(u, v) \quad (v \in V)$$

where $a(u, v)$ is the Dirichlet form for the operator A, i.e.

$$a(u, v) = \sum_{|\alpha| \leq m} (F_\alpha(\cdot, \xi_m(u)), D^\alpha v),$$

and $b(u,v)$ is the Dirichlet form for the operator B, i.e.

$$b(u, v) = \sum_{|\beta| \leq m-1} (H_\beta(\cdot, \xi_{m-1}(u)), D^\beta v).$$

Let us note that for smooth solutions u of the eigenvalue problem defined in the above definition, the *natural* boundary conditions which are imposed upon u through the fact that $a(u, v) - \lambda b(u, v) = (A(u), v) - \lambda(B(u), v)$, $v \in V$, are in general nonlinear and involve the eigenvalue λ, except for the special case of the Dirichlet problem in which $V = W_0^{m,p}(G)$, the closure in $W^{m,p}(G)$ of the testing functions in G.

By our remarks above, for all u and v in V,

$$a(u, v) = (f'(u), v); \qquad b(u, v) = (h'(u), v)$$

at least if our calculations for f' and h' can be verified under precise analytical hypotheses (as is done below). The equality $a(u, v) = \lambda b(u, v)$, $v \in V$, is therefore equivalent to the functional equation in V, $f'(u) = \lambda h'(u)$. Here, f' is the operator from V to V^* which plays the same role for A as the operator T did in the discussion of §1 for nonselfadjoint operators, and similarly, h' plays the same role for B.

The equation $f'(u) = \lambda h'(u)$ which characterizes the eigenvalue problems in the sense of the definition above suggests an extremely useful and conceptually important method for the interpretation of such eigenvalue problems, namely they are the solution by Lagrange multipliers of the constrained variational problem of minimizing $f(u)$ with $h(u)$ held constant or the dual problem of minimizing $h(u)$ with $f(u)$ held constant. We have already spoken in the very last part of §1 of constrained variational problems, i.e. minimizing functionals over convex subsets of V. We now consider a related but different problem, minimizing a functional over a level set of another functional.

Let us consider this problem in a general form.

LEMMA 6. *Let V be a Banach space, f and h two real functionals on V which are once continuously differentiable in the sense of Fréchet. If c is a real number, the level surface of the functional f at level c is*

$$M_c(f) = \{u | u \in V, f(u) = c\}.$$

If we suppose that $f'(u) \neq 0$ *for any* u *in* $M_c(f)$, *then* $M_c(f)$ *is a manifold of codimension* 1 *in* V, *and we can consider the function* h_0 *on* $M_c(f)$ *obtained by restricting the function* h *to* $M_c(f)$. *Then the critical points of* h_0, *the restriction* h *to* $M_c(f)$, *are solutions of the equation*

$$h'(u) = \lambda f'(u).$$

PROOF OF LEMMA 6. $M_c(f)$ is locally a C^1 submanifold of V and h_0, being the restriction of a C^1 function, itself is once continuously differentiable on $M_c(f)$. A critical point u of h_0 is by definition a point at which $(h_0)'(u) = 0$.

The tangent space to $M_c(f)$ at a point u in that manifold can be identified with a subspace of V, namely the subspace $V_u = \{v | (f'(u), v) = 0\}$. For v in V_u, we have $((h_0)'(u), v) = (h'(u), v)$ since h and h_0 coincide on $M_c(f)$. We choose an element v_0 of V such that $(f'(u), v_0) = 1$, which is possible because we assume that $f'(u) \neq 0$. Then any element w of V can be written uniquely in the form

$$w = c(w)v_0 + v, \; c(w) \in R^1, \; v \in V_u.$$

Indeed, we set $c(w) = (f'(u), w)$ and note that

$$(f'(u), w - c(w), v_0) = (f'(u), w) - c(w)(f'(u), v_0) = 0$$

so that $v = w - c(w)v_0$ lies in V_u. The uniqueness of the decomposition follows similarly by applying $f'(u)$ to both sides of the decomposition.

If u is a critical point of h_0, then $(h'(u), v) = (h_0)'(u), v) = 0$ for all v in V_u, and conversely the vanishing of $(h'(u), v)$ for all v in V_u implies that u is a critical point of h_0. For such a point u and any w in V, we see that

$$(h'(u), w) = (h'(u), c(w)v_0 + v) = c(w)(h'(u), v_0)$$

where $v \in V_u$ and $(h'(u), v) = 0$. Here $c(w) = (f'(u), w)$. Hence for all w in V,

$$(h'(u), w) = \lambda(f'(u), w); \quad \lambda = (h'(u), v_0) \in R^1$$

i.e. $h'(u) = \lambda f'(u)$. Q.E.D.

From the formal point of view, the relation between f and h in Lemma 6 is symmetric, and if we know that on a level set of h, $M_c(h)$, $h'(u) = 0$, then the critical points of the functional f^0 obtained by restricting f to $M_c(h)$ are solutions u of the eigenvalue problem $f'(u) = \lambda h'(u)$, $\lambda \in R^1$. In the concrete context of elliptic eigenvalue problems, the hypotheses which we impose on the character of f and h are very far from symmetric and the topological properties of their level surfaces as subsets of V and of their derivatives f' and h' are quite different for the two functions. Nevertheless, we shall show that under both types of normalization, it is possible to obtain results on critical points of the pair (A, B) and we shall obtain normalized eigenfunctions lying on the level surface $M_c(f)$ as well as normalized eigenfunctions lying on $M_c(h)$.

Let us now describe in detail the hypotheses which we impose upon the functions defining the elliptic eigenvalue problem:

ASSUMPTIONS (C). (1) *F* and *H* *are two functions,* $F: G \times R^{s_m} \to R^1, H: G \times R^{s_{m-1}} \to R^1$. *For each fixed* ξ *in* R^{s_m}, $F(\cdot, \xi)$ *is measurable on G. For each fixed x in the complement of a null set in G,* $F(x, \cdot)$ *is once continuously differentiable in G. For each fixed* η *in* $R^{s_{m-1}}$, $H(\cdot, \eta)$ *is measurable on G. For each fixed x in the complement of a null set in G,* $H(x, \cdot)$ *is once continuously differentiable on* $R^{s_{m-1}}$.

For a given exponent p with $1 < p < \infty$, *the functions F and H satisfy the following inequalities:*

$$|F(x, \xi_m)| \leq c(\xi_b)(x) + c_1(\xi_b) \sum_{m-n/p \leq |\alpha| \leq m} |\xi_\alpha|^{s_\alpha},$$

$$|H(x, \eta_{m-1})| \leq c(\eta_b)(x) + c_1(\eta_b) \sum_{m-n/p \leq |\beta| \leq m-1} |\eta_\beta|^{t_\beta},$$

where $s_\alpha^{-1} \geq p^{-1} - n^{-1}(m - |\alpha|)$, $s_\beta < +\infty$, $t_\beta < s_\beta$, *b is the greatest integer less than* $m - n/p$, c_1 *is a continuous function from* R^{s_b} *to* R^1, *and c is a continuous function from* R^{s_b} *to* $L^p(G)$.

(2) *If* $F_\alpha = \partial F/\partial \xi_\alpha$ *for* $|\alpha| \leq m$, $H_\beta = \partial H/\partial \eta_\beta$ *for* $|\beta| \leq m - 1$, *then* F_α *should satisfy the three conditions of Assumptions* (B) *of* §1 *corresponding to the exponent p.* $H_\beta(x, \eta)$ *should satisfy the following inequality:*

$$|H_\beta(x, \eta)| \leq c_\beta(\eta_b)(x) + c_1(\eta_b) \sum_{m-n/p \leq |\alpha| \leq m-1} |\eta_\alpha|^{q_{\alpha\beta}}$$

where c_1 *is a continuous function from* R^{s_b} *to* R^1, c_β *is a continuous function from* R^{s_b} *to* $L^{q_\beta}(G)$, *and the exponents* q_β *and* $q_{\alpha\beta}$ *satisfy the inequalities:*

$$q_\beta > (s_\beta)', \text{ where } s_\beta^{-1} = p^{-1} - (m - |\beta|)n^{-1}, \quad \text{for } |\beta| \geq m - n/p,$$

$$q_\beta = 1, \quad \text{for } |\beta| < m - n/p,$$

$$q_{\alpha\beta} < s_\alpha(s'_\beta)^{-1}; \quad |\beta| \geq m - n/p,$$

$$q_{\alpha\beta} \leq s_\alpha; \quad |\beta| < m - n/p.$$

LEMMA 7. *Let G be a bounded, smoothly bounded open subset of* R^n, *F and H two functions satisfying the conditions of Assumptions* (C), *for a given exponent p,* $1 < p < \infty$, *and set*

$$f(u) = \int_G F(x, \xi_m(u)(x))dx,$$

$$h(u) = \int_G H(x, \xi_{m-1}(u)(x))dx,$$

$$a(u, v) = \sum_{|\alpha| \leq m} (F_\alpha(\cdot, \xi_m(u)), D^\alpha v),$$

$$b(u, v) = \sum_{|\beta| \leq m-1} (H_\beta(\cdot, \xi_{m-1}(u)), D^\beta v).$$

Then

(a) *If V is a closed subspace of* $W^{m,p}(G)$, *then f and h are once differentiable functionals on V, and their derivatives* f' *and* h' *satisfy the equations*

$$(f'(u), v) = a(u, v), \qquad (h'(u), v) = b(u, v) \quad (u, v \in V).$$

(b) *h is a weakly continuous functional on bounded subsets of V and h' is completely continuous from V to V^*, i.e. if u_n converges weakly to u in V, then $h'(u_n)$ will converge strongly to $h(u)$ in V^*.*

(c) *f' is a continuous mapping of V into V^* which maps bounded sets into bounded sets and satisfies the condition:*

(S) *If $\{u_n\}$ is a weakly convergent sequence in V with limit u and if $(f'(u_n) - f'(u), u_n - u) \to 0$, then u_n converges strongly to u in V.*

(d) *f is bounded on bounded sets and $f(u) \to +\infty$ as $||u|| \to +\infty$. f' is coercive and $(f'(u), u) \to +\infty$ as $||u|| \to +\infty$.*

PROOF OF LEMMA 7. Using the Sobolev imbedding theorem (Theorem 3 of §1) along the lines of the proof of Lemmas 3 and 3′ in the Appendix to §1, we verify directly that under the conditions given in Assumptions (C), $f(u)$ and $h(u)$ are well defined and continuous real valued functionals on $W^{m,p}(G)$ and hence on V and that h is weakly continuous on bounded subsets of V. Furthermore, f is bounded on bounded sets and so is h. Similarly, we verify along the lines of Lemma 3 that $a(u, v)$ and $b(u, v)$ are well defined for all u and v in V, and that there exist continuous mappings T and C from V to V^* such that

$$a(u, v) = (Tu, v), \; b(u, v) = (Cu, v) \quad (u, v \in V).$$

T is a continuous mapping of V into V^* which carries bounded sets into bounded sets, is coercive, and satisfies condition (S). C is a completely continuous mapping of V into V^* which carries bounded sets into bounded sets. To complete the proof of Lemma 7, we must show that $f' = T$, $h' = C$, and $f(u) \to +\infty$ as $||u|| \to +\infty$.

To prove that $f' = T$, we consider u and v in V, and note that

$$f(u + tv) - f(u) = \int_G [F(x, \xi(u)(x) + t\xi(v)(x)) - F(x, \xi(u)(x))]$$

$$= t \int_G \int_0^1 \sum_{|\alpha| \le m} F_\alpha(x, \xi(u) + st\xi(v)(x)) D^\alpha v(x) ds\, dx.$$

Hence as $t \to +\infty$, $t^{-1}(f(u + tv) - f(u)) \to a(u, v)$ by dominated convergence. Similarly $t^{-1}(h(u + tv) - h(u)) \to b(u, v)$ as $t \to 0$.

To show that $f(u) \to +\infty$ as $||u|| \to +\infty$, we note that since f' is coercive, there exists $R > 0$ and $c > 0$ such that for all u in V with $||u|| \ge R$, $(f'(u), u) \ge c > 0$. Let u be a point in the exterior of the ball of radius R about the origin in V. On the ball of radius R, there exists a constant M such that $|f(v)| \le M$, for $||v|| = R$. The element u may be written in the form $u = rv$, $r \ge 1$, $||v|| = R$. Then

$$f(u) - f(v) = f(rv) - f(v) = \int_1^r (f'(sv)v) dx$$

$$= \int_1^r s^{-1}(f'(sv), sv) dx \ge c \int_1^r s^{-1} ds \ge c \log(r),$$

i.e.

$$f(u) \geq \log(\|u\|/R) - M \to \infty, \quad \|u\| \to \infty. \text{ Q.E.D.}$$

We shall apply Lemma 7 together with a number of abstract theorems about eigenfunctions to obtain results for nonlinear elliptic eigenvalue problems.

For the existence of a single normalized eigenfunction, we have the following relatively simple abstract theorems.

THEOREM 14. *Let V be a separable reflexive Banach space, f and h two C^1 functions on V with real values such that h is weakly continuous on bounded subsets of V, h' is compact and f' satisfies condition* (S) *and maps bounded sets into bounded sets. Suppose that for a given constant c, the level set $M_c(f) = \{u | u \in V, f(u) = c\}$ is bounded and that for u in $M_c(f)$, $(f'(u), u) \neq 0$. Suppose further that there exists a point u_0 in $M_c(f)$ and a constant $c > 0$ such that for all u in $M_c(f)$ for which $h(u) \geq h(u_0)$, $(h'(u), u) \geq c$.*

Then h assumes its maximum on $M_c(f)$ at a point u which is a solution of the equation $f'(u) = \lambda h'(u)$ for some real number λ.

THEOREM 15. *Let V be a separable reflexive Banach space, f and h C^1 functions from V to the real line with h weakly continuous, h' a compact mapping of V into V^*, and f' satisfying condition* (S). *Suppose that f' maps bounded sets into bounded sets, that $f(u) \to +\infty$ as $\|u\| \to +\infty$, and that for a given constant c and points u in the level set $M_c(h) = \{u | u \in V, h(u) = c\}$, $(h'(u), u) > 0$ and for each $R > 0$, there exists $c(R) > 0$ such that $(h'(u), u) \geq c(R)$ for u in $M_c(h)$ with $\|u\| \leq R$.*

Then f assumes its minimum on the set $M_c(h)$ at a point u which is a solution of the eigenvalue equation $f'(u) = \lambda h'(u)$ for some real number λ.

Translating Theorems 14 and 15 into results on nonlinear elliptic eigenvalue problems, we have the following:

THEOREM 16. *Let G be a bounded, smoothly bounded open set in R^n, $A(u)$ and $B(u)$ two qusilinear partial differential operators of order $2m$ and $s \geq 2m - 2$ respectively on G in the divergence form*

$$A(u) = \sum_{|\alpha| \leq m} (-1)^{|\alpha|} D^\alpha F_\alpha(\cdot, \xi_m(u)),$$

$$B(u) = \sum_{|\beta| \leq m-1} (-1)^{|\beta|} D^\beta H_\beta(\cdot, \xi_{m-1}(u)),$$

where $a(u, v)$, $b(u, v)$ are the Dirichlet forms of A and B, respectively, $F_\alpha = \partial F/\partial \xi_\alpha$, $H_\beta = \partial H/\partial \eta_\beta$ and $F(x, \xi)$ and $H(x, \eta)$ are two functions satisfying the conditions of Assumptions (C) *for a given exponent p, $1 < p < \infty$. Let V be a closed subspace of the corresponding Sobolev space $W^{m,p}(G)$ with V containing the testing functions in G, and let f and h be the functionals defined by*

$$f(u) = \int_G F(x, \xi_m(u)(x))dx, \quad h(u) = \int_G H(x, \xi_{m-1}(u)(x))dx.$$

Let c be a real number such that for $f(u) = c$, $a(u, u) \neq 0$ *and suppose that there exists* $c_0 > 0$ *and* u_0 *with* $f(u_0) = c$ *such that for u in* $M_c(f)$ *with* $h(u) \geq h(u_0)$, $b(u, u) \geq c_0$.

Then h assumes its maximum on the set $M_c(f)$ *at a point v which is an eigenfunction for the eigenvalue problem* $A(v) = \lambda B(v)$, λ *real, with null variational boundary conditions corresponding to V, i.e.* $a(u, w) = \lambda b(u, w)$, $(w \in V)$.

THEOREM 17. *Let G be a bounded, smoothly bounded open set in* R^n, $A(u)$ *and* $B(u)$ *two quasilinear partial differential operators of order* $2m$ *and* $s \leq 2m - 2$, *respectively, in the divergence form*

$$A(u) = \sum_{|\alpha| \leq m} (-1)^{|\alpha|} D^\alpha F_\alpha(\cdot, \xi_m(u)),$$

$$B(u) = \sum_{|\beta| \leq m-1} (-1)^{|\beta|} D^\beta H_\beta(\cdot, \xi_{m-1}(u)).$$

Suppose that $F_\alpha = \partial F/\partial \xi_\alpha$, $H_\beta = \partial H/\partial \eta_\beta$, *where* $F(x, \xi)$ *and* $H(x, \eta)$ *are two functions satisfying the conditions of Assumptions* (C) *for a given exponent p with* $1 < p < \infty$. *Let* $a(u, v)$ *and* $b(u, v)$ *be the Dirichlet forms corresponding to the divergence representations of A and B, respectively, i.e.*

$$a(u, v) = \sum_{|\alpha| \leq m} (F_\alpha(\cdot, \xi_m(u)), D^\alpha v),$$

$$b(u, v) = \sum_{|\beta| \leq m-1} (H_\beta(\cdot, \xi_{m-1}(u)), D^\beta v).$$

Let V be a closed subspace of $W^{m,p}(G)$ *containing* $W_0^{m,p}(G)$, *and let f and h be the functionals on V defined by*

$$f(u) = \int_G F(x, \xi_m(u)(x))dx, \qquad h(u) = \int_G H(x, \xi_{m-1}(u)(x))dx.$$

Suppose that for a given real number c, and all u in $M_c(h) = \{u | u \in V, h(u) = c\}$, $b(u, u) > 0$, and $a(u, u) \geq k(\|u\|_{m,p}) > 0$ *for all u in* $M_c(h)$.

Then f assumes its minimum on $M_c(h)$ *at a point u which is a solution of the eigenvalue problem* $A(u) = \lambda B(u)$ *with null boundary conditions of variational type with respect to V.*

We remark that Theorems 16 and 17 are translations of Theorems 14 and 15 in terms of elliptic eigenvalue problems, using Lemma 7. It therefore suffices to prove Theorems 14 and 15.

PROOF OF THEOREM 14. Since V is separable, we may construct an increasing sequence $\{V_n\}$ of finite dimensional subspaces of V such that the union of the V_n is dense in V and if $M_{c,n} = M_c(f) \bigcap V_n$, then the union of the $M_{c,n}$ is dense in $M_c(f)$.

Let f_n be the restriction of f to V_n, h_n the restriction of h to V_n. Then $M_{c,n} = \{u | u \in V_n, f_n(u) = c\} = M_c(f_n)$. Let ψ_n be the injection mapping of V_n in V, ψ_n^* the projection mapping of V^* on V_n^*. Then it follows easily that $f'_n = \psi_n^* f' \psi_n$, $h'_n = \psi_n^* h' \psi_n$, where f'_n, h'_n are the derivatives of f_n and h_n, respectively, and are continuous mappings of V_n into V_n^*.

Let S be the supremum of the values of h on $M_c(f)$, S_n the supremum of the values of h_n on $M_{c,n}$. Since h_n is the restriction of h to V_n and the union of $M_{c,n}$ is dense in $M_c(f)$, $\{S_n\}$ is an increasing sequence converging to S.

For u in $M_{c,n}$, $(f'_n(u), u) = (f'(u), u) \neq 0$. This shows that each $M_{c,n}$ is a manifold of codimension 1 in V_n. Since f is continuous, $M_c(f_n)$ is a closed subset of V_n. $M_{c,n}$ is contained in $M_c(f)$ which is bounded by hypothesis. Thus each $M_{c,n}$ is a compact manifold. In particular, h_n assumes its maximum on $M_{c,n}$ on at least one point u_n, and for any such u_n, $h'_n(u_n) = \lambda_n f'_n(u_n)$, $h(u_n) = S_n$, $f(u_n) = c$. We may assume without any loss of generality that u_0 of the hypothesis lies in V_n for all n. Then it follows since $h(u_n) \geq h(u_0)$ that $(h'(u_n), u_n) \geq c > 0$, with c independent of n.

Since $(h'_n(u_n), u_n) = (h'(u_n), u_n)$ and $(f'_n(u_n), u_n) = (f'(u_n), u_n)$, we have $(h'(u_n, u_n) = \lambda_n(f'(u_n), u_n)$. The u_n lies in $M_c(f)$ and are uniformly bounded, $(h'(u_n), u_n) \geq c$, and since f' maps bounded sets into bounded sets, $|(f'(u_n), u_n)| \leq k_0$. Hence $|\lambda_n| \geq c_1 > 0$ for all n. In particular,

$$f'_n(u_n) = k_n h'_n(u_n), \quad |k_n| \leq K \text{ for all } n.$$

Since V is reflexive and $\{u_n\}$ is bounded, we may pass to an infinite subsequence and assume that u_n converges weakly to u in V. Since h is weakly continuous on bounded sets, $h(u_n) = S_n \to S$, $h(u) = S$. We wish to prove that u lies in $M_c(f)$ and that u is an eigenfunction of the pair (f', h').

Suppose that v is a point of the space V_m for some m and consider $n \geq m$. Since h' is a compact mapping, we may assume that $h'(u_n)$ converges strongly in V^* to some element w. Since the numbers k_n are bounded, after passing to a subsequence, we may assume that $k_n \to k$ as $n \to \infty$. Now consider

$$(f'(u_n), u_n - v) = (f'_n(u_n), u_n - v) = k_n(h'_n(u_n), u_n - v) = k_n(h'(u_n), u_n - v)$$

where $(h'(u_n), u_n - v) \to (w, u - v)$ since $h'(u_n)$ converges strongly to w and $u_n - v$ converges weakly to $u - v$. Thus $(f'(u_n), u_n - v) \to k(w, u - v)$ for any v in the dense union of the spaces V_m.

Let v be any element of V. Then we can find v^j in some V_m such that $\|v - v^j\| < \varepsilon$. Then

$$(f'(u_n), u_n - v) - k(w, u - v) = (f'(u_n), u_n - v^j) - k(w, u - v^j) + (f'(u_n) - kw, v^j - v).$$

The last term can be estimated by

$$|(f'(u_n) - kw, v^j - v)| \leq [\|f'(u_n)\| + |k|\|w\|] \quad \|v^j - v\| \leq M\varepsilon.$$

The other term goes to zero. Hence

$$\overline{\lim} |f'(u_n), u_n - v) - k(w - v)| \leq \varepsilon$$

and therefore $(f'(u_n), u_n - v) \to k(w, u - v)$, $v \in V$.

In particular, we may set $v = u$. Then we obtain $(f'(u_n) - f'(u), u_n - u) \to 0$ since $(f'(u), u_n - u) \to 0$ by the weak convergence of $u_n - u$ to 0, and $(f'(u_n), u_n - u) \to k(w, u - u) = 0$.

Applying the property (S) for f', it follows that u_n converges strongly to u. Hence $f(u_n) \to f(u)$ and since $f(u_n) = c$, $f(u)$ must equal c and $u \in M_c(f)$. We already know that $h(u) = S$, so that u is a maximum point of h on $M_c(f)$. Since $M_c(f)$ is a manifold with $f'(u) \neq 0$, it follows that

$$h'(u) = \xi f'(u), \quad \xi \in R^1.$$

Since $(h'(u), u) = \xi(f'(u), u)$, $\xi \neq 0$, and setting $\lambda = \xi^{-1}$, we obtain $f'(u) = \lambda h'(u)$. Q.E.D.

PROOF OF THEOREM 15. We again construct an increasing sequence $\{V_n\}$ of finite dimensional subspaces of V whose union is dense in V. If $M_{c,n} = V_n \cap M_c(h)$, in this case, we may assume that the union of the $M_{c,n}$ is dense in $M_c(h)$. Let f_n and h_n be the restrictions of f and h to V_n. Then $M_{c,n} = M_c(h_n) = \{u | u \in V_n, h_n(u) = h(u) = c\}$. For each u in V_n, $(h_n'(u), u) = (h'(u), u) \neq 0$. It follows that each $M_{c,n}$ is a manifold as well as $M_c(h)$.

Let m be the infimum of f on $M_c(h)$, and for each n, let m_n be the inf of f on $M_{c,n}$. If v_0 is a point in $M_{c,1}$, then since $f(u) \to +\infty$ as $||u|| \to \infty$, there exists $R > 0$ such that for $||u|| \geq R$, $f(u) > f(v_0)$. Hence the infimum of f on $M_{c,n}$ coincides with its infimum on $M_{c,n} \cap B_R(0)$, and since $M_{c,n}$ is closed in V_n, this latter set is compact. Hence there exists at least one point u_n in $M_{c,n}$ such that $f(u_n) = m_n$. We note that since the union of the $M_{c,n}$ is dense in M_c, m_n is a decreasing sequence which converges to m. Since $M_{c,n}$ is a manifold on which $h_n'(u) \neq 0$, then at the extremum point u_n for f_n on $M_{c,n} = M_c(h_n)$, we have $f_n'(u_n) = \lambda_n h_n'(u_n)$.

By the hypothesis, $(h'(u_n), u_n) \geq c(R) > 0$ for all n, since $||u_n|| \leq R$. Since $(f'(u_n), u_n) = \lambda_n(h'(u_n), u_n)$, and since $(f'(u_n), u_n)$ is bounded because f' maps bounded sets into bounded sets, $|\lambda_n| \leq M_0$ for some constant M_0 and all n. If we pass to an infinite subsequence and use the compactness of h', we may assume that $\lambda_n \to \lambda$ and $h'(u_n)$ converges strongly to w. By the reflexivity of V, we may assume that u_n converges weakly to some u in V. Since h is weakly continuous on bounded sets, $h(u_n) \to h(u)$, so that $h(u) = c$ and u lies in $M_c(h)$.

Let v be an element of some space V_m. Then for $n > m$, we have

$$\begin{aligned}(f'(u_n), u_n - v) &= (f_n'(u_n), u_n - v) = \lambda_n(h_n'(u_n), u_n - v) \\ &= \lambda_n(h'(u_n), u_n - v) \to \lambda(w, u_n - v).\end{aligned}$$

Arguing as in the proof of Theorem 14, we note from the uniform boundedness of $||f'(u_n||$ that for all v in V, $(f'(u_n), u_n - v) \to \lambda(w, u - v)$. In particular, $(f'(u_n), u_n - u) \to 0$, and hence $(f'(u_n) - f'(u), u_n - u) \to 0$. Applying the property (S) for f', it follows that u_n converges strongly to u. Hence $f(u_n) = m_n \to f(u)$, so that $f(u) = m = \min\{f(u) : u \in M_c(h)\}$. Hence u is an eigenfunction of $f'(u) = \lambda h'(u)$. Q.E.D.

We now come to the qualitatively most interesting part of the nonlinear eigenvalue theory of variational type, the application of the ideas of the Lusternik-Schnirelman theory for the existence of critical points of functions on manifolds. In Theorems 14, 15, 16, and 17 we established, first in a general Banach space context and then in a specialization to classes of partial differential operators, results on the existence of critical points of functionals which were extrema of the functional and thereby correspond to the lowest order eigenvalues of the linear theory. If we wish to establish the existence of more eigenfunctions in a systematic way, we must consider techniques for getting hold of higher order eigenfunctions or unstable critical points. Such a technique is provided by the topological theory of critical points initiated by Lusternik and Schnirelman in the early 1920's and whose abstract form has been thoroughly discussed in the lectures of Palais.

The connection of the Lusternik-Schnirelman theory with the problem of constructing higher order eigenfunctions of nonlinear operators was clear from the inception of the theory. In one of his earliest publications, Lusternik (Monatsh. Math., 1923) brought out in an explicit way the connection of the concept of *category in the Lusternik-Schnirelman sense* and the classical *minimax* principle of Poincaré-Fisher-Weyl-Courant *et. al.* which serves in the linear cases as the most convenient tool for deriving and studying the higher order eigenfunctions by a variational method.

In the context of eigenvalue problems connected with quasilinear elliptic partial differential operators, we can only apply the results of the general Lusternik-Schnirelman theory on infinite dimensional manifolds after some additional reworking to put them in a suitably general form. In addition, to get the results for elliptic problems in their full generality it is necessary for us in addition to reconstruct the passage from the finite to the infinite dimensional situation by a method different from that which Palais has employed in the general theory on Finsler manifolds. Thus the discussion which we give below contains two relatively independent approaches for applying the Lusternik-Schnirelman principle to nonlinear eigenvalue theory. (1) The verification in certain cases of the basic assumptions of the general theory for functions on Finsler manifolds satisfying the Palais-Smale Condition (C) and the discussion for the eigenvalue theory of how to establish Condition (C) for the functions involved. (2) The derivation of a Lusternik-Schnirelman theory for submanifolds of a Banach space directly by a Galerkin approximation argument of the type which we have employed in the preceding discussion to establish existence theorems for solutions of boundary value problems and for critical points assuming extrema.

Let us now restate the conclusions of the general Lusternik-Schnirelman theory of Finsler manifolds as Palais has derived them in his lectures and point out what features of these results and of the technical arguments used in their derivation create difficulties for their application to the study of nonlinear elliptic eigenvalue problems. We consider a C^2 Finsler manifold M, modelled on a Banach space V_0. The Finsler structure on such a manifold is given by assigning a norm to the tangent space at each point which is equivalent to the given norm on the Banach space V^0 and where the equivalence varies continuously over the points of the manifold.

The Finsler norm on the tangent space $T_x(M)$ at each point x of M induces the dual norm on $T_x^*(M) = (T_x(M))^*$, the cotangent space of M at x. In particular, if h is a real-valued function on M, then for each x, $h'(x)$, $(=dh(x)$ in Palais' notation), is an element of $T_x^*(M)$ and $||h'(x)||$ is determined by the Finsler structure. In addition, we define a metric on M by defining the length of a C^1 curve given by $\{x = x(t), 0 \leq t \leq a\}$ to be $\int_0^a ||x'(t)||dt$ and letting the distance between two points be the infimum of the length of curves in M which join the two points.

DEFINITION. *Let h be a C^1 function from M to the reals, where M is a C^1 manifold with Finsler structure. Then h is said to satisfy Condition* (C) *on M if for any sequence $\{u_n\}$ in M for which $h(u_n)$ is bounded from above and for which $||h'(u_n)|| \to 0$, we can extract an infinite subsequence $\{u_n\}$ converging in the metric of M.*

Using this definition, the generalized Lusternik-Schnirelman principle in the form given by Palais asserts the following:

(I) *Let M be a C^2 manifold with a Finsler structure, h a real valued C^1 function on M such that h is bounded from below, h satisfies Condition* (C), *and for any real constant c, the subspace*

$$N_c(h) = \{u \mid u \partial M, h(u) \leq c\}$$

is complete in the metric of M. Then if $n(h)$ is the number of critical points of h and if cat (M) *is the Lusternik-Schnirelman category of M,*

$$n(h) \geq \operatorname{cat}(M).$$

(II) *If h is the function described in* (I) *and m is any integer* $\leq$ cat (M), *let*

$$c_m(h) = \inf_{\substack{K \text{ compact} \\ \operatorname{cat}_M(K) \geq m}} \max_{u \in K} h(u).$$

(*We remind the reader that the* cat (M) *is the least number of closed sets C, in M which are contractible in M such that $M = \bigcup_j C_j$, while $\operatorname{cat}_M(K)$ is the least number of closed subsets K_j of K, each contractible in M such that $K = \bigcup_j K_j$.*)

Then there exists a critical point u_m of h such that

$$h(u_m) = c_m(h).$$

The assertion of (II) is the generalized minimax principle of Lusternik, can be shown to include the minimax principle in the ordinary sense in the linear case for finite dimensional and compact selfadjoint operators, and, as we see in detail below, provides an extremely important tool for carrying through perturbation and limit arguments on the higher order eigenfunctions. (This is of course its primary importance in the linear case, as was first shown clearly by Hermann Weyl in 1912.)

The difficulty in applying the above results to the manifolds which we construct in the eigenvalue theory is the requirement that the manifold M be of class C^2. Though this can be assured in some cases, it requires additional and extraneous hypotheses on both the regularity of the functions defining the elliptic eigenvalue problems and, more important, a complete new set of hypotheses on the growth of these functions for large arguments in ξ_m. Even if one imposes such requirements, it seems impossible to assure this fact for the M involved if $p \leq 2$, though for $p = 2$, one can modify the above result slightly by changing the condition that M be of Class C^2 to the slightly weaker condition that M be of class C^{2-}, where a mapping in a Banach space is of class C^{2-} if it is of class C^1 and its first derivative is locally Lipschitzian. (For the eigenvalue theory done in this style, see the author's paper in Ann. of Math., 1965.)

The requirement that M be C^2 or C^{2-} seems to be firmly rooted in the general result for Finsler manifolds because of the technical problems involved in the proof of the Lusternik-Schnirelman principle, namely the construction of pseudo-gradient vector fields on M which define a flow on M. We know that such vector fields in general must be locally Lipschitzian if they are to define an unique flow, and barring special hypotheses which produce continuous vector fields on the infinite dimensional manifold M for which an unique flow exists, the existence of locally Lipschitzian pseudo-gradient vector fields makes sense only if M is of class C^{2-} so that the tangent bundle to M has a well-defined locally Lipschitzian structure.

There are two ways to get past this obstacle which we present in the present discussion, one argument in a relatively abstract form which suffices for our eigenvalue problems if $p \geq 2$, the second a reconstruction of the infinite dimensional theory for submanifolds of a Banach space which works for all p with $1 < p < \infty$.

The first direction of argument is based upon the following simple general theorem.

THEOREM 18. *Let M be a C^1 manifold with Finsler structure modelled upon a Banach space V_0, h a C^1 function from M to R^1 such that h is bounded from below on M and h satisfies Condition* (C) *on M. Suppose that there exists a C^1 diffeomorphism g of M onto another Finsler manifold M_1 of class C^{2-} with M_1 complete in its Finsler metric and there exists a constant $c > 0$ such that for any v in $T_x(M)$, $x \in M$, $||dg_x(v)||_{g(x)} \leq c||v||$.*

Then the Lusternik-Schnirelman principle (I) *and* (II) *above holds for h on M.*

In particular, it follows from Theorem 18 that the Lusternik-Schnirelman principle holds for all C^1 functions on all compact C^1 manifolds.

PROOF OF THEOREM 18. We form a new function h_1 on M_1 by composing h with the diffeomorphism g_1 of M_1 on M, $g_1 = g^{-1}$, $h_1 = hg_1$. This function h_1 is of class C^1 on the C^2 manifold M_1, and x_1 in M_1 is a critical point of h_1 in M_1 with $h_1(v_n) \leq K_0$ and $\|h_1'(v_n)\| \to 0$. Let $u_n = g_1(v_n)$. Then $h(u_n) = h_1(v_n) \leq K_0$, logical invariant, cat $(M) =$ cat (M_1), and the class of compact subsets K such that $\text{cat}_K(M) \geq m$ is mapped by g on the class of compact subsets K_1 of M_1 such that $\text{cat}_{M_1}(K_1) \geq m$. To prove the Lusternik-Schnirelman principle for h on M, it therefore suffices to show that the new function h_1 on the C^{2-} manifold M_1 satisfies Condition (C).

By hypothesis, h satisfies Condition (C) on M. Suppose $\{v_n\}$ is a sequence in M_1 with $h_1(v_n) \leq K_0$ and $||h_1'(v)_n|| \to 0$. Let $u_n = g_1(v_n)$. Then $h(u_n) = h_1(v_n) \leq K_0$, and since $h = h_1 g$, $h'(u_n) = g_{v_n}^*(h_1'(v_n))$. Here g_v^* is the linear mapping of $T_v^*(M_1)$ onto $T_{g_1(v)}^*(M)$ which is dual to the mapping $dg_{g_1(v)}$ mapping $T_{g_1(v)}(M)$ on $T_v(M_1)$. By our hypothesis $||dg_u|| \leq c$ for all points u on M. Hence $||g_v^*|| \leq c$ for all points v of M_1. On the other hand,

$$||h'(u_n)|| \leq ||g_{v_n}^*||\,||h_1'(v_n)|| \leq c||h_1'(v_n)|| \to 0.$$

Since h satisfies Condition (C) on M, we can extract an infinite subsequence $\{u_{n_j}\}$ which converges to some u in M. Since g is continuous, the corresponding subsequence $v_{n_j} = g(u_{n_j})$ converges to $v = g(u)$ in M_1.

Thus Condition (C) holds for h_1 on M_1, and the proof of Theorem 18 is complete.

We apply Theorem 18, using the following simple prototype of a C^1 diffeomorphism g:

THEOREM 19. *Let V be a Banach space whose norm is C^1 on its unit sphere, M a C^1 submanifold in V which is the level surface $M_c(f)$ of a C^1 function f on V. Suppose that $(f'(u), u) > 0$ for u on M. Suppose further that each ray from the origin hits M in exactly one point. Let g be the mapping of M on the unit sphere given by $g(u) = u/||u||$.*

Then g is a C^1 diffeomorphism of M on the unit sphere $S_1(V)$ of V and there exists a constant $c > 0$ independent of f such that for any u in M and any v in $T_u(M)$, then for the Finsler structures on M and $S_1(V)$ induced by the norm in V, $||dg_u(v)|| \leq c||v||$.

It follows from Theorem 19 that if V has a unit sphere which is a manifold of class C^{2-}, we can use Theorem 19 together with Theorem 18 to get rid of the difficulty in applying the Lusternik-Schnirelman theory to functions on C^1 manifolds.

PROOF OF THEOREM 19. By hypothesis, the norm function $g(u)$ is of class C^1 on the complement of 0 in V. Hence the unit sphere $S_1(V)$ is a C^1 submanifold of V. Let v be a point of $S_1(V)$, and let $K(u)$ be the gradient of the norm function at v, so that $K(v) \in V^*$,

$$(K(v), v) = ||v||, \quad ||K(v)|| = 1.$$

The tangent space to $S_1(V)$ at v is the subspace $T_v = \{u|\,(K(v), u) = 0\}$. For any element w of V, w may be written uniquely in the form $w = cv + t_v$, $t_v \in T_v$, where $c = (K(v), w)/||v|| = (K(v), w)$.

The mapping g may be defined as a continuous mapping of $V - \{0\}$ onto $S_1(V)$, and is obviously C^1 on the complement of the origin in V. The derivative $g_u' = dg_u$ of g at a point u of V maps the tangent space to $V - \{0\}$ at u into T_v, $v = g(u)$, by setting

$$g_u'(w) = t_{g(u)} = w - (K(g(u)), w)v = w - (K(u), w)u/||u||.$$

Hence $\|g'(w)\| \leq \|w\| + \|K(u)\|\,\|w\| \leq 2\|w\|$. Hence the same inequality holds for all tangent vectors to M at u.

To complete the proof of Theorem 19, it suffices to show that under the hypothesis $(f'(u), u) \neq 0$ at each point u of $M = M_c(f)$, g is a diffeomorphism of M on $S_1(V)$. We know already that g maps M injectively onto $S_1(V)$ and that g is continuous. It suffices therefore to show that $g'_u = dg_u$, at each u of M, is an isomorphism of $T_u(M)$ with $T_{g(u)}(S_1(V))$.

We identify each of these tangent spaces with its image in V under the imbedding maps of M and $S_1(V)$ into V. Then $T_u(M) = \{w | w \in V, \ (f'(u), w) = 0\}$ while $T_{g(u)}(S_1(V)) = \{z | \, z \in V, (K(u), z) = 0\}$. Let w be an element of $T_u(M)$. Then

$$w = (K(u), w)^u/\|u\| + t_u(w), \quad (t_u(w) \in T_{g(u)}(S_1(V))),$$

i.e. $w = (K(u), w)u/\|u\| + g'_u(w)$.

We note first that g'_u is an injective map of $T_u(M)$ into $T_{g(u)}(S_1(V))$ since if for a given w, $g'_u(w) = 0$, we should have $w = cu$, $(c \in R^1)$, while $0 = (f'(u), w) = c(f'(u), u)$ implies that $c = 0$ and $w = 0$. Hence the kernel of $g'_u = \{0\}$.

To show that g'_u is a surjective mapping of $T_u(M)$ on $T_{g(u)}(S_1(V))$, let z be an element of $T_{g(u)}(S_1(V))$. We seek to construct an element w of $T_u(M)$ such that $g'_u(w) = z$. Such an element w may be taken in the form $w = cu + z$. In order that w lies in $T_u(M)$, it suffices that $(f'(u), w) = 0$, i.e. $c(f'(u), u) + (f'(u), z) = 0$. Since $(f'(u), u) \neq 0$, we may choose $c = (f'(u), z)/(f'(u), u)$, and the proof of surjectivity is complete.

Thereby, the proof of Theorem 19 is complete. Q.E.D.

In order to obtain nontrivial applications of the Lusternik-Schnirelman theory to the existence of eigenfunctions we must construct a function $\tilde{h}$ which satisfies Condition (C) on a manifold $\tilde{M}$ having large Lusternik-Schnirelman category and for which the Lusternik-Schnirelman Principle holds as stated above, while the critical points of $\tilde{h}$ correspond to eigenfunctions of our given eigenvalue problem. We already know that if we consider the C^1 function h on the level manifold $M_c(f)$ of the C^1 function f in V and if $f'(u) \neq 0$ for all u on $M_c(f)$, then the critical points u of the function h_0 obtained by restricting h to $M_c(f)$ are solutions of the eigenvalue problem

$$h'(u) = \lambda f'(u) \quad (\lambda \in R^1).$$

In most cases, however, the topological structure of the level surface $M_c(f)$ is relatively trivial and permits no conclusion of any significant kind to be obtained from the Lusternik-Schnirelman theory without basic further assumptions. Such an assumption is the following which we put in the following form for emphasis:

ASSUMPTION (E). *Let M be a submanifold of the Banach space V, h a function from M to the reals. Then*

(1) *M is invariant under the involution π, $\pi(u) = -u$. M does not contain 0 and each closed ray emanating from the origin hits M in exactly one point. The mapping $g: M \to S_1(V)$ given by $g(u) = u/\|u\|$ is bicontinuous from M to $S_1(V)$.*

(2) *h is invariant under the involution π on M, i.e. $h(-u) = +h(u)$ for all u in M.*

If $M = M_c(f)$ for a given function f on V, then the condition that M is invariant under π is simply the condition that $f(u) = c$ implies $f(-u) = c$ and follows if f is a function which is even on some symmetric set S in V containing $M_c(f)$, i.e. $f(u) = f(-u)$, $u \in S$.

THEOREM 20. *Let V be an infinite dimensional Banach space with C^{2-} norm on the complement of the origin, f and h two C^1 functions on V, c a real number with the manifold $M = M_c(f)$ and h_0 the restriction of h to M. Suppose that M and h_0 are even in the sense of Assumption* (E), *h_0 is bounded from below on $M(f)$, and $(f'(u), u) > 0$ for u in $M_c(f)$. Let $\tilde{M}$ be the quotient space of $M_c(f)$ under the involution π, $(\pi u = -u)$, $\tilde{h}$ the function on $\tilde{M}$ induced by h_0. Suppose that h_0 satisfies the Condition* (C) *on $M_c(f)$. Then*

(a) *$\tilde{M}$ is a C^1-manifold with a canonical Finsler structure inherited from M. $\tilde{M}$ is homeomorphic to the ∞-dimensional projective space obtained by taking the quotient space of $S_1(V)$ by π and has Lusternik-Schnirelman category $+\infty$.*

(b) *$\tilde{h}$ satisfies Condition* (C) *on M and has an infinite number of critical points on $\tilde{M}$. Each critical point of $\tilde{h}$ is the image under the identification map $M \to M$ of exactly two critical points of h on M, u and $(-u)$.*

(c) *The Lusternik-Schnirelman minimax principle holds for h on M in the following sense: For a compact symmetric subset K of M, let $p - \operatorname{cat}_M(K) =$ the category of K/π relative to $\tilde{M}$. Then if*

$$c_m(h) = \inf_{p-\operatorname{cat}_M(K)\geq m} \max_{u\in K} h(u),$$

then for each integer $m \geq 1$, there exists a critical point u_m of the pair (f', h') such that $h(u_m) = c_m$, i.e. $f(u_m) = \lambda_m h'(u_m)$.

PROOF OF THEOREM 20. Proof of (a). By Theorem 19, M is diffeomorphic with $S_1(V)$ with the diffeomorphism given by $g(u) = u/\|u\|$. Since g commutes with π, g induces a diffeomorphism g of $\tilde{M}$ with $S_1(V)/\pi$, which is the infinite dimensional projective space in question. By standard results (which essentially are based upon the principle that $S_1(V)/\pi$ has the same homotopy type as the inductive limit of an increasing sequence of finite dimensional projective spaces of increasing dimensions going to infinity), $\operatorname{cat}(S_1(V)/\pi) = +\infty$. (For a detailed argument by elementary methods, see the writer's paper in Ann. of Math. 1965.) The existence of the canonical Finsler structure on $\tilde{M}$ follows from the fact that π is an isometry on V.

PROOF OF (b). Since h satisfies Condition (C) by hypothesis, it follows from the construction of $\tilde{h}$ and the Finsler structure on $\tilde{M}$ that for each u in M, the mapping of $T_u(M)$ into $T_{\pi u}(\tilde{M})$ induced by π is an isometric isomorphism and yields the Finsler norm defined on $T_{\pi u}(\tilde{M})$. It follows immediately from the definition of Condition (C) that $\tilde{h}$ must satisfy this condition if h does. Since h_0 is bounded from below on M, it follows from Theorem 18 that the Lusternik-Schnirelman Principle is valid for $\tilde{h}$ on $\tilde{M}$ and hence h has an infinite number of critical points on $\tilde{M}$. It is obvious from the fact that $d\pi$ is an isomorphism that u is a critical point of h on M if and only if πu is a critical point of $\tilde{h}$ on $\tilde{M}$. Hence h has an infinite number of critical points on M.

PROOF OF (c). $\tilde{M}$ is mapped diffeomorphically onto $S_1(V)/\pi$ by a diffeomorphism $\tilde{g}$ such that $\|d\tilde{g}\|$ is uniformly bounded. By Theorem 18, the Lusternik Principle (II) is valid for $\tilde{h}$ on $\tilde{M}$. Translating this fact into terms of the function h on M, we obtain the conclusions of assertion (c). Q.E.D.

It is evident from Theorem 20 that it is essential to determine sufficient conditions for h_0, the restriction of h to $M_c(f)$, to satisfy Condition (C) on $M_c(f)$. The simplest such condition for our purposes is given by the following theorem.

THEOREM 21. *Let V be a Banach space with C^1 norm on the complement of the origin, f and h two C^1 functions on V, c a real number and $M_c(f) = \{u \mid u \in V, f(u) = c\}$ the level set of f at level c. Suppose that all of the following conditions hold.*

(1) *There exists a constant c_0 for all u in $M_c(f)$, $(f'(u), u) \geq c_0 > 0$ so that $M_c(f)$ is a submanifold of codimension 1 in V.*

(2) *f' maps bounded sets into bounded sets. $M_c(f)$ is bounded. h' is compact on bounded sets on which h is bounded.*

(3) *If S is a subset of $M_c(f)$ on which h is bounded, there exists a positive constant $c_S > 0$ such that $(h'(u), u) \geq c_S$ for u in S.*

(4) *f' is a proper mapping from $M_c(f)$ into V^*, i.e. the inverse image of a compact set in V^* is compact in $M_c(f)$.*

PROOF OF THEOREM 21. By condition (1), $M_c(f)$ is a submanifold of V and its tangent space at a given point u can be considered as a closed subspace of V, where

$$T_u(M_c(f)) = \{w \mid w \in V, (f'(u), w) = 0\}.$$

For a given u in $M_c(f)$ and any v in V, v can be written in one and only one way in the form

$$v = c_u(v)u + w_u(v); \qquad w_u(v) \in T_u(M_c(f)),$$

where $c_u(v) = (f'(u), v)/(f'(u), u)$. Since $(f'(u), u) \geq c_0 > 0$ for all u in $M_c(f)$, $M_c(f)$ is bounded, and f' maps bounded subsets of V into bounded subsets of V^*, it follows that

$$|c_u(v)| = |(f'(u), v)| \cdot |f'(u), u)|^{-1} \leq c_0^{-1}\|f'(u)\| \cdot \|v\| \leq K\|v\|$$

with a constant K independent of u in $M_c(f)$. It follows that $\|w_u(v)\| = \|v - c_u(v)u\| \leq K_1\|v\|$ for all u in $M_c(f)$.

Suppose $\{u_n\}$ is a sequence in $M_c(f)$ with $h(u_n)$ bounded and $\|h_0'(u_n)\| \to 0$. For any w in $T_u(M_c(f))$, $(h_0'(u), w) = (h'(u), w)$. Let v be any element of V. Then

$$(h'(u_n), v) - c_{u_n}(v)(h'(u_n), u_n) = (h'(u_n, w_{u_n}(v))$$

where

$$|(h'(u_n), w_{u_n}(v))| = |(h_0'(u_n), w_{u_n}(v))| \leq \|h_0'(u_n)\| \, \|w_{u_n}(v)\| \leq K\|h_0'(u_n)\| \cdot \|v\|,$$

with K independent of n. Hence

$$|(h'(u_n), v) - c_{u_n}(v)(h'(u_n), u_n)|$$

$$= \left|\left(h'(u_n) - \frac{(h'(u_n), u_n)}{(f'(u_n), u_n)} f'(u_n), v\right)\right| \leq \varepsilon_n \|v\| \quad (\varepsilon_n \to 0 \text{ as } n \to \infty).$$

Thus

$$h'(u_n) - \frac{(h'(u_n), u_n)}{(f'(u_n), u_n)} f'(u_n) \to 0 \quad (n \to \infty).$$

Since $h(u_n)$ is bounded, there exists $c_1 > 0$ such that for all n, $(h'(u_n), u_n) \geq c_1 > 0$. Since u_n is a bounded set, $(f'(u_n), u_n)$ is uniformly bounded for all n. Hence

$$f'(u_n) - \frac{(f'(u_n), u_n)}{(h'(u_n), u_n)} h'(u_n) \to 0.$$

Since h' is compact on $M_c(f)$, we can pass to a subsequence and assume that $(f'(u_n), u_n)(h'(u_n), u_n)^{-1} h'(u_n)$ converges strongly to w in V^*. Since f' is proper on $M_c(f)$ and since the closure of the elements of a convergent sequence is compact in V^*, it follows that u_n is relatively compact in $M_c(f)$. Hence we can choose an infinite subsequence which converges strongly in $M_c(f)$ and Condition (C) for h_0 has been established. Q.E.D.

LEMMA 8. *Let V be a reflexive Banach space, f a C^1 function on V such that f' satisfies Condition* (S). *Then f' is proper on any bounded closed subset of V.*

PROOF OF LEMMA 8. Let K be a compact subset of V^*, S a bounded closed subset of V. We wish to show that $(f')^{-1}(K) \cap S$ is compact. Let $\{u_n\}$ be an infinite sequence in $(f')^{-1}(K) \cap S$. Then $w_n = f'(u_n)$ lie in the compact set K. We wish to show that u_n has a convergent subsequence (since the limit of that subsequence must then lie in the closed set $(f')^{-1}(K) \cap S$). Since S is bounded and K is compact, we may pass to a subsequence and assume that u_n converges weakly to u in V and that $w_n = f'(u_n)$ converges strongly to some w in V^*. By hypothesis, f' satisfies Condition (S). We consider $(f'(u_n) - f(u), u_n - u)$. We know that $f'(u_n) - f(u)$ converges strongly in V to $w - f(u)$, while $u_n - u$ converges weakly to 0 in V^*. Hence

$$(f'(u_n) - f(u), u_n - u) \to (w - f(u), 0) = 0.$$

By Condition (S), we may find a strongly convergent infinite subsequence of the sequence u_n. Q.E.D.

Collecting the results of the preceding theorems together with Lemma 8, we obtain the following general form of the Lusternik-Schnirelman principle for eigenvalue problems in a Banach space V whose norm is of class C^{2-} on the complement of the origin in V:

THEOREM 22. *Let V be an infinite dimensional reflexive Banach space with norm of class C^{2-} on the complement of the origin in V, f and h two C^1 functions from V to the reals, c a real number such that $M_c(f) = \{u \mid u \in V, f(u) = c\}$ is a bounded subset of V and the following conditions are satisfied.*

(1) *There exists $c_0 > 0$ such that for all u in $M_c(f)$, $(f'(u), u) \geq c_0 > 0$. $M_c(f)$ is invariant under the involution π (where $\pi u = -u$), and for u in $M_c(f)$, $h(u) = h(-u)$. Each ray through the origin intersects $M_c(f)$ in exactly one point.*

(2) *For each subset S of $M_c(f)$ for which h is bounded, there exists a constant $c_2 > 0$ such that $(h'(u), u) \geq c_2 > 0$ for u in S. For each such subset S, $h'(S)$ is relatively compact in V^*. h is bounded from below on $M_c(f)$.*

(3) *f' maps bounded sets into bounded sets and satisfies Condition* (S). *Then:*

(a) *There exists an infinite number of distinct elements u_n of $M_c(f)$ such that*

$$f'(u_n) = \lambda_n h'(u_n), \quad \lambda_n \in R^1.$$

(b) *Let*

$$c_m(h) = \inf_{\substack{p-\mathrm{cat}_{M_c(f)}(K) \geq m \\ K \text{ compact in } M}} \max_{u \in K} h(u).$$

Then there exists u_m in $M_c(f)$ with $h(u_m) = c_m$ which satisfies the equation $f'(u_m) = \xi_m h'(u_m)$, $\xi_m \in R^1$.

PROOF OF THEOREM 22. It follows from Theorem 20 that under our present hypotheses, it suffices to show that h_0 (the restriction of h to $M_c(f)$) satisfies Condition (C) on $M_c(f)$. By Theorem 21, it suffices under our hypotheses to show that f' is proper on bounded closed subsets of $M_c(f)$. Since V is reflexive and f' satisfies Condition (S), it follows from Lemma 8 that f is proper on bounded closed subsets of V. Q.E.D.

We can now translate Theorem 22 into an existence theorem for an infinite number of normalized eigenfunctions for a nonlinear elliptic eigenvalue problem.

THEOREM 23. *Let $A(u)$ and $B(u)$ be a pair of quasilinear partial differential operators of order $2m$ and $s \leq 2m - 2$ defined on a bounded smoothly bounded open subset G of R^n and in generalized divergence form*

$$A(u) = \sum_{|\alpha| \leq m} (-1) D^\alpha F_\alpha(\cdot, \xi_m(u)),$$

and

$$B(u) = \sum_{|\beta| \leq m-1} (-1)^{|\beta|} D^\beta H_\beta(\cdot, \xi_{m-1}).$$

Let the functions F_α and H_β satisfy the conditions of Assumptions (C) *and A and B be the Euler-Lagrange operators of the multiple integrals for a given exponent p with $2 \leq p < \infty$,*

$$f(u) = \int_G F(x, \xi_m(u)(x))dx, \qquad h(u) = \int_G H(x, \xi_{m-1}(u)(x))dx,$$

while $a(u, v)$ and $b(u, v)$ are the Dirichlet forms corresponding to the divergence representation of A and B, respectively, i.e.

$$a(u, v) = \sum_{|\alpha| \leq m} (F_\alpha(\cdot, \xi_m(u)), D^\alpha v),$$

$$b(u, v) = \sum_{|\beta| \leq m-1} (H_\beta(\cdot, \xi_{m-1}(u)), D^\beta v).$$

Suppose that for each x, ξ, and η

$$F(x, -\xi) = F(x, \xi); \qquad H(x, -\eta) = H(x, \eta).$$

Suppose further that $h(u) > 0$ for each u in V, a closed subspace of $W^{m,p}(G)$, with $u \neq 0$ and that $(h'(u), u)$ is bounded from below by a positive constant on any bounded set on which $h(u) \geq d_0 > 0$ (d_0 an arbitrary positive constant).

Then, for any sufficiently large positive c, there exist an infinite number of distinct u_n in V with $f(u_n) = c$ such that $A(u_n) = \lambda_n B(u_n)$ in the variational sense with respect to the subspace V of $W^{m,p}(G)$, or more precisely; $u_n \in V$, and

$$a(u_n, v) = \lambda_n b(u_n, v), \quad (v \in V).$$

We may show in addition that u_n can be chosen with $h(u_n) = c_n(h, c)$ where $c_n(h, c)$ is the Lusternik-Schnirelman minimax value:

$$c_n(h, c) = \sup_{\substack{K \text{ compact, symmetric in } M_c(f) \\ p-\text{cat}(K) \geq n}} \min_{u \in K} h(u).$$

Proof of Theorem 23. We will apply Theorem 22 to the proof of the given theorem but to the pair of functions f and $h_1 = 1/h$, not to the pair f and h. Let V be the given closed subspace of $W^{m,p}(G)$. Since we assume $p \geq 2$, the norm of V is of class C^{2-} on the complement of the origin in V. V is reflexive.

By the results of §1, it follows from our assumptions on F_α that f' is a continuous mapping of V into V^* which maps bounded sets into bounded sets and satisfies Condition (S). Moreover, $(f'(u), u) \to \infty$ as $||u|| \to +\infty$. Moreover, we know that under these hypotheses, $f(u) \to +\infty$ as $||u|| \to +\infty$. It follows therefore that for any constant c, the level set $M_c(f)$ is bounded while for c sufficiently large, there exists $d_1 > 0$ such that $(f'(u), u) \geq d_1$ for all u in $M_c(f)$.

Since the functions F and H are odd in ξ and η, respectively, $f(-u) = f(u)$, $h(-u) = h(u)$ for all u in V. Moreover, $h_1(-u) = h_1(u)$. Thus for all c, $M_c(f)$ is invariant under the involution π and for u in $M_c(f)$, $h_1(-u) = h_1(u)$. To complete the verification of the Assumptions (E) for the pair $(M_c(f), h_1)$, we must show that each ray through the origin hits $M_c(f)$ in exactly one point. However, this follows for sufficiently large c by the following argument: There exists $R > 0$ such that $(f'(u), u) \geq d_1 > 0$ for all u in V with $||u|| \geq R$. Since f maps bounded sets into bounded sets, there exists c_R such that for $c \geq c_R$, $M_c(f)$ is entirely contained in the set $\{u | u \in V, ||u|| \geq R\}$. Consider the ray $u = ty$ through the origin, $0 \leq t < +\infty$. For $c \geq c_R$, such a ray can hit $M_c(f)$ in at most one point since

$$f(ty) - f(t_1 y) = \int_{t_1}^{t} \frac{d}{ds} f(sy) ds = \int_{t_1}^{t} (f'(sy), y) ds \geq d_1 \int_{t_1}^{t} s^{-1} ds > 0$$

if $t > t_1$ and $||t_1 y|| \geq R$. Since $f(ty) \to +\infty$ as $t \to \infty$, it follows moreover that the ray must actually intersect $M_c(f)$ exactly once for $c \geq c_R$.

The function h is bounded from above on $M_c(f)$, h is weakly continuous, and h' is compact on bounded sets in V. Hence $h_1 = 1/h$ is bounded from below on $M_c(f)$. Since $h'_1 = (h^{-2})h'$, it follows that h'_1 is compact on any bounded set S on which h_1 is bounded from above, i.e. on which h is bounded from below by a positive constant. Similarly,

$$(h'_1(u), u) = h(u)^{-2}(h'(u), u) \geq d_0 > 0$$

on any closed set S in $M_c(f)$ for which $h(u)$ is bounded from below by a positive constant, i.e. on which $h_1(u)$ is bounded from above.

Thus we see that all the hypotheses of Theorem 22 are satisfied for such values of c for the pair f and h_1. The conclusions of that theorem for f and h_1 are equivalent to the conclusions stated for the differential operators A and B in Theorem 23. Q.E.D.

We now turn to another method of obtaining the existence of an infinite number of eigenfunctions based upon the use of Galerkin approximations rather than the application of arguments about infinite dimensional manifolds. This second method has two basic advantages over the first: (1) It works for all values of the exponent p determining the order of growth of the top order terms of $A(u)$ with $1 < p < +\infty$, and (2) it is both more elementary than the general Lusternik-Schnirelman theory on infinite dimensional Finsler manifolds and somewhat more concrete in that it obtains the desired eigenfunctions as the limit of eigenfunctions of corresponding problems on finite dimensional subspaces of V. Indeed, it would be more descriptive to call this method, the extension of the Lusternik-Schnirelman theory by the approximation method of Rayleigh-Ritz since it consists of finding critical points of functionals in an infinite dimensional space by taking limits of critical points of the same functionals on finite dimensional subspaces (a procedure which in concrete equations of mathematical physics is usually called the Rayleigh-Ritz method).

For this phase of the discussion of our general eigenvalue problem, we shall not proceed by the cumulative collection of parts and subresults leading up to the final statement of the existence theorem for eigenfunctions, as we did in the preceding argument using the LS-theory on Finsler manifolds. We begin rather with the statement of the general abstract result, we shall give the proof of this result, and after that proof is complete, we shall briefly mention the application of the abstract result to the case of elliptic eigenvalue problems.

THEOREM 24. *Let V be a separable reflexive Banach space of infinite dimension, f and h two C^1 functions on V, c a real number such that for u in the level set $M_c(f) = \{u | u \in V, f(u) = c\}$, $(f'(u), u) \geq c_0 > 0$. We assume that f' maps bounded sets into bounded sets and satisfies Condition (S), that for the given constant c, $M_c(f)$ is bounded, h is weakly continuous on bounded subsets of V, $h(u) > 0$ on $M_c(f)$ and for each $\delta > 0$, $h^{-1}(\delta, \infty)$ is compact on $M_c(f)$. We assume also that $M_c(f)$ is invariant under the involution π which sends u into $(-u)$ and that for u in $M_c(f)$, $h(-u) = h(u)$. Suppose also that each ray through the origin in V intersects $M_c(f)$ in exactly one point. Suppose that on any subset S of $M_c(f)$ on which $h(u)$ is bounded from below by a positive constant, h' is compact on S and $(h'(u), u)$ is bounded from below by a positive constant on S.*

Then: (a) *There exists an infinite number of distinct elements u_n of $M_c(f)$ such that $f'(u_n) = \xi_n h'(u_n)$.*

(b) *We may obtain the eigenfunctions u_n by the following minimax criterion: Let c_m for each $m \geq 1$ be given by*

$$c_m = \inf_{\substack{K \text{ compact in } M_c(f) \\ p-\mathrm{cat}_M(K) \geq m}} \max_{u \in K} h(u).$$

Then there exists v_m in $M_c(f)$ such that $h(v_m) = c_m$ and $f'(v_m) = \xi_m h'(v_m)$ for real numbers ξ_m such that $\xi_m \to +\infty$. The minimax levels $c_m \to 0$ as $m \to \infty$.

(c) *If $\{V_j\}$ is an increasing sequence of finite dimensional subspaces of dimension j of V whose union is dense in V, there exist $v_{j,m}$ for $j \geq m$ such that $v_{j,m} \in V_j \cap M_c(f)$, $h(u_j) = c_{j,m}$ (f_j and h_j are the restrictions of f and h to V_j),*

$$c_{j,m} = \inf_{\substack{K \text{ compact in } M_c(f) \cap V_j \\ p-\mathrm{cat}_M K > m}} \max_{u \in K} h(u),$$

$$f'_j(v_{j,m}) = \xi_{j,m} h'_j(v_{j,m})$$

and one can extract strongly convergent subsequences of $\{v_{j,m}\}$ and $\{\xi_{j,m}\}$ for a given m which converge to v_m and ξ_m, respectively, which satisfy the conditions of (b).

(d) *In the situation of part* (c), *$c_{j,m} \to c_m$ as $j \to \infty$.*

We have formulated Theorem 24 in such a fashion that it includes as part of its statement most of the subresults used in carrying through its proof. We can thus carry through the different parts of the proof of the theorem by referring to the various assertions made in its statement.

PROOF OF THEOREM 24. By the hypotheses which we have made upon $M_c(f)$ it follows that the projection mapping g which carries u into $u/\|u\|$ is a C^1 diffeomorphism of $M_c(f)$ on the unit sphere $S_1(V)$. Since g commutes with π, and π leaves $M_c(f)$ invariant, it follows as before that $M_c(F)/\pi$ is homeomorphic to the infinite dimensional projective space $S_1(V)/\pi$ and therefore has LS-category $= +\infty$. The projected category of a compact subset K of $M_c(F)$ with respect to $M_c(f)$ is defined to be the category of its image in $M_c(f)/\pi$ with respect to that space. The existence of compact subsets of arbitrarily high category follows from the corresponding property for the infinite dimensional projective space.

Let $\{V_j\}$ be a given increasing sequence of finite dimensional subspaces of V whose union is dense in V, and suppose for simplicity as we have done in assertion (c) of the theorem, that $\dim(V_j) = j$ for each integer $j \geq 1$. We let f_j and h_j be the restriction of the functions f and h to the subspaces V_j. Let $M = M_c(f)$, $M_j = M_c \cap V_j$. Then $M_j = M_c(f_j)$. If ψ_j is the injection mapping of V_j onto V and if ψ_j^* is the dual projection mapping of V^* onto V_j^*, then by an obvious calculation,

$$f'_j = \psi_j^* f' \psi_j, \qquad h'_j = \psi_j^* h' \psi_j.$$

In particular, for u in V_j, $(h'_j u, u) = (h'u, u)$; $(f'_j u, u) = (f'u, u)$.

By hypothesis, $(f'(u), u) \geq d_0 > 0$ for all u in $M_c(f)$. It follows that $(f'_j u, u) > 0$ for all j and all u in $M_c(f_j) = M_j$ so that M and all the M_j are submanifolds of codimension 1 in V and the V_j, respectively. Since each M_j and M itself are closed and $M_j \subset M$, which is bounded by hypothesis, each M_j is a compact $(j-1)$-dimensional manifold. Since M and V_j are invariant under π, so is their intersection M_j and each ray through the origin in V_j intersects M_j in exactly one point. Hence M/π is homeomorphic to the $(j-1)$-dimensional real projective space $S_1(V_j)/\pi$ and therefore has LS-category j. Since these manifolds are compact C^1 manifolds, we may apply the Lusternik-Schnirelman theory to the C^1 even functions h_j on these manifolds (where h_j automatically satisfy Condition (C) since M_j is compact). Hence we may find $v_{j,m} (m \leq j)$ in M_j such that $h(v_{j,m}) = c_{j,m}$, the mth Lusternik-Schnirelman minimax value for $m \leq p - \mathrm{cat}_M M_j = \mathrm{cat}(M_j/\pi)$, and these $v_{j,m}$ will satisfy relations of the form

$$h'_j(v_{j,m}) = \zeta_{j,m} f'_j(v_{j,m}); \quad \zeta_{j,m} \in R^1.$$

Taking the value of both sides at the point $v_{j,m}$ we see that

$$(h'(v_{j,m}), v_{j,m}) = \zeta_{j,m}(f'(v_{j,m}), v_{j,m}),$$

with the coefficients of $\zeta_{j,m}$ different from zero. Dividing, we can rewrite these equations in our canonical form

$$f'_j(v_{j,m}) = \xi_{j,m} h'(v_{j,m}); \quad \xi_{j,m} \in R^1.$$

Since $V_j \cap S_1(V)$ has dense union in $S_1(V)$, it follows that the union of the increasing sequence of subsets M_j of $M_c(f) = M$ is dense in M. It follows from this fact that for each fixed m, $c_{j,m} \to c_m$ as $j \to \infty$. Note that we are considering in the definition of $c_{j,m}$, compacts K in M_j whose p-category relative to M is at least m. Since this is a subfamily of those for a larger M_k or for M itself, it follows that $c_{j,m} \leq c_{k,m}$ for $j < k$, and for all j, $c_{j,m} \leq c_m$. On the other hand, for $\varepsilon > 0$ given, consider a compact subset K of M of p-category $\geq m$ such that $\min_{u \in K} h(u) \geq c_m - \varepsilon$. For j sufficiently large, this can be deformed by a small deformation through M into a compact subset K of some M_j, and the deformation can be made to commute with π. Since $p\text{-}\mathrm{cat}_M(K_1) \geq p\text{-}\mathrm{cat}_M(K) \geq m$, it follows that

$\min_{u \in K_1} h(u) \le c_{j,m}$ for the given j, and $\min_{u \in K_1} h(u) \ge \min_{u \in K} h(u) - \varepsilon$ from the smallness of the deformation. Hence $c_{j,m} \ge c_m - 2\varepsilon$ for j sufficiently large, i.e. $c_{j,m} \to c_m$ as $j \to \infty$. Thus the proof of part (d) is complete.

The minimax levels c_m decrease as $m \to \infty$. We assert that $c_m \to 0$ as $m \to \infty$. Suppose $\varepsilon > 0$ is given. Then by our hypothesis, the subset of $M_c(f)$ for which $h(u) \ge \varepsilon$ is compact. If j is chosen sufficiently large, there exists a deformation of this set in M into M_j which commutes with π. Hence any compact subset K of M for which $\inf_{u \in K} h(u) \ge \varepsilon$ must be deformable through M into M_j by a deformation commuting with π and has p-category in $M \le j$. Hence $c_m \le \varepsilon$ for $m \ge j$, i.e. $c_m \to 0$.

Since $h(v_{j,m}) = c_{j,m} \to c_m > 0$, $h'(v_{j,m})$ is relatively compact in V^* and $(h'(v_{j,m}), v_{j,m}) \ge d_1 > 0$ for all j. Since $f'_j(v_{j,m}) = \xi_{j,m} h'_j(v_{j,m})$, we obtain

$$(f'(v_{j,m}), v_{j,m}) = (f'_j(v_{j,m}), v_{j,m}) = \xi_j(h'_j(v_{j,m}), v_{j,m}) = \xi_j(h'(v_{j,m}), v_{j,m}).$$

Thus

$$|\xi_{j,m}| = |(f'(v_{j,m}), v_{j,m})(h'(v_{j,m}), v_{j,m})^{-1}| \le k_m$$

independently of j. Passing to a subsequence, we may assume that $\xi_{j,m}$ converge as $j \to \infty$ to some constant ξ_m and that $h'(v_{j,m})$ converges strongly in V^* to some element w. It follows that if we assume (by again passing to an infinite subsequence) that $v_{j,m}$ converges weakly in V to an element v_m of V, then for v in any subspace V_k, $j \ge k$,

$$\begin{aligned}(f'(v_{j,m}), v_{j,m} - v) &= (f'_j(v_{j,m}), v_{j,m} - v) = \xi_{j,m}(h'_j(v_{j,m}), v_{j,m} - v) \\ &= \xi_{j,m}(h'(v_{j,m}), v_{j,m} - v) \to \xi_m(w, v_m - v).\end{aligned}$$

Since $f'(v_{j,m})$ is bounded, it follows by an easy argument (as earlier in this section) that $(f'(v_{j,m}), v_{j,m} - v) \to \xi_m(w, v_m - v)$ for all v in V. If we take $v = v_m$, we have then $(f'(v_{j,m}), v_{j,m} - v_m) \to 0$, $j \to \infty$. Hence

$$(f'(v_{j,m}) - f'(v_m), v_{j,m} - v_m) \to 0, \quad (j \to \infty).$$

By hypothesis, f' satisfies Condition (S), and it follows from the last limit that $v_{j,m} \to v_m$ strongly in V. Since f is continuous in the strong topology, $f(v_{j,m}) \to f(v_m)$ so that $f(v_m) = c$, i.e. $v_m \in M_c(f)$. Since $h(v_{j,m}) = c_{j,m} \to c_m$ and h is continuous, $h(v_m) = c_m$.

Finally for v in V_k, $j \ge k$

$$0 = (f'(v_{j,m}) - \xi_{j,m} h'(v_{j,m}), v) \to (f'(v_m) - \xi_m h'(v_m), v).$$

Since the union of the V_k is dense in V, it follows that $f'(v_m) = \xi_m h'(v_m)$. Q.E.D.

For the sake of brevity, we can state our final result for elliptic problems in the following abbreviated form:

THEOREM 25. *The result of Theorem* 23 *for nonlinear elliptic problems is valid for any exponent* p *with* $1 < p < \infty$.

Details of the application of Theorem 24 to more general elliptic eigenvalue problems are given in a forthcoming paper by the writer.

PART III

3. **Nonlinear parabolic problems and semigroups of nonexpansive operators in Banach spaces.** In the present section, we shall consider parabolic initial boundary value problems for nonlinear parabolic operators of the form

$$\partial u/\partial t = -A(u), \tag{1}$$

where $A(u)$ is a quasilinear elliptic partial differential operator of the form considered in §1. It is our purpose in this discussion to prove the existence of strong solutions of such problems by applying the recently developed theory of semigroups of nonlinear nonexpansive operators in Hilbert and Banach spaces.

We consider $A(u)$ as in §1 to be given in generalized divergence form

$$A(u) = \sum_{|\alpha| \le m} (-1)^{|\alpha|} D^\alpha A_\alpha(\cdot, \xi_m(u)), \tag{2}$$

with a corresponding Dirichlet form

$$a(u, v) = \sum_{|\alpha| \le m} (A_\alpha(\cdot, \xi_m(u)), D^\alpha v).$$

If V is a closed subspace of the Sobolev space $W^{m,p}(G)$, and if the coefficient functions A_α satisfy the conditions of either Assumptions (A) or Assumptions (B) of §1 with $p \ge 2$, there exists a continuous mapping T of V into V^* such that for all u and v of V, $a(u, v) = (Tu, v)$. We say that u in V is a solution of the elliptic equation $A(u) = w$ with null variational boundary conditions with respect to V whenever $a(u, v) = (w, v)$ for all v in V, i.e. if and only if $T(u) = w$. For the parabolic problem, we make the following definition:

DEFINITION. *Let G be a bounded, smoothly bounded domain, p an exponent with $2 \le p < \infty$, V a closed subspace of the Sobolev space $W^{m,p}(G)$. Let $A(u)$ be a quasilinear elliptic operator of order $2m$ on G in the generalized divergence form*

$$A(u) = \sum_{|\alpha| \le m} (-1)^{|\alpha|} D^\alpha A_\alpha(\cdot, \xi_m(u))$$

whose coefficient functions A_α satisfy the conditions of Assumptions (B). *If $a(u, v)$ is the corresponding Dirichlet form, let T be the mapping of V into V^* given by $(T(u), v) = a(u, v)$.*

We define a nonlinear operator A with domain $D(A)$ in $L^2(G)$ and with values in $L^2(G)$ by setting

$$D(A) = \{u | \, u \in V, T(u) \in L^2(G)\}$$

$$A(u) = T(u), \quad u \in D(A).$$

Note that since $V \subset L^2(G)$, $(L^2(G))^* = L^2(G) \subset V^*$, and it makes sense to speak of an element of V^* lying in $L^2(G)$.

DEFINITION. *If* $u: R^+ \to D(A)$, *u is said to be a strong solution of the equation*

$$du/dt = -A(u)(t), \quad t > 0 \tag{3}$$

with initial value $u(0) = u_0 \in D(A)$, *if* $u(t)$ *is strongly differentiable in t from the right in* $L^2(G)$, *the equation* (3) *is satisfied for all t in* R^+, $u(t)$ *is weakly continuous in V and* $A(u(t))$ *is weakly continuous in* $L^2(G)$.

THEOREM 26. *Let G be a bounded, smoothly bounded domain,* $A(u)$ *an elliptic operator as in the definition above which satisfies the conditions of the Assumptions* (A) *or* (B) *for an exponent* $p \geq 2$. *Suppose further that for a given subspace V of* $W^{m,p}(G)$,

$$a(u, u - v) - a(v, u - v) \geq 0,$$

(*i.e. the corresponding operator T from V to V is monotone*). *Then for each* u_0 *in* $D(A)$, *there exists one and only one strong solution of the differential* $du(t)/dt = -A(u(t))$, $t > 0$, *with* $u(0) = u_0$.

We shall derive the result of Theorem 26 from the following more general theorem:

THEOREM 27. *Let H be a Hilbert space, A an operator* (*in general nonlinear*) *with domain* $D(A)$ *in H and values in H. Suppose that A is monotone and that* $R(A + I) = H$, *Then there exists one and only one strong solution of the differential equation*

$$du/dt(t) = -A(u(t)), \quad t \geq 0$$

with initial value $u(0) = u$ *for each given element* u_0 *of* $D(A)$.

Theorem 27 is due to Y. Komura (J. Math. Soc. Japan 1967) and generalizes earlier weaker results on monotone differential equations in Hilbert space due to the writer and to T. Kato. The writer first proved Theorem 26 in 1965 without making use of Theorem 27 but applying arguments similar to the proof of Theorem 27 under more concrete hypotheses (see the writer's paper in Proc. Bratislava Symposium on Differential Equations, 1966).

PROOF OF THEOREM 26 FROM THEOREM 27. By our earlier results, under the given assumptions, T is a monotone coercive continuous mapping of V into V^*. Hence $R(T) = V^*$. If we set $H = L^2(G)$, $V \subset H \subset V^*$, in the sense that the injection mappings are continuous with dense images and the pairing between V and V^* is an extension of the H-inner product, it follows that the identity mapping I of H into H when restricted to V becomes a monotone continuous linear mapping from V into V^*. Hence $T + I$ is monotone, continuous and coercive, and once more $R(T + I) = V^*$.

Consider the nonlinear operator $A + I$ with domain $D(A)$ in H and values in H. Since A is a restriction of T, it follows that A is a monotone mapping from $D(A)$ in H to H. By the definition of $D(A)$, the element u of V lies in $D(A)$ if and only if Tu lies in H. Suppose $(T + I)u = w$ lies in H, where u is an element of V.

Since $V \subset H$, and $Tu = w - u$, where $w - u$ lies in H, it follows that $u \in D(A)$, $Tu = Au$, and $(A + I)u = w$. Hence $R(A + I) = H$.

Thus Theorem 27 may be applied to give us the existence of a strong solution of our parabolic problem. Q.E.D.

The crucial feature of the differential equation $du/dt = -A(u(t))$ for a monotone operator A in a Hilbert space which makes the existence proof for Theorem 27 possible is the characteristic property of the solutions if they exist. Consider two solutions $u(t)$ and $v(t)$ of this differential equation, and note that

$$\begin{aligned} \frac{d}{dt}||u(t) - v(t)||^2 &= 2\left(\frac{du}{dt}(t) - \frac{dv}{dt}(t), u(t) - v(t)\right) \\ &= -2(A(u(t)) - A(v(t)), u(t) - v(t)) \\ &\leq 0, \end{aligned}$$

i.e. for $s < t$

$$||u(t) - v(t)||^2 \leq ||u(s) - v(s)||^2.$$

If we define the transition operator $U(t)$ for our differential equation by setting $U(t)(u(0)) = u(t)$ for any solution, then the last fact that we derived can be expressed intuitively by saying that the transition operators are nonexpansive operators in H,

$$||U(t)u - U(t)v|| \leq ||u - v||, \quad (t \geq 0),$$

or in other terms, $-A$ generates a semigroup $U(t)$ of nonexpansive operators.

The generalization to more general Banach spaces of this particular property of monotone operators in Hilbert spaces gives rise to the general class of *accretive* operators in Banach spaces, which are defined as follows.

DEFINITION. *Let X be a Banach space, X^* its conjugate space. By a duality mapping J (J single valued), we mean a mapping J (not necessarily continuous) of X into X^* such that for all u in X,*

$$||Ju|| = ||u||, \quad (Ju, u) = ||u||^2.$$

Then a mapping A with domain $D(A)$ in X and values in X is said to be accretive (with respect to J) if for all u and v of $D(A)$,

$$(Au - Av, J(u - v)) \geq 0,$$

(where we use (u, w) to denote the pairing between u in X and w in X^*).

For $X = H$, I is a duality mapping, and A is accretive if and only if it is monotone. For L^p spaces, the duality mapping is the familiar mapping, $u \to ||u||^{2-p}|u(x)|^{p-2}u(x)$.

The property that a mapping be accretive with respect to a duality mapping J is independent of the choice of J (if any real choice exists), in any Banach space and therefore we can speak of *accretive* operators without any modifying adjectives. The existence of at least one duality mapping in any Banach space X follows easily from the Hahn-Banach theorem.

We shall obtain Theorem 27 as a special case of the following more general theorem whose proof is substantially the same as that of Theorem 27.

THEOREM 28. *Let X be a Banach space whose norm is continuously differentiable on the complement of the origin, and let J be the duality mapping of X into X^* which is the derivative of the function $g(u) = \frac{1}{2}||u||^2$. Suppose that J is uniformly continuous on bounded subsets of X. Let A be an accretive mapping from a domain $D(A)$ in X with values in X and suppose that $R(A + I) = X$.*

Then for every u_0 in $D(A)$, there exists one and only one solution $u: R^+ \to X$ of the differential equation

$$du(t)/dt = -A(u(t)), \quad t \geq 0$$

with $u(0) = u_0$.

By a solution in Theorem 28, we mean a strongly continuous function $u: R^+ \to X$ which is weakly once differentiable to X, $u(t) \in D(A)$ for all $t \geq 0$, $A(u(t))$ is weakly continuous, and

$$du(t)/dt = -A(u(t)), \quad t \geq 0.$$

We also require that $||A(u(t))||$ be nonincreasing in t on R^+. If X is uniformly convex, a weakly continuous function $v(t)$ from R^+ to X with $||v(t)||$ nonincreasing with t must be strongly continuous from the right. It follows that if X is uniformly convex, every solution in the sense just described is strongly differentiable from the right for each $t \geq 0$. This remark connects up the solutions described in Banach spaces with uniformly continuous J as in Theorem 28 and the strong solutions of Theorem 27 in Hilbert spaces (and for uniformly convex Banach spaces X).

Theorem 28 is equivalent to a theorem first proved by T. Kato (J. Math. Soc. Japan, 1967) as a generalization of Komura's theorem. Weaker results for accretive operators had previously been obtained by the writer (Bull. Amer. Math. Soc., March 1967) when A is continuous or the sum of an unbounded linear operator and a continuous one. Further extensions of Kato's theorem have been given by the writer, Kato, and others. (See the writer's forthcoming paper in Proc. Amer. Math. Soc. Symposium on Nonlinear Functional Analysis, April 1968.)

PROOF OF THEOREM 28. By hypothesis, A is an accretive mapping and $R(A + I) = X$. For any $\lambda > 0$, if we set $A_\lambda = \lambda A + I$, for u, v in $D(A)$,

$$(A_\lambda u - A_\lambda v, J(u - v)) = \lambda(Au - Au, J(u - v)) + (u - v, J(u - v)) \geq ||u - v||^2.$$

Hence

$$||u - v||^2 \leq (A_\lambda u - A_\lambda v, u - v) \leq ||A_\lambda u - A_\lambda v|| \cdot ||u - v||,$$

so that $||A_\lambda u - A_\lambda v|| \geq ||u - v||$. Thus if $R(A_\lambda) = X$, $(A_\lambda)^{-1}$ is well defined and is Lipschitzian with Lipschitz constant one, i.e. A_λ^{-1} in that case is a nonexpansive mapping of X into X. Since $A + \xi I = \xi(\xi^{-1}A + I)$ for $\xi > 0$, we see that if

$R(A + \xi I) = X$ for $\xi > 0$, then $(A + \xi I)^{-1}$ is a Lipschitzian operator with $||(A + \xi I)^{-1}||_{\text{Lip}} \leq 1/\xi$.

ASSERTION 1. *For all* $\lambda > 0$, $R(A_\lambda) = X$.

PROOF. It suffices to show $R(A + \xi I) = X$ for $\xi > 0$. Suppose this is true for one value ξ_0. To solve the equation $Au + \xi u = y$, we set $u = (A + \xi_0 I)^{-1}x$. Then, we obtain the following equation as a sufficient condition for u to be a solution of the original equation:

$$x + (\xi - \xi_0)(A + \xi_0 I)^{-1}x = y.$$

This equation will have a solution x for each y if $||(\xi - \xi_0)(A + \xi_0 I)^{-1}||_{\text{Lip}} < 1$, i.e. if $\xi_0^{-1}(\xi - \xi_0) < 1$. This will be true if $0 < \xi < 2\xi_0$. Repeating this argument n times, we see that $R(A + \xi I) = X$ for $0 < \xi < 2^n\xi_0$. Thus $R(A + \xi I) = X$ for all $\xi > 0$. Q.E.D.

ASSERTION 2. *If we set* $R_\lambda = A_\lambda^{-1}$, $T_\lambda = AR_\lambda$, *then* T_λ *is a Lipschitzian accretive mapping of* X *into* X. By definition, $u = R_\lambda v$ if and only if $\lambda Au + u = v$, so that

$$T_\lambda u = AR_\lambda u = \lambda^{-1}(I - R_\lambda)u.$$

Since I and R_λ are nonexpansive, T_λ is Lipschitzian with constant at most $2\lambda^{-1}$. For u and v in X,

$$\begin{aligned}(T_\lambda u - T_\lambda v, J(u - v)) &= \lambda^{-1}\{||u - v||^2 - (R_\lambda u - R_\lambda v, J(u - v))\}\\ &\geq \lambda^{-1}\{||u - v||^2 - ||R_\lambda u - R_\lambda v||\cdot||u - v||\}\\ &\geq 0.\end{aligned}$$

Using Assertion 2, we can construct a family of approximating equations for the equation we wish to solve, namely,

$$du_\lambda(t)/dt = -T_\lambda(u_\lambda(t)), \quad t \geq 0,$$

$$u_\lambda(0) = u_0 \in D(A).$$

The existence and uniqueness of the solutions of these approximating equations follows easily from the Lipschitzian character of T_λ. Any two solutions of the differential equation for a fixed value of λ will have the property that $||u(t) - v(t)||$ does not increase with t.

ASSERTION 3. *If* $u \in D(A)$, $||T_\lambda u|| \leq ||Au||$ *for all* $\lambda > 0$.

PROOF. $T_\lambda u = AR_\lambda u = \lambda^{-1}(I - R_\lambda u)$. Since $u = R_\lambda(\lambda Au + u)$, it follows from the fact that R_λ is nonexpansive that

$$||u - R_\lambda u|| = ||R_\lambda(u + \lambda Au) - R_\lambda u|| \leq ||\lambda Au|| = \lambda||Au||.$$

Hence, $||T_\lambda u|| \leq ||Au||$, $\lambda > 0$, $u \in D(A)$. Q.E.D.

ASSERTION 4. $||du_\lambda(t)/dt|| \leq ||Au_0||$, $t \geq 0$.

PROOF. Since $u_\lambda(t + h)$ is a solution of the same equation as u_λ, we obtain

$$||u_\lambda(t + h) - u_\lambda(t)|| \leq ||u_\lambda(h) - u_\lambda(0)||.$$

Dividing by $h > 0$ and letting $h \to 0+$, we obtain $||du_\lambda(t)/dt|| \leq ||du_\lambda(0)/dt|| = ||T_\lambda u_0|| \leq ||Au_0||$. Q.E.D.

It follows from Assertion 4 that the functions $u_\lambda(t)$ are uniformly Lipschitzian. Since $T_\lambda(u_\lambda(t)) = -du_\lambda/dt$, it follows also that $||T_\lambda(u_\lambda(t))|| \leq ||Au_0||$. Since $u_\lambda(t) - R_\lambda(u_\lambda(t)) = \lambda T_\lambda(u_\lambda(t))$, it follows that if we set $v_\lambda(t) = R_\lambda(u_\lambda(t))$, then $||u_\lambda(t) - v_\lambda(t)|| \leq \lambda||Au_0||$.

ASSERTION 5. *As $\lambda \to 0$, $u_\lambda(t)$ converges strongly to $u(t)$ in X for some continuous function $u(t)$, where the convergence is uniform on bounded intervals.*

PROOF. Let $\lambda, \xi > 0$. Then

$$\begin{aligned} d/dt\{||u_\lambda(t) - u_\xi(t)||^2\} &= -2(T_\lambda(u_\lambda(t)) - T_\xi(u_\xi(t)), J(u_\lambda(t) - u_\xi(t)) \\ &= -2(Av_\lambda(t) - Av_\xi(t), J(u_\lambda(t) - u_\xi(t)) \\ &= -2(Av_\lambda(t) - Av_\xi(t), J(v_\lambda(t) - v_\xi(t)) + R_{\xi,\lambda}, \end{aligned}$$

where

$$R_{\xi,\lambda}(t) = +2(Av_\lambda(t) - Av_\xi(t), J(v_\lambda(t) - v_\xi(t)) - J(u_\lambda(t) - u_\xi(t)).$$

Hence

$$d/dt||u_\lambda(t) - u_\xi(t)||^2 \leq |R_{\xi,\lambda}(t)|,$$

where

$$|R_{\xi,\lambda}(t)| \leq 4||Au_0|| \; ||J(v_\xi(t) - v_\lambda(t)) - J(u_\xi(t) - u_\lambda(t))||.$$

Since the difference of the arguments of J in the two terms inside the norm can be made uniformly small by making λ and ξ sufficiently small and since J is uniformly continuous on bounded sets,

$$d/dt\{||u_\lambda(t) - u_\xi(t)||^2\} < \zeta$$

for any $\zeta > 0$ if $\lambda, \xi < \lambda_0(\zeta)$. Hence $||u_\lambda(t) - u_\xi(t)||^2 < \zeta t$, for such λ, ξ. Q.E.D.

ASSERTION 6. *A is maximal accretive, i.e. if $(Av - w, J(v - u)) \geq 0$ for all v in $D(A)$, then $u \in D(A)$, $w = Au$.*

PROOF. We add $||v - u||^2$ to the left-hand side of the inequality and obtain

$$(Av + v - w - u, J(v - u)) \geq 0.$$

For fixed z in X, $t > 0$, set $v_t = (A + I)^{-1}(u + w + tz)$. Then $Av + v - u - w = tz$, and we find after cancelling t that $(z, J(v_t - u)) \geq 0$. Letting $t \to 0+$, $v_t - u \to (A + I)^{-1}(u + w)$. Since J is continuous

$$(z, J((A + I)^{-1}(w + u) - u)) \geq 0.$$

Since this is true for all z in X, $(A + I)^{-1}(u + w) = u$. Hence u lies in $D(A)$, and $(A + I)u = w + u$, i.e. $Au = w$. Q.E.D.

ASSERTION 7. *If u_j converges strongly to u and Au_j converges weakly to w, then $u \in D(A)$, $Au = w$.*

PROOF. For each j, $v \in D(A)$, $(Av - Au_j, J(v - u_j)) \geq 0$. Taking the limit as $j \to \infty$ and using the fact that J is continuous, we obtain $(Av - w, J(v - u)) \geq 0$. Hence by Assertion 6, $u \in D(A)$, $Au = w$. Q.E.D.

ASSERTION 8. *$u(t) \in D(A)$ for all $t > 0$ and $A(u(t))$ is weakly continuous.*

PROOF. Since $\|v_\lambda(t) - u_\lambda(t)\| \leq \lambda \|Au_0\|$, $v_\lambda(t) \to u(t)$ strongly in X, uniformly on bounded intervals of t. $T_\lambda(u(t)) = A(v_\lambda(t))$ is uniformly bounded in norm for all $\lambda > 0$ and all t in R^+. For fixed t and any subsequence of λ such that $A(v_\lambda(t))$ converges weakly, the limit must equal $A(u(t))$. The hypotheses of Theorem 28 imply the space is reflexive, so that $A(v_\lambda(t))$ converges weakly to $A(u(t))$ where $u(t) \in D(A)$ for all $t \geq 0$. Since $u(t)$ is strongly continuous and $\|A(u(t))\| \leq \|Au(0)\|$, it follows similarly that $A(u(t))$ is weakly continuous.

To complete the proof of Theorem 28, we must show that u is weakly differentiable in X on R^+ and satisfies the desired differential equation. For each $\lambda > 0$,

$$u_\lambda(t) = u_0 - \int_0^t A(v_\lambda(s))ds.$$

Taking weak limits on both sides and using bounded convergence, we obtain

$$u(t) = u_0 - \int_0^t A(u(s))ds.$$

Hence u is once weakly differentiable and

$$du(t)/dt = -A(u(t)).$$

We know moreover that $\|Au(t)\| \leq \|Au(0)\|$.

Finally, let $u_s(t)$ be the corresponding solution on the interval $t \geq s$ with $u(s) = u_s(s)$. Then by uniqueness, $u_s(t) = u(t)$, $t \geq s$. However

$$\|Au(t)\| = \|Au_s(t)\| \leq \|A(u(s))\|, \quad t \geq s. \qquad \text{Q.E.D.}$$

UNIVERSITY OF CHICAGO.

A FIXED POINT THEOREM FOR HOLOMORPHIC MAPPINGS[1]

CLIFFORD J. EARLE AND RICHARD S. HAMILTON

1. Throughout this note, E and F will be complex Banach spaces of any dimension, and $X \subset E$, $Y \subset F$ will be bounded regions (= bounded connected open sets). A map $f: X \to F$ is holomorphic if and only if its Fréchet derivative $Df(x): X \to F$ exists at each $x \in X$ and is complex linear. Equivalently [**5**], f can be represented locally by absolutely convergent power series with complex homogeneous terms.

We say that a subset X' of X lies *strictly inside* X if there exists $\varepsilon > 0$ such that $\|x' - y\| > \varepsilon$ for all $x' \in X'$ and $y \in E \backslash X$.

THEOREM 1. *If $f: X \to X$ is holomorphic and $f(X)$ lies strictly inside X, then f has a unique fixed point.*

For infinite dimensional E, our condition on $f(X)$ is much weaker than the usual condition of compact closure in X. If $X \subset \mathbf{C}^n$, the two conditions coincide, and Theorem 1 is already known (see [**4**, p. 83] and [**8**, p. 322]).

We prove Theorem 1 in §4. Our proof uses a Finsler metric on X which satisfies the Schwartz-Pick Lemma that holomorphic maps do not increase distances. We define our metric in §2 and prove the Schwarz-Pick Lemma in §3. The last two sections, §§6 and 7, are devoted to a closer look at the metric. In the finite dimensional case it turns out to be the infinitesimal form of Carathéodory's metric [**1**]. That metric was already used by H.-J. Reiffen to prove a fixed point theorem for holomorphic maps in a finite dimensional complex analytic space [**8**, p. 322].

2. Denote by $H^\infty(X)$ the Banach space of bounded holomorphic functions on X, and by $H^\infty(X)^*$ its dual space. The evaluation map $\phi: X \to H^\infty(X)^*$ is defined by $\phi(x)(f) = f(x)$ for $x \in X$ and $f \in H^\infty(X)$. The following lemma is well known; we shall give its proof in §5.

LEMMA 1. *The map ϕ is a holomorphic embedding of X as a submanifold of $H^\infty(X)^*$. Further,*

(2.1) $(D\phi(x)e)f = Df(x)e$

(2.2) $\|\phi(x) - \phi(y)\| \geqq \delta\|x - y\|$

for all $x, y \in X$, $e \in E$, $f \in H^\infty(X)$ and some $\delta > 0$.

The *Carathéodory-Reiffen Finsler metric* on X is the function α on $X \times E$ defined by

$$\alpha(x, e) = \|D\phi(x)e\|$$

[1] Both authors were partially supported by NSF Grant GP-8219.

for all $x \in X$ and $e \in E$. Since $D\phi(x)$ is an isomorphism of E onto a closed subspace of $H^\infty(X)^*$, and since holomorphic maps are C^∞, α satisfies these conditions:

(a) $\alpha: X \times E \to \mathbf{R}$ is a locally Lipschitz function,

(b) for each $x \in X$, the function $\alpha(x, \cdot)$ is a norm equivalent to the original norm $\|\cdot\|$ on E.

Thus, α is a Finsler metric on X. The CRF-length $L(\gamma)$ of a C^1 curve γ in X and the CRF-distance $\rho(x, y)$ between two points in X are defined in the usual way. Since α is induced by the embedding ϕ, $L(\gamma)$ is merely the length of the curve $\phi \circ \gamma$ measured in $H^\infty(X)^*$, and $\rho(x, y)$ is the infimum of the lengths of C^1 curves joining $\phi(x)$ and $\phi(y)$ *in the manifold* $\phi(X)$. From that observation and (2.2) we get the rather crude estimate

(2.3) $\rho(x, y) \geqq \|\phi(x) - \phi(y)\| \geqq \delta\|x - y\|$ for all x and y in X and some $\delta > 0$.

3. We will now show that the CRF metric satisfies the Schwarz-Pick Lemma.

THEOREM 2. *For any bounded regions $X \subset E$ and $Y \subset F$ and any holomorphic map $f: X \to Y$, the induced map of tangent vectors does not increase lengths in the* CRF *metrics, i.e.*

$$\alpha(f(x), Df(x)e) \leqq \alpha(x, e)$$

for all $x \in X$ and $e \in E$.

PROOF. The holomorphic map $f: X \to Y$ induces a linear map $H^\infty(f): H^\infty(Y) \to H^\infty(X)$ defined by composition with f, and $\|H^\infty(f)\| \leqq 1$. Therefore, the adjoint map $H^\infty(f)^*: H^\infty(X)^* \to H^\infty(Y)^*$ also has norm $\leqq 1$.

Let $\phi: X \to H^\infty(X)^*$ and $\psi: Y \to H^\infty(Y)^*$ be the respective evaluation maps. Then the diagram

$$\begin{array}{ccc} X & \xrightarrow{\phi} & H^\infty(X)^* \\ {\scriptstyle f}\downarrow & & \downarrow{\scriptstyle H^\infty(f)^*} \\ Y & \xrightarrow{\psi} & H^\infty(Y)^* \end{array}$$

commutes. By the Chain Rule

$$\alpha(f(x), Df(x)e) = \|D\psi(f(x))Df(x)e\| = \|H^\infty(f)^* D\phi(x)e\| \leqq \|D\phi(x)e\| = \alpha(x, e).$$

Theorem 2 is proved.

For a discussion of some other metrics which satisfy the Schwarz-Pick Lemma, the reader should consult [2] and [6].

4. We shall next prove Theorem 1. By hypothesis, $f: X \to X$ is holomorphic, and there is a number $\varepsilon > 0$ such that $\|f(x) - y\| > \varepsilon$ for all $x \in X$ and $y \in E \ X$. There is also a number M such that $\|x\| \leqq M$ for all $x \in X$. Put $\mu = \varepsilon/2M$, fix $x \in X$, and define $g: X \to E$ by

$$g(y) = (1 + \mu)f(y) - \mu f(x) = f(y) + \mu(f(y) - f(x))$$

for all $y \in X$. Then g is holomorphic and $g(X) \subset X$, so

$$\alpha(g(x), Dg(x)e) \leqq \alpha(x, e)$$

for all $e \in E$, by Theorem 2. But $g(x) = f(x)$ and $Dg(x) = (1 + \mu)Df(x)$, so

$$\alpha(f(x), Df(x)e) = (1 + \mu)^{-1}\alpha(g(x), Dg(x)e)$$

and

$$\alpha(f(x), Df(x)e) \leqq (1 + \mu)^{-1}\alpha(x, e)$$

for all $x \in X$ and $e \in E$.

Put $\theta = (1 + \mu)^{-1} < 1$. Then $\rho(f(x), f(y)) \leqq \theta\rho(x, y)$ for all x and $y \in X$. The sequence $(f^n(x_0))$ is therefore ρ-Cauchy, for any $x_0 \in X$. Because of (2.3), ρ-Cauchy sequences are norm-Cauchy, and $(f^n(x_0))$ has a limit in E. Since that limit is in the closure of $f(X)$, it belongs to X and is the unique fixed point of f.

5. Now we prove Lemma 1. Let $x \in X$ and choose $\eta > 0$ so that the open ball $N_\eta(x)$ of radius η around x in E lies in X. For each $e \in E$ and $n \geqq 0$ we define a linear functional $P_n(e)$ on $H^\infty(X)$ by

$$P_n(e)(f) = (1/n!)D^nf(x)(e, \ldots, e)$$

for $f \in H^\infty(X)$. Since f is bounded in $N_\eta(x)$, it is represented there by its Taylor series; thus

$$\sum_{n=0}^{\infty} P_n(e)(f) = f(x + e) = \phi(x + e)(f)$$

for all $f \in H^\infty(X)$ and $x + e \in N_\eta(x)$. In other words

(5.1) $$\phi(y) = \sum_{n=0}^{\infty} P_n(y - x)$$

for all $y \in N_\eta(x)$. We claim that (5.1) defines an absolutely convergent power series. That is immediate from

LEMMA 2 (CAUCHY'S DERIVATIVE ESTIMATES). *$P_n(e) \in H^\infty(X)^*$ and $\|P_n(e)\| \leqq \eta^{-n}\|e\|^n$.*

PROOF. If $\|e\| < 1$, then

$$P_n(e)(f) = \frac{1}{2\pi i}\int_{|\zeta|=\eta} f(x + \zeta e)\zeta^{-(n+1)}d\zeta,$$

and $\|P_n(e)(f)\| \leqq \eta^{-n}\|f\|$. Lemma 2 follows, because $P_n(\lambda e) = \lambda^n P_n(e)$ for all $\lambda \in \boldsymbol{C}$.

We conclude that ϕ is holomorphic, since it is represented locally by an absolutely convergent power series with complex homogeneous terms [5]. Moreover

$$(D\phi(x)e)(f) = P_1(e)(f) = Df(x)e,$$

proving (2.1). That ϕ is an embedding follows from (2.2), which we shall now prove. Choose $\delta > 0$ so that $\|x\| < \delta^{-1}$ for all $x \in X$. For any x and y in X choose (by the Hahn-Banach theorem) a linear functional l of norm one on E with

$l(x - y) = \|x - y\|$. Then l is bounded by δ^{-1} in X, so $l \in H^\infty(X)$ and

$$\|\phi(x) - \phi(y)\| \geq \delta\|\phi(x)(l) - \phi(y)(l)\| = \delta\|l(x) - l(y)\| = \delta\|x - y\|.$$

That completes the proof of Lemma 1. We remark without proof that the range of $D\phi(x)$ will split in $H^\infty(X)^*$ if and only if the embedding of E in its double dual E^{**} splits.

6. If X is the unit ball $B = \{x \in E;\ \|x\| < 1\}$ in E, some explicit information about the CRF metric can be obtained. The finite dimensional case of our lemma was proved by Carathéodory [**1**].

LEMMA 3. *If B is the unit ball of E, the* CRF *metric in E satisfies*
(a) $\alpha(0, e) = \|e\|$
(b) $\rho(0, x) = \tanh^{-1}\|x\|$
for all $x \in B$ and $e \in E$.

PROOF. First we note, with Carathéodory, that the CRF metric on the unit disk $\Delta = \{\lambda \in \mathbf{C}; |\lambda| < 1\}$ is the Poincaré metric of constant curvature -1. In general, for any unit vector $u \in E$, the map $\lambda \to \lambda u$ is a CRF isometry from Δ onto its image in B (because the Hahn-Banach theorem gives a holomorphic left inverse map from B to Δ). That proves the lemma.

Lemma 3 determines $\alpha(x, e)$ completely if B is homogeneous, since biholomorphic automorphisms of B are length preserving. For example

PROPOSITION. *Suppose B is the unit ball of the Banach algebra $C(S)$ of continuous functions on the compact space S. Then*

$$\alpha(x, e) = \|e(1 - |x|^2)^{-1}\|$$

for all $x \in B$ and $e \in C(S)$.

PROOF. Put $T_x(y) = (y - x)(1 - \bar{x}y)^{-1}$ for x and y in B. Then T_x is a biholomorphic automorphism of B and $T_x(x) = 0$. Hence

$$\alpha(x, e) = \alpha(0, DT_x(x)e) = \|DT_x(x)e\|.$$

The proposition follows by computation.

We remark that many Banach algebras have unit balls with a transitive group of biholomorphic automorphisms. L. Harris [**3**] has studied the unit balls of J^*-algebras (these include all C^*-algebras). Their CRF metric is

$$\alpha(x, e) = \|(1 - xx^*)^{-1/2}e(1 - x^*x)^{-1/2}\|.$$

7. Finally we want to point out the connection between the CRF metric on X and the Gleason parts of the maximal ideal space $\mathcal{M}$ of the Banach algebra $H^\infty(X)$. Recall that $\mathcal{M}$ is the set of $\psi \in H^\infty(X)^*$ such that $\psi(1) = 1$ and $\psi(fg) = \psi(f)\psi(g)$ for all f and $g \in H^\infty(X)$. Then $\|\psi\| = 1$ for all $\psi \in \mathcal{M}$, and the evaluation map $\phi: X \to H^\infty(X)^*$ maps X into $\mathcal{M}$. The *hyperbolic metric on* $\mathcal{M}$ was defined by J. Lewittes [**7**] by the formula

$$d(\psi_1, \psi_2) = \sup\{\rho(\psi_1(f), \psi_2(f)); f \in H^\infty(X), \|f\| < 1\},$$

where of course ρ is the Poincaré (=CRF) distance in the unit disk Δ. d is clearly a (sometimes infinite) metric on $\mathscr{M}$. The property of having finite hyperbolic distance defines an equivalence relation on $\mathscr{M}$; the equivalence classes are the *Gleason parts.* It turns out that ψ_1 and ψ_2 are equivalent if and only if $\|\psi_1 - \psi_2\| < 2$; that is an immediate consequence of

LEMMA 4 (LEWITTES [7, p. 1090]). For all ψ_1 and ψ_2 in $\mathscr{M}$,

$$d(\psi_1, \psi_2) = 2\tanh^{-1}(\|\psi_1 - \psi_2\|/2).$$

The proof of Lemma 4 is geometric. It amounts to the assertion that if $\rho(a, b)$ is fixed, with a and b in Δ, then the norm $|a - b|$ is maximized when $a + b = 0$.

Of course, the hyperbolic metric induces a metric on X through the embedding $\phi: X \to \mathscr{M}$. Explicitly,

$$d(x, y) = 2\tanh^{-1}(\|\phi(x) - \phi(y)\|/2) \tag{7.1}$$

for all x and y in X. Since we also have

$$d(x, y) = \sup\{\rho(f(x), f(y)); f \in H^\infty(X), \|f\| < 1\},$$

Theorem 2 implies that $d(x, y) \leqq \rho(x, y)$. Hence the metric d is finite on X, and $\phi(X)$ lies in one Gleason part of $\mathscr{M}$. d is the *Carathéodory metric* on X [1]. Formula (7.1) shows that the CRF metric α is the infinitesimal form of the Carathéodory metric. In fact, for any C^1 curve γ in X we have

$$\lim_{t \to 0^+} d(\gamma(0), \gamma(t))/t = \|(\phi \circ \gamma)'(0)\| = \alpha(\gamma(0), \gamma'(0)).$$

REFERENCES

1. C. Carathéodory, *Über das Schwarzsche Lemma bei analytischen Funktionen von zwei komplexen Veränderlichen,* Math. Ann. **97** (1926), 76–98.

2. S. S. Chern, H. I. Levine and L. Nirenberg, *Intrinsic norms on a complex manifold,* (to appear).

3. L. A. Harris, *Schwarz lemmas for Jordan algebras of operators,* (to appear).

4. M. Hervé, *Several complex variables, local theory,* Oxford Univ. Press and Tata Institute of Fundamental Research, 1963.

5. E. Hille and R. S. Phillips, *Functional analysis and semigroups,* Amer. Math. Soc. Colloq. Publ., Vol. 31, Amer. Math. Soc., Providence, R.I., 1957.

6. S. Kobayashi, *Invariant distances on complex manifolds and holomorphic mappings,* J. Math. Soc. Japan **19** (1967), 460–480.

7. J. Lewittes, *A note on parts and hyperbolic geometry,* Proc. Amer. Math. Soc. **17** (1966), 1087–1090.

8. H.-J. Reiffen, *Die Carathéodorysche Distanz und ihre zugehörige Differentialmetrik,* Math. Ann. **161** (1965), 315–324.

CORNELL UNIVERSITY

VARIATION INTEGRALS IN FIBER BUNDLES

HALLDOR I. ELIASSON

Introduction. Throughout this paper N will denote a compact, connected manifold of class C^∞ and dimension n and $\pi: W \to N$ will denote a fiber bundle of class C^∞ over N. We will be concerned with variation integrals $J: C^\infty(W) \to \boldsymbol{R}$, which extend to a function on the manifold $H^k(W)$, of sections of class H^k (square integrable derivatives up to order k) in W. The main subject of our investigation is to characterize those variation integrals, which can be described by a C^∞ function J satisfying condition (C) of Palais and Smale [**10**]: Let f_j be a sequence in $H^k(W)$, such that $J(f_j)$ is bounded and $\|dJ(f_j)\|$ converges to zero, then f_j has a convergent subsequence. Moreover, we like to have a regularity theorem, which asserts, that the critical points of J are sections of class C^∞.

This problem, which can be described as a nonlinear Dirichlet problem, has been studied by Palais and his students [**8**], [**12**] and [**13**]. See also [**2**] for a discussion of energy functions. In the work of Palais and Saber W is embedded in a vector bundle E and the differential operators in question are linear operators on E restricted to sections in W. Here we consider a broader class of differential operators (also introduced in [**8**]) and treat the nonlinear Dirichlet problem in a completely intrinsic manner. Mrs. K. Uhlenbeck has also studied the same type of operators in a partially intrinsic treatment [**13**] and obtained results quite parallel to those presented here.

The variation integral J is of the form $J(f) = \int_N P(f)$, where P is a differential operator of order k, mapping sections in W to functions on N. In local coordinates

$$P: C^\infty(U, V) \to C^\infty(U, \boldsymbol{R}), \quad U \subset \boldsymbol{R}^n, V \subset \boldsymbol{R}^m,$$

can be written in the form

$$P(f)(x) = \sum_\alpha A_\alpha(x, f(x)) \cdot (D^{\alpha_1} f(x), \ldots, D^{\alpha_j} f(x)),$$

where the coefficient A_α is a C^∞ map

$$A_\alpha: U \times V \to L(L^{\alpha_1}(\boldsymbol{R}^n, \boldsymbol{R}^m), \ldots, L^{\alpha_j}(\boldsymbol{R}^n, \boldsymbol{R}^m); \boldsymbol{R})$$

and $D^s f: U \to L^s(\boldsymbol{R}^n, \boldsymbol{R}^m)$ is the derivative of f of order s. We sum over $\alpha = (\alpha_1, \ldots, \alpha_j)$ with $1 \leqq \alpha_\nu \leqq k$, $|\alpha| \leqq w$. This type of operators "quasi-polynomial operators of order k and weight w" is the simplest type, which is invariant under coordinate transformations of both the domain U and the range V and thus admits a generalization to fiber bundles. We call P an energy operator (and J an energy integral), if $w = 2k$ and the principal part of P is of the form

$$A_{(k,k)}(x, y) \cdot (\xi, \eta) = \langle G(x, y) \cdot \xi, \eta \rangle,$$

where $G(x, y)$ is a positive definite linear isomorphism. This property is again invariant under coordinate transformations.

Our main result (Theorem 16) states, that an energy integral J restricted to a subset of $H^k(W)$ relatively compact in $C^0(W)$ satisfies condition (C) and we prove an interior regularity theorem for the Euler-Lagrange equation $dJ(f) = 0$ (Theorem 8).

The problem to obtain conditions on J, such that boundedness of J on a subset A of $H^k(W)$ implies, that A is relatively compact in $C^0(W)$, seems to be very difficult. We do obtain a particular result in this direction (Theorem 17) and as a consequence we prove that the energy function

$$E_k : H^k(W) \to \mathbf{R}, \qquad E_k(f) = \sum_{i=1}^{k} \int_N \|\nabla^i f\|^2, \quad k > n/2,$$

satisfies condition (C).

We construct an intrinsic Riemannian metric for the manifold $H^k(W)$ (Theorem 13) and prove, that $H^k(W)$ is a complete Riemannian manifold in this metric. Moreover, we prove that a bounded subset in $H^k(W)$ is relatively compact in $C^0(W)$ (Theorem 19), so that any energy integral J of order $k > n/2$ restricted to a bounded subset of $H^k(W)$, satisfies condition (C). In particular, the set of critical points of J is locally compact in $H^k(W)$.

In the last paragraph we construct, as an example, a variation integral $J : H^2(N, M) \to \mathbf{R}$ satisfying condition (C) and having exactly the harmonic maps $N \to M$ as critical points. Here $n = 2$ or 3 and M has nonpositive sectional curvature. This improves the results of Eells and Sampson [**3**], in the case $\dim N = 2$ or 3.

1. **Global linear theory.**

1.1. *RMC structures on vector bundles.* N will denote a compact, connected manifold of class C^∞ and dimension $n \geqq 1$. ∂N will denote the boundary of N, $\tau : TN \to N$ the tangent bundle of N and $VB(N)$ the category of finite dimensional vector bundles of class C^∞ over N. $C^k(E)$ with E in $VB(N)$, $0 \leqq k \leqq \infty$, will denote the linear space of sections of class C^k in E and $C_c^\infty(E)$ the linear space of C^∞ sections having compact support in the interior of N.

A Riemannian metric for a vector bundle E in $VB(N)$ is a C^∞ section g in $L^2_s(E, \mathbf{R})$, such that $g(x)$ is an inner product for the fiber E_x, for all x in N. A connection for E is e.g. given by a map $K : TE \to E$, the connection map, which is locally of the form

$$K(x, \xi, y, \eta) = (x, \eta + \Gamma(x)\cdot(y, \xi)),$$

where $\Gamma : U \to L(\mathbf{R}^n, \mathbf{E}; \mathbf{E})$ is a C^∞ map, see [**4**]. The covariant derivative of a section ξ in $C^k(E)$ is then defined as a section $\nabla\xi$ in $C^{k-1}(L(TN, E))$ by the formula

$$\nabla\xi(x)\cdot v = K(T\xi(v)), \quad v \in T_xN.$$

This means locally:

$$\nabla\xi(x)\cdot y = D\xi(x)\cdot y + \Gamma(x)\cdot(y, \xi(x)).$$

In fact ∇ can be characterized as a linear differential operator

$$\nabla : C^\infty(E) \to C^\infty(L(TN, E))$$

of order one and with symbol $\sigma_\nabla(\omega)\cdot\xi = \omega \otimes \xi$.

We say that the covariant differentiation ∇ is Riemannian with respect to a Riemannian metric g for E, iff for any sections $\xi, \eta \in C^\infty(E)$ the derivative of the function $g(\xi, \eta)$ in the direction of some $v \in T_xN$, is given by

$$dg(\xi, \eta)(x)\cdot v = g(x)\cdot(\nabla\xi(x)\cdot v, \eta(x)) + g(x)\cdot(\xi(x), \nabla\eta(x)\cdot v).$$

A RMC structure for E consists then of a Riemannian metric together with a Riemannian connection. The existence of a RMC structure for any E in $VB(N)$ is well known and in particular, there is a unique symmetric Riemannian connection, corresponding to a given Riemannian metric, for the tangent bundle TN (the so-called Levi-Civita connection, symmetric means without torsion or locally: $\Gamma(x)\cdot(y, z) = \Gamma(x)\cdot(z, y)$).

We will now assume, that $E, F \in VB(N)$ have each a RMC structure and we give the vector bundle $L(F, E)$, the RMC structure, which is induced as follows. The Riemannian metric is defined by

$$\langle A, B\rangle = \mathrm{Trace}\,(B^*A) = \textstyle\sum_i \langle A\cdot e_i, B\cdot e_i\rangle,$$

where $\{e_i\}$ is an orthonormal base for F_x and $A, B \in L(F_x, E_x)$. A corresponding Riemannian connection for $L(F, E)$ is defined in [**4**] and its covariant differentiation is characterized by $\nabla(A\cdot\xi) = \nabla A \otimes \xi + A\cdot\nabla\xi$, for any $A \in C^\infty(L(F, E))$ and $\xi \in C^\infty(F)$. Taking $F = TN$, a RMC structure for TN and E induces a RMC structure on $L^j(TN, E)$ for all $j \geqq 1$, defined inductively using $L^{j+1}(TN, E) = L(TN, L^j(TN, E))$. The jth covariant derivative of a section ξ in E is then a section $\nabla^j\xi$ in $L^j(TN, E)$.

We have an inner product defined in $C^\infty(E)$ by $\langle\xi, \eta\rangle_0 = \int_N \langle\xi, \eta\rangle * 1$, where $*1$ is the volume element of N. We define the divergence of a section X in TN by $\mathrm{div}X = \mathrm{Trace}\,\nabla X$ and of a section A in $L(TN, E)$ by

$$\langle\eta, \mathrm{Div}A\rangle = \mathrm{Trace}(\nabla A^* \otimes \eta) \quad \text{for all } \eta \in C^\infty(E).$$

LEMMA 1. *We have for any* $X \in C_c^\infty(TN)$: $\int_N \mathrm{div}X * 1 = 0$ *and for any* $\eta \in C_c^\infty(E)$, $A \in C_c^\infty(L(TN, E))$, $\langle\eta, \mathrm{Div}A\rangle_0 = -\langle\nabla\eta, A\rangle_0$.

PROOF. The first formula is classic [**11**] and follows from $\mathrm{div}X * 1 = d(*1\cdot X)$. The second follows from

$$\begin{aligned}\mathrm{div}(A^*\cdot\eta) &= \mathrm{Trace}(\nabla(A^*\cdot\eta)) \\ &= \mathrm{Trace}(\nabla A^* \otimes \eta) + \mathrm{Trace}(A^*\cdot\nabla\eta) \\ &= \langle\eta, \mathrm{Div}A\rangle + \langle A, \nabla\eta\rangle.\end{aligned}$$

We will review briefly some of the classical facts on linear differential operators, in the language of covariant differentiation. We refer to [**9**] for a detailed treatment.

Any linear differential operator $P:C^\infty(E) \to C^\infty(F)$ can be written in the form

$$P(\xi) = \sum_{j=0}^{k} A_j \cdot \nabla^j \xi, \quad \text{with } A_j \in C^\infty(L(L^j(TN, E), F)),$$

where k is the order of P. The coefficient A_k of the principal part is independent of the RMC structures of TN, E, F. The symbol σ_P of P is defined as a fiber map:

$$\sigma_P : T^*N \to L(E, F), \qquad T^*N = L(TN, \mathbf{R}),$$

by

$$\sigma_P(\omega) \cdot \xi = A_k(\omega^k \otimes \xi), \ \omega^k = \omega \otimes \omega \otimes \ldots \otimes \omega \ k\text{-times}.$$

P is called elliptic, iff $\sigma_P(\omega)$ is injective for every $\omega \neq 0$. If P is of order $2k$ and $E = F$, then P is called strongly elliptic, iff $(-1)^k \sigma_P(\omega)$ is positive definite for all $\omega \neq 0$. This means, since N is compact

$$\langle (-1)^k \sigma_P(\omega) \cdot \xi, \xi \rangle \geqq \lambda \|\omega\|^{2k} \|\xi\|^2, \quad \lambda > 0.$$

for all $(\omega, \xi) \in T^*N \oplus E$.

The adjoint operator P^* of P is uniquely defined by the property

$$\langle P^*(\xi), \eta \rangle_0 = \langle \xi, P(\eta) \rangle_0 \quad \text{for all } \xi \in C_c^\infty(F), \eta \in C_c^\infty(E).$$

From the formulas

$$\sigma_{PQ}(\omega) = \sigma_P(\omega) \sigma_Q(\omega), \qquad \sigma_{P^*}(\omega) = \sigma_P(\omega)^*,$$

it is easily seen, that P^*P is strongly elliptic of order $2k$ if P is elliptic of order k. If $P = \sum A_j \nabla^j$, then $P^* = \sum (-\operatorname{Div})^j A_j^*$, since $-\operatorname{Div}$ is the adjoint of ∇ by Lemma 1. ∇^k is a linear operator of order k and has the symbol

$$\sigma_{\nabla^k}(\omega) \cdot \xi = \omega^k \otimes \xi.$$

∇^k is therefore elliptic and $(-\operatorname{Div})^k \nabla^k$ is strongly elliptic with symbol $(-1)^k \|\omega\|^{2k} \xi$. $\Delta = \operatorname{Div} \nabla$ is called the Laplacian on E and $-\Delta$ is strongly elliptic. More generally $(-1)^k \nabla^{2k}$ followed by any complete contraction is strongly elliptic, in fact its principal part is independent of the art of contraction, as is seen from the formula in [4] for the curvature tensor R

$$\nabla^2 \xi \cdot (X, Y) - \nabla^2 \xi(Y, X) = R(X, Y)\xi.$$

1.2. *Spaces of sections.* We now introduce norms in the linear space $C^\infty(E)$, $E \in VB(N)$, as follows. We choose some RMC structure for TN and E and define

$$|\xi| = \sum_{j=0}^{k} \sup_{x \in N} \|\nabla^j \xi(x)\|,$$

$$\|\xi\|_{(p,k)} = \sum_{j=0}^{k} \left(\int_N \|\nabla^j \xi\|^p * 1\right)^{1/p},$$

for a nonnegative integer k and some real number $p \geqq 1$. Then $C^k(E)$ is a Banach space with norm $|\ |_k$ and we define the Banach spaces $C_0^k(E)$, $L_k^p(E)$ to be the completions of $C_c^\infty(E)$ in the norms $|\ |_k$, $\|\ \|_{(p,k)}$. $L_k^p(E)$ is the well-known Sobolev space and $H^k(E) = L_k^2(E)$ is a Hilbert space with inner product

$$\langle \xi, \eta \rangle_k = \sum_{j=0}^{k} \langle \nabla^j \xi, \nabla^j \eta \rangle_0,$$

and the corresponding norm is $\|\xi\|_k = \langle \xi, \xi \rangle_k^{1/2}$.

Two different RMC structures for TN, E would induce equivalent norms both for $C^k(E)$ and $L^p_k(E)$, due to the fact that N is compact and if ∇, ∇' denote the two covariant derivations, then $\nabla^j - \nabla'^j$ is a linear differential operator of order $j - 1$.

For $p > 1$, $1/p + 1/p' = 1$, and k a positive integer, we define

$$\|\xi\|_{(p',-k)} = \sup_\eta \frac{|\langle \xi, \eta \rangle_0|}{\|\eta\|_{(p,k)}}, \quad \eta \in C^\infty_c(E)$$

and define $L^{p'}_{-k}(E)$ to be the completion of $C^\infty_c(E)$ in this norm. $L^p_k(E)$ and $L^{p'}_{-k}(E)$, $p > 1$, are then dual Banach spaces with $\langle \xi, \eta \rangle_0$ as a pairing.

THEOREM 1. *Let r be an integer, k a positive integer and $1 \leqq p$ $(1 < p$ if $r < k)$. Then*

$$\nabla^k : L^p_r(E) \to L^p_{r-k}(L^k(TN, E))$$

is a continuous linear map and

$$\|\nabla^k \xi\|_{(p,r-k)} \leqq \|\xi\|_{(p,r)}, \quad \xi \in L^p_r(E).$$

PROOF. It is sufficient to prove the norm inequality for C^∞ sections. For $r \geqq k$ it is obvious from the definition of the norms and else we use

$$\begin{aligned} |\langle \nabla^k \xi, \eta \rangle_0| &= |\langle \nabla^{k-j}\xi, (-\mathrm{Div})^j \eta \rangle_0 \\ &\leqq \|\nabla^{k-j}\xi\|_{(p,i)} \|\mathrm{Div}^j \eta\|_{(p',-i)}, \end{aligned}$$

which holds for any i and $0 \leqq j \leqq k$. In case $0 \leqq r < k$ we take $j = k - r$ and $i = 0$, then we obtain

$$\begin{aligned} |\langle \nabla^k \xi, \eta \rangle_0| &\leqq \|\nabla^r \xi\|_{(p,0)} \|\mathrm{Div}^{k-r}\eta\|_{(p',0)} \\ &\leqq \|\xi\|_{(p,r)} \|\eta\|_{(p',k-r)}, \end{aligned}$$

using $\|\mathrm{Div}^s \eta\| \leqq \|\nabla^s \eta\|$ and the Theorem for $r \geqq k$. In case $r < 0$ we put $j = k$ and $i = r$, then

$$\begin{aligned} |\langle \nabla^k \xi, \eta \rangle_0| &\leqq \|\xi\|_{(p,r)} \|\mathrm{Div}^k \eta\|_{(p',-r)} \\ &\leqq \|\xi\|_{(p,r)} \|\eta\|_{(p',k-r)}, \end{aligned}$$

which completes the proof.

COROLLARY 1. *Let P be a linear differential operator of order k from E to F. Then there is a unique extension of $P : C^\infty_c(E) \to C^\infty_c(F)$ to a continuous linear map $P : L^p_r(E) \to L^p_{r-k}(F)$, for any integer r and $p \geqq 1$ $(p > 1$ if $r < k)$.*

We quote here the following fundamental results on Sobolev spaces (see e.g. [1] and [7]).

Sobolev embedding theorems. Let $E \in VB(N)$, then (1) We have a completely continuous linear inclusion

$$L^p_r(E) \subset C^s_0(E) \quad \text{if } r - n/p > s.$$

(2) We have a continuous linear inclusion

$$L^p_r(E) \subset L^q_s(E) \quad \text{if } r \geqq s, r - n/p \geqq s - n/q,$$

which is completely continuous if the inequalities are strict.

We will sometimes denote $|\ |_k$ by $\|\ \|_{(\infty,k)}$ and C^k by L_k^∞ for convenience.

THEOREM 2. *Let $1 \leqq p, q \leqq \infty$, $0 \leqq m \leqq k$ and fix some RMC structure for N. Then there is a constant c depending only on p, q, m, k and the Riemannian manifold N, such that for any $E \in VB(N)$ with some RMC structure, we have*

$$\|\xi\|_{(q,m)} \leqq c\|\xi\|_{(p,k)} \quad \text{for all } \xi \in L_k^p(E),$$

provided $k - n/p \geqq m - n/p$ with $>$ if $q = \infty$.

REMARK. The existence of a c depending on E is granted by the Sobolev embedding Theorem, but it is of fundamental interest to us to be able to choose c independently of E. This has also been worked out independently by K. Uhlenbeck [**13**].

PROOF. We have a constant $c = c(q, r, N)$, such that

$$\|f\|_{(q,0)} \leqq c\|f\|_{(r,1)} \quad \text{for all } f \in L_1^r(N, \boldsymbol{R}),$$

provided $1 - n/r \geqq -n/q$ with $>$ if $q = \infty$. Now with $\xi \in C_c^\infty(E)$

$$d\|\xi\|\cdot v = \|\xi\|^{-1}\langle \xi, \nabla\xi\cdot v\rangle, \ \|d\|\xi\|\ \| \leqq \|\nabla\xi\|$$

which implies $\|(\|\xi\|)\|_{(r,1)} \leqq \|\xi\|_{(r,1)}$. Then $\|\xi\|_{(q,0)} = \|(\|\xi\|)\|_{(q,0)} \leqq c\|\xi\|_{(r,1)}$, which proves the theorem for $m = 0$, $k = 1$. Now suppose $m = 0$, $k > 1$ and choose $r_1, \ldots, r_{k-1}$, such that $-n/q \leqq 1 - n/r_1 \leqq \ldots \leqq v - n/r_\mu \leqq \ldots \leqq k - n/p$, then

$$\begin{aligned} \|\xi\|_{(r_v,\mu)} &= \sum_{j=0}^{\mu} \|\nabla^j\xi\|_{(r_v,0)} \leqq c_\mu \sum_{j=0}^{\mu} \|\nabla^j\xi\|_{(r_{v+1},1)} \\ &\leqq 2c_\mu\|\xi\|_{(r_{v+1},\mu+1)} \end{aligned}$$

and we obtain $\|\xi\|_{(q,0)} \leqq c\|\xi\|_{(p,k)}$. For $m > 0$, we have

$$\begin{aligned} \|\xi\|_{(q,m)} &= \sum_{j=0}^{m} \|\nabla^j\xi\|_{(q,0)} \leqq \sum_{j=0}^{m} c_j\|\nabla^j\xi\|_{(p,k-j)} \\ &\leqq c\|\xi\|_{(p,k)}. \end{aligned}$$

which completes the proof.

The following theorem, in the form of multiplication of functions on N, is due to Palais [**8**].

THEOREM 3. *Let $s, \alpha_0, \ldots, \alpha_j$ be integers and $1 \leqq q, p < \infty$ with $q > 1$ if $s < 0$. Let $F, E_1, \ldots, E_j \in VB(N)$ and put $E_0 = L(E_1, \ldots, E_j; F)$. Then the evaluation map $ev: E_0 \times E_1 \times \ldots \times E_j \to F$, $ev(\xi_0, \ldots, \xi_j) = \xi_0\cdot(\xi_1, \ldots, \xi_j)$ induces a continuous multilinear map*

$$L_{\alpha_0}^{p_0}(E_0) \times \ldots \times L_{\alpha_j}^{p_j}(E_j) \to L_s^q(F)$$

and consequently a continuous linear inclusion

$$L_{\alpha_0}^{p_0}(L(E_1, \ldots, E_j; F)) \to L(L_{\alpha_1}^{p_1}(E_1), \ldots, L_{\alpha_j}^{p_j}(E_j); L_s^q(F)),$$

provided one of the following conditions is satisfied:

(1) $\alpha_\nu \geqq n/p_\nu$ *for all* $\nu = 0, \dots, j$ *and* $s - n/q \leqq \min_\nu(\alpha_\nu - n/p_\nu)$ *with* $s - n/q < 0$, *if* $\alpha_\nu = n/p_\nu$ *for more than one* ν.

(2) $\alpha_\nu \geqq 0$ *for all* ν *and*

$$\max(-n, s - n/q) \leqq - \sum_{\nu=0}^{j} (n/p_\nu - \alpha_\nu)^+$$

(*where* $x^+ = x$ *if* $x \geqq 0$ *and* $= 0$ *if* $x < 0$) *with strict inequality if* $\alpha_\nu = n/p_\nu$ *for some* ν.

PROOF. Suppose first, that s is nonnegative. With $\xi_\nu \in L^{p_\nu}_{\alpha_\nu}(E_\nu)$ we know that $\nabla^t(\xi_0 \cdot (\xi_1, \dots, \xi_j))$ is a sum of terms $\nabla^{\beta_0}\xi_0 \cdot (\nabla^{\beta_0}\xi_1, \dots, \nabla^{\beta_j}\xi_j)$ with $|\beta| = t \leqq s$. With numbers q_ν, such that $1 \leqq q_\nu \leqq \infty$ and $\sum 1/q_\nu \leqq 1/q$ we have

$$\begin{aligned} \|\nabla^{\beta_0}\xi_0 \cdot (\nabla^{\beta_1}\xi_1, \dots, \nabla^{\beta_j}\xi_j)\|_{(q,0)} &\leqq \text{const.} \|\nabla^{\beta_0}\xi_0\|_{(q_0,0)} \cdots \|\nabla^{\beta_j}\xi_j\|_{(q_j,0)} \\ &\leqq \text{const.} \|\xi_0\|_{(q_0,\beta_0)} \cdots \|\xi_j\|_{(q_j,\beta_j)}, \end{aligned}$$

where if $q_\nu = \infty$ we mean the sup norm. Moreover, we have

$$\|\xi_\nu\|_{(q_\nu,\beta_\nu)} \leqq \text{const.} \|\xi_\nu\|_{(p_\nu,\alpha_\nu)} \text{ if } \beta_\nu - n/q_\nu \leqq \alpha_\nu - n/p_\nu$$

with $q_\nu < \infty$ if $\beta_\nu = \alpha_\nu - n/p_\nu$. We are done if we find $q_0, \dots, q_j$ with these properties. We proceed as follows.

If $\beta_\nu - \alpha_\nu + n/p_\nu > 0$, put $n/q_\nu = \beta_\nu - \alpha_\nu + n/p_\nu$.

If $\beta_\nu - \alpha_\nu + n/p_\nu < 0$, put $q_\nu = \infty$.

Then it is easily seen, that the q_ν with $\beta_\nu - \alpha_\nu + n/p = 0$ can be chosen so large, that $\sum 1/q_\nu \leqq 1/q$ holds, provided either (1) or (2) is satisfied.

Suppose now, that s is a negative integer. Then we must show, that

$$L^{p_0}_{\alpha_0}(L(E_1, \dots, E_j, F; \mathbf{R})) \subset L(L^{p_1}_{\alpha_1}(E_1), \dots, L^{p_j}_{\alpha_j}(E_j), L^{q'}_{-s}(F); L^1_0(N, \mathbf{R}))$$

since then $\langle \xi_0 \cdot (\xi_1, \dots, \xi_j), \eta \rangle \in L^1_0(N, \mathbf{R})$ with $\eta \in L^{q'}_{-s}(F)$. Using the previous result, we see easily that the inclusion is continuous. Q.E.D.

THEOREM 4. *Let* $|s| \leqq r$, $1 \leqq p, q < \infty$, $r > n/p$ *and* $q > 1$ *if* $s < 0$. *Then we have a continuous linear inclusion*

$$L^p_r(L(E, F)) \subset L(L^q_s(E), L^q_s(F)), \qquad E, F \in VB(N)$$

provided $-n - r + n/p \leqq s - n/q \leqq r - n/p$.

PROOF. If $s \geqq n/q$, then condition (1) in Theorem 3 requires only $s - n/q \leqq r - n/p$ and if $0 \leqq s < n/q$, then condition (2) in Theorem 3 is satisfied. Thus for $s \geqq 0$ we need only $s - n/q \leqq r - n/p$. Let $s < 0$. Then by duality we must have

$$L^p_r(L(F, E)) \subset L(L^{q'}_{-s}(F), L^{q'}_{-s}(E))$$

or $-s - n/q' \leqq r - n/p$ by our previous result. Q.E.D.

REMARK. In the terminology of [**4**], Theorem 4 asserts, that property $L^p_r L \subset L(L^q_s, L^q_s)$ holds for these section functors.

THEOREM 5. *Let $E, F \in VB(N)$ and let V be an open subset of E, which projects onto N. Let $f: V \to F$ be a C^∞ fiber map, then*

$$L_k^p(f): L_k^p(V) \to L_k^p(F), \qquad L_k^p(f)(\xi) = f \circ \xi$$

is a continuous map, provided $k > n/p$.

PROOF. It suffices to prove that

$$\nabla^k \circ L_k^p(f): L_k^p(V) \to L_0^p(F)$$

is continuous, using any RMC structure. Now using Lemma 2.3 [**4**], $\nabla^k(f \circ \xi)$ is a sum of terms of the form

$$(\nabla_1^r D_2^j f \circ \xi) \cdot (\nabla^{\alpha_1}\xi, \ldots, \nabla^{\alpha_j}\xi), \quad r + |\alpha| = k.$$

$\nabla_1^r D_2^j f$ is a C^∞ fiber map and induces a continuous map

$$C_0^0(V) \to C_0^0(L(L^{\alpha_1}(TN, E), \ldots, L^{\alpha_j}(TN, E); L^k(TN, F))) = X,$$

since C_0^0 is a manifold model [**4**]. Moreover we have a continuous inclusion $L_k^p(V) \subset C_0^0(V)$ by the Sobolev embedding Theorem and the map

$$\nabla^{\alpha_\nu}: L_k^p(V) \to L_{k-\alpha_\nu}^p(L^{\alpha_\nu}(TN, E))$$

is continuous by Theorem 1. We are therefore done, if we have a continuous linear inclusion

$$X \subset L(L_{k-\alpha_1}^p(L^{\alpha_1}(TN, E)), \ldots, L_{k-\alpha_j}^p(L^{\alpha_j}(TN, E)); L_0^p(L^k(TN, E))).$$

This is the case, if condition (2) in Theorem 2 is satisfied (the proof being the same, replacing $L_{\alpha_0}^p$ by C_0^0). We have $\sum_{\nu=1}^j(-k + \alpha_\nu + n/p)^+ \leqq |\alpha| - j(k - n/p) \leqq |\alpha| - k + n/p \leqq n/p$, with strict inequality if some $-k + \alpha_\nu + n/p = 0$, which completes the proof.

THEOREM 6. *If $k > n/p$, then the section functor L_k^p is a manifold model. In particular the map $L_k^p(f)$ in Theorem 5 is of class C^∞ and its derivative is: $DL_k^p(f) = L_k^p(D_2 f)$.*

PROOF. That L_k^p is a manifold model means, that the three conditions: $L_k^p \subset C_0^0$, $L_k^p L \subset L(L_k^p, L_k^p)$ and Theorem 5 holds [**4**]. The last statement is proved in Lemma 4.1 [**4**].

1.3. *Quasi-polynomial differential operators.* Let $E, F \in VB(N)$ and let V be an open subset of E, which projects onto N. A quasi-polynomial differential operator from V to F of order k and weight w is a map

$$P: C^\infty(V) \to C^\infty(F),$$

which can be written, using some RMC structures, as

$$P(\xi) = \textstyle\sum_\alpha (A_\alpha \circ \xi) \cdot (\nabla^{\alpha_1}\xi, \ldots, \nabla^{\alpha_j}\xi),$$

$\alpha = (\alpha_1, \ldots, \alpha_j), j \geqq 0, 1 \leqq \alpha_\nu \leqq k, |\alpha| = \sum \alpha_\nu \leqq w$ and

$$A_\alpha: V \to L(L^{\alpha_1}(TN, E), \ldots, L^{\alpha_j}(TN, E); F)$$

is a fiber preserving map of class C^∞.

The form of the representation is independent of the RMC structure used, although the coefficients A_α depend upon it. However, if $w = pk$, where p is an integer, then the coefficient A_α with $\alpha = (\alpha_1, \ldots, \alpha_p)$, $\alpha_v = k$, is independent of the RMC structure and the corresponding term $(A_\alpha \circ \xi)\cdot(\nabla^k\xi, \ldots, \nabla^k\xi)$ is called the principal part of P. Note, that a zero weight term $A_0 \circ \xi$ $(j = 0)$ can occur. We will denote the set of quasi-polynomial differential operators of order k and weight w from V to F by $PD_k^w(V, F)$.

It is easily seen, that if $f: V' \to V$, $V' \subset E'$, is a C^∞ fiber map, then

$$P': C^\infty(V') \to C^\infty(F), \qquad P'(\xi') = P(f \circ \xi')$$

is in $PD_k^w(V', F)$. The formula for P in local coordinates is of the same form as above, replacing ∇ by ordinary differentiation D.

Part 2 of the next theorem has been proven by Palais [**8**], but the extensions 1 and 3 are necessary for the regularity theorem that follows.

THEOREM 7. *Let $1 \leqq p, q < \infty$ and r, k, s, w be integers with $r > n/p$, $1 \leqq k \leqq w$, $s \leqq r - k$ and let $q > 1$ if $s < 0$. If $P \in PD_k^w(V, F)$, $V \subset E$, then we have a unique continuous extension of P to a C^∞ map*

$$P: L_r^p(V) \to L_s^q(F),$$

provided

(1) *If $r \geqq k + n/p$, then $s - n/q \leqq r - k - n/p$ with strict inequality if $r - k - n/p = 0$ and $w \geqq 2k$.*

(2) *If $k + n/p > r \geqq k$,* then

$$\max(-n, s - n/q) \leqq -w/k(k + n/p - r).$$

(3) *If $k > r \geqq w/2$, then*

$$s - n/q \leqq r - n/p - k$$

with strict inequality if $w - k = r - n/p$, and

$$\max\left(-n, s - \frac{n}{q}\right) \leqq -\frac{w}{r}\frac{n}{p}$$

with strict inequality if $2r = w$ and $s - n/q = -n$.

PROOF. Let $P(\xi) = \sum_\alpha (A_\alpha \circ \xi)\cdot(\nabla^{\alpha_1}\xi, \ldots, \nabla^{\alpha_j}\xi)$, $1 \leqq \alpha_v \leqq k$, $|\alpha| \leqq w$. The map $L_r^p(A_\alpha): L_r^p(V) \to L_r^p(L(L^{\alpha_1}(TN, E), \ldots, L^{\alpha_j}(TN, E); F)) = X$ is of class C^∞ by Theorem 6, since $r > n/p$. Then as

$$\nabla^{\alpha_v}: L_r^p(V) \to L_{r-\alpha_v}^p(L^{\alpha_v}(TN, E))$$

is of class C^∞, we are done if we have a continuous linear inclusion

$$X \subset L(L_{r-\alpha_1}^p(L^{\alpha_1}(TN, E)), \ldots, L_{r-\alpha_j}^p(L^{\alpha_j}(TN, E)); L_s^q(F)).$$

(1) Suppose $r \geqq k + n/p$, then $r - \alpha_v \geqq r - k \geqq n/p$ and by Theorem 3 we must have $s - n/q \leqq \min(r - \alpha_v - n/p)$ with strict inequality if $r - k = n/p$ and $\alpha_v = k$ for more than one v. This is a consequence of our condition in 1.

(2) Suppose $k + n/p > r \geqq k$. Now $r - \alpha_\nu \geqq k - \alpha_\nu \geqq 0$ for all ν and the condition is by Theorem 3:

$$\max(-n, s - n/q) \leqq -\sum(\alpha_\nu - r + n/p)^+,$$

with strict inequality if $r - \alpha_\nu = n/p$ for some ν. An elementary computation shows the maximum, over all possible α, of $\sum(\alpha_\nu - r + n/p)^+$ to be equal to $(w/k)(k - r + n/p)$ and taken only if $jk = w$ and all $\alpha_\nu = k$, but in that case $r - \alpha_\nu \neq n/p$, so our condition in 2 is sufficient.

(3) Let $k > r \geqq w/2$ and suppose first that all $\alpha_\nu \leqq r$. Then P is of order r and using condition 2 in Theorem 7, which we just proved, we must have

$$\max\left(-n, s - \frac{n}{p}\right) \leqq -\frac{w}{r}\frac{n}{p}.$$

Now suppose some $\alpha_\nu > r$, e.g. $\alpha_j > r$. Then we have for $\nu \neq j : \alpha_\nu \leqq w - \alpha_j \leqq 2r - r = r$. We must show

$$L_r^p L \subset L(L_{r-\alpha_1}^p, \ldots, L_{r-\alpha_{j-1}}^p, L_{-s}^{q'}; L_{\alpha_j - r}^{p'}).$$

Since $r - \alpha_\nu > 0$ and $-s \geqq 0$, this holds by Theorem 3, if

$$\alpha_j - r - \frac{n}{p'} \leqq -\sum_{\nu<j}\left(-r + \alpha_\nu + \frac{n}{p}\right)^+ - \left(s + \frac{n}{q'}\right)^+$$

with strict inequality if $r - \alpha_\nu - n/p = 0$ for some $\nu < j$ or if $s + n/q' = 0$. Suppose $w - \alpha_j - r + n/p > 0$, then

$$\sum_{\nu<j}(\alpha_\nu - r + n/p)^+ \leqq w - \alpha_j - r + n/p$$

with equality only if $\alpha_\nu = w - \alpha_j$ for some $\nu < j$ and $\alpha_\nu = 0$ for all other $\nu < j$. But in that case $r - \alpha_\nu \neq n/p$ for all $\nu < j$. Using this estimate, we are done if $s - n/q \leqq w + 2(r - n/p)$ with strict inequality if $s - n/q = -n$. This holds, since $w < 2r$ implies $-wn/rp < -w + 2(r - n/p)$.

Suppose now that $w - \alpha_j - r + n/p \leqq 0$. Then $\alpha_\nu - r + n/p \leqq 0$ for all $\nu < j$ and in case $s + n/q' > 0$ the first condition in (3) Theorem 7 implies

$$\alpha_j - r - n/p' \leqq -s - n/q'$$

with strict inequality if necessary, which is the condition above. If $s + n/q' \leqq 0$ we use (1) Theorem 3 and the condition is now

$$\alpha_j - r - \frac{n}{p'} \leqq \min\left(r - \alpha_\nu - \frac{n}{p} - s - \frac{n}{q'}\right), \quad \nu < j.$$

This follows from $s - n/q \leqq r - k - n/p$ and $\alpha_j + \alpha_\nu \leqq w$,

$$-n \leqq -\frac{w}{r}\frac{n}{p} \leqq -w + 2\left(r - \frac{n}{p}\right).$$

COROLLARY 7.1. *Let $P \in PD_k^k(V, F)$, $V \subset E$, $r > n/p$, $r \geqq k/2$ and $k \leqq pr$. Then $P : L_r^p(V) \to L_{r-k}^p(F)$ of class C^∞.*

COROLLARY 7.2. *Let $P \in PD_k^{pk}(V, F)$, $V \subset E$, $r > n/p$ and $r \geqq k$. Then $P: L_r^p(V) \to L_{r-k}^1(F)$ of class C^∞.*

THEOREM 8 (REGULARITY). *Let N be without boundary and suppose $P \in PD_m^m(V, F)$, $V \subset E$, is of the form*

$$P(\xi) = (G \circ \xi) \cdot L\xi + Q(\xi),$$

where L is a linear elliptic operator of order m, $G: V \to L(F, F)$ is a C^∞ fiber map, such that $G(v)$ is a linear isomorphism for all $v \in V$ and $Q \in PD_{m-1}^m(V, F)$. If $\xi \in L_r^p(V)$ with $pr > n$, $pr \geqq m$, $2r \geqq m$, $p(n-1) \geqq n$ if $n \geqq 2$, and if $P(\xi) = 0$, then $\xi \in C^\infty(V)$.

PROOF. We show that $\xi \in L_r^p(V)$ implies $\xi \in L_{r+1}^p(V)$ for any integer r satisfying the conditions in the theorem. The theorem then follows by induction and the Sobolev embedding Theorem.

The proof goes as follows. We take $s = r + 1 - m$ and choose the largest q such that $Q(\xi) \in L_s^q(F)$, using Theorem 7. Then

$$L\xi = -(G\xi)^{-1} \cdot Q(\xi) \in L_s^q(F),$$

since $(G \circ \xi)^{-1} \in L_r^p(L(F, F))$ by Theorem 5 and $L_r^p L \subset L(L_s^q, L_s^q)$ by Theorem 4, provided $-n - r + n/p \leqq s - n/q \leqq r - n/p$, which is to be shown. Since we will be able to choose $q > 1$ (for $n > 1$), it now follows from the regularity theorems for linear elliptic operators, that $\xi \in L_{r+1}^q(V)$. If $q < p$, then we repeat the same cycle and obtain a sequence $1 < q_1 < q_2 < \ldots$ with $\xi \in L_{r+1}^{q_\nu}(V)$ and get $q_\nu \geqq p$ for finite ν, which implies that $\xi \in L_{r+1}^p(V)$. The regularity theorem for linear elliptic operators used here, is well known, but possibly not available in the literature for $s < 0$. There is an easy proof using pseudo differential operators, as L. Nirenberg has kindly pointed out to me.

The choice of q goes as follows.

(1) If $r \geqq m - 1 + n/p$, we can take $q = p$.

(2) $m - 1 + n/p > r \geqq m - 1 > 0$. Take $q = q_1$:

$$\frac{n}{q_\nu} = r + 1 - \left(\frac{m}{m-1}\right)^\nu \left(r - \frac{n}{p}\right), \quad \nu = 1, 2, \ldots.$$

Then if $n \geqq 2$: $n \geqq 1 + n/p > n/q_1 > n/q_2 > \ldots$ and

$$s - \frac{n}{q_\nu} \geqq s - \frac{n}{q} = -m + \frac{m}{m-1}\left(r - \frac{n}{p}\right) \geqq -n,$$

as seen by checking for $r = \max(m - 1, m/p, n/p)$.

(3) $m - 1 > r \geqq m/2$. If $r - n/p - m + 1 < -mn/rp$, then we can take $q = p$. If $r - n/p - m + 1 \geqq -mn/rp$, then we take

$$\frac{n}{q} = r + 1 - \frac{m}{r}\left(r - \frac{n}{p}\right) = 1 + \frac{n}{p} - \left(\frac{m}{r} - 1\right)\left(r - \frac{n}{p}\right)$$

$$\frac{n}{q_\nu} = r + 1 - \left(\frac{m}{r+1}\right)^{\nu-1} \frac{m}{r}\left(r - \frac{n}{p}\right).$$

2. Global nonlinear theory.

2.1. *RMC structure in fiber bundles.* M will denote a C^∞ manifold without boundary and of dimension m. $\pi: W \to N$ is a C^∞ fiber bundle with M as a standard fiber. We then have the following vector bundles and maps:

$$\begin{array}{ccccccc} T_2W & \to & TW & \overset{T\pi}{\to} & \pi^*TN & \to & TN \\ \downarrow & & \downarrow & & \downarrow & & \downarrow \tau \\ W & = & W & = & W & \overset{\pi}{\to} & N \end{array}$$

T_2W is the kernel of the bundle map $T\pi$ and is called the vertical tangent bundle of W, since the vectors in T_2W at $a \in W$ are tangent to curves in the fiber W_x, $x = \pi(a)$. A local trivialization for π is given by a C^∞ diffeomorphism $\Phi: \pi^{-1}(U) \to U \times M$, where U is an open subset of N, such that $\Phi(a) = (x, \Phi_x(a))$, $x = \pi(a)$, where $\Phi_x: W_x \to M$ is a C^∞ diffeomorphism. The maps

$$T\Phi: T\pi^{-1}(U) \to TU \times TM$$
$$T_2\Phi: T_2\pi^{-1}(U) \to U \times TM$$

are then the tangent trivializations for the vector bundles TW and T_2W.

DEFINITION. A connection map for π is a C^∞ section H in $L(TW, T_2W)$, which at each point is a surjective projection. A RMC structure for π consists of a connection map for π and RMC structures for the vector bundles $\tau: TN \to N$ and $T_2W \to W$ (see 1.1).

In particular, if $W = N \times M$, then the projection $(\tau, id): TW = TN \times TM \to N \times TM = T_2W$ is a connection map. If we have a RMC structure for $TW \to W$, then the orthogonal projection $TW \to T_2W$ would give us a connection map for π and a RMC structure for $T_2W \to W$ would also be induced.

We will now suppose, that $\pi: W \to N$ has some RMC structure. We see easily, that in a local trivialization the local representative $H_0 = T_2\Phi \circ H \circ (T\Phi)^{-1}$ of the the connection map H is of the form

$$H_0(v, \xi) = \xi + \Lambda(x, y)\cdot v, \quad v \in T_xN, \xi \in T_yM$$

with $\Lambda \in C^\infty(L(T_1(U \times M), T_2(U \times M)))$. This local property is of course invariant and characterizes the connection map H.

Let $f: X \to W$ be a differential map. We define the covariant derivative ∇f of f by

$$\nabla f = H \circ Tf: TX \to T_2W.$$

The map $(T\pi, H): TW \to \pi^*TN \oplus T_2W$ is a bundle equivalence (in a local trivialization it maps (v, ξ) at (x, y) to $(v, \xi + \Lambda(x, y)\cdot v)$ at (x, y)) and $Tf = (T\pi, H)^{-1}(Th, \nabla f)$, $h = \pi \circ f$. In particular, if f is a section in W, then ∇f contains all the informations about the tangent Tf. We will usually interpret the covariant derivative ∇f of a section f in W as a section in $L(TN, T_fW)$, $T_fW = f^*T_2W$. Now we have a RMC structure in T_fW, the pull-back by f of the RMC structure in T_2W (see [**4**]). Therefore, we have an induced RMC structure on $L^j(TN, T_fW)$ for all j. In particular, we have higher covariant derivatives of f and more generally, if ξ is a section in $L^j(TN, T_fW)$, then $\nabla\xi$ is a section in $L^{j+1}(TN, T_fW)$.

LEMMA 2. *We have the following local formulas for a section ξ in $L^j(TN, T_hW)$ and $h \in C^\infty(W)$, where Γ_N, Γ_2 and Λ denote the local connectors for TN, T_2W and W:*

$$(\nabla\xi)(x)\cdot(y, y_1, \ldots, y_j) = D\xi(x)\cdot(y, y_1, \ldots, y_j) + \Gamma_2(x, h(x))\cdot(\xi(x)\cdot(y_1, \ldots, y_j), (y, Dh(x)\cdot y)) - \sum_{\nu=1}^{j} \xi(x)\cdot(y_1, \ldots, y_{\nu-1}, \Gamma_N(x)\cdot(y, y_\nu), y_{\nu+1}, \ldots, y_j),$$

$\nabla h(x) = Dh(x) + \Lambda(x, h(x))$, $\nabla^j h = D^j h + P_j(h)$, *where P_j is a quasi-polynomial differential operator of order $j - 1$ and weight j.*

PROOF. The first formula follows from the local formulas in [**4**], the second from the local representation of H given here and the third follows from the two first by induction on j.

Let a, b be two points in W, we define the vertical distance between a and b by

$$d_2(a, b) = \inf_\alpha \int_0^1 \|\nabla\alpha(t)\| dt,$$

where the infimum is over all C^1 curves $\alpha:[0, 1] \to W$, with $\alpha(0) = a$ and $\alpha(1) = b$. Then $d_2(a, b) + d_N(\pi(a), \pi(b))$, is an admissible metric for the space W, if d_N denotes the distance function for N. The following lemma is a generalization of a well-known norm inequality for functions on a domain satisfying the cone property [**6**].

LEMMA 3. *Let $n < p < \infty$, then there is a constant l depending only on p and the Riemannian manifold N, such that*

$$d_2(f(x), f(y)) \leqq l d_N(x, y)^{1-n/p}\left(\int_N \|\nabla f\|^p * 1\right)^{1/p}$$

for all $f \in C^\infty(W)$ and $x, y \in N$.

PROOF. Let c be a geodesic in N of minimal length connecting x and $y:c:[0, 1] \to N$, $\nabla\partial c = 0$, $c(0) = x$, $c(1) = y$ and $r = \|\partial c\| = d_N(x, y)$. Let $Y_1, \ldots, Y_{n-1}$ be parallel fields along c such that $(1/r)\partial c, Y_1, \ldots, Y_{n-1}$ form an orthonormal base for the tangent space of N at every point on c. We put

$$\phi(s, t) = \exp\left(rt(1 - t)\sum_{\nu=1}^{n-1} s_\nu Y_\nu(t)\right),$$

for $s = (s_1, \ldots, s_{n-1}) \in \boldsymbol{R}^{n-1}$ and $0 \leqq t \leqq 1$. Now we can find a cone K (or a ball if N is without boundary) in $\boldsymbol{R}^{n-1}$ of a finite positive volume (depending on the RMC structure of N) such that ϕ restricted to $K \times (0, 1)$ is a diffeomorphism onto its image in N for a suitable choice of $Y_1, \ldots, Y_{n-1}$ and such that we have

$$\left\|\frac{d}{dt}\phi(s, t)\right\| \leqq Cr$$

$$|\det \partial\phi(s, t)| \geqq C'r^n(t(1 - t))^{n-1}$$

with some positive constants C, C', for all $s \in K$, $0 < t < 1$. Then since $\phi(s, 0) = x$ and $\phi(s, 1) = y$ for all $s \in K$, we get

$$d_2(f(x), f(y)) \leqq \int_0^1 \left\| H \frac{d}{dt} f(\phi(s, t)) \right\| dt$$

$$\leqq \int_0^1 \|\nabla f(\phi(s, t))\| \left\| \frac{d}{dt} \phi(s, t) \right\| dt.$$

Integration over K and Hölder inequality give

$$\mathrm{Vol}(K) d_2(f(x), f(y)) \leqq r^{1-n/p} \left(\int_N \|\nabla f\|^p\right)^{1/p} \left(C'' \int_0^1 (t(1-t))^\gamma dt\right)^{1/p'}$$

with constant $C'' > 0$ and $\gamma = -p'/p(n-1)$. The integral over t is finite, since $p > n$ implies $\gamma > -1$. Q.E.D.

The RMC structure for $T_2 W \to W$ induces a RMC structure for $TW_x \to W_x$ (pulling back by the inclusion $W_x \subset W$) for all $x \in N$. Therefore, we have an exponential map $\exp: T_2 W \to W$, such that $\exp: (T_2 W)_a = T_a W_x \to W_x$, $x = \pi(a)$, is the exponential map for the Riemannian manifold W_x.

If $h \in C^0(W)$ we define a map

$$\exp_h : C^0(T_h W) \to C^0(W) \quad \text{by } \exp_h(\zeta) = \exp \circ \zeta.$$

By a well-known property of the exponential map, there is an open neighborhood V_h of the set of zero vectors in $T_h W$, such that exp restricted to any fiber of V_h is injective. Then $\exp_h$ restricted to $C^0(V_h)$ is injective and provides the Banach manifold $C^0(W)$ with a chart.

2.2. *Manifolds and bundles of sections.* As before $\pi: W \to N$ will denote a C^∞ fiber bundle over a compact C^∞ manifold N. We will moreover assume, that W does admit a continuous section, so that $C^\infty(W)$ is not empty.

THEOREM 9. *Let $k > n/p$, $1 \leqq p < \infty$. Then there is a unique Banach manifold $L_k^p(W)$ characterized as follows. $L_k^p(W)$ is a subset of $C^0(W)$, such that given $h \in C^\infty(W)$ and some RMC structure for π, then*

$$\exp_h : L_k^p(V_h) \to L_k^p(W),$$

is a chart for $L_k^p(W)$, where $V_h \subset T_h W$ is chosen such that $\exp_h : C^0(V_h) \to C^0(W)$ is a chart for $C^0(W)$.

The proof is given by Palais in [7] and follows also from Theorem 5.1 [**4**], since L_k^p is a manifold model by Theorem 6.

If $h \in C^\infty(W)$, then we will denote by $L_k^p(W)_h$ the connected component of $L_k^p(W)$, which contains h. Then any section in $L_k^p(W)_h$ has the same restriction to the boundary of N as h.

THEOREM 10. *Let $k > n/p$, $1 \leqq p, q < \infty$, $|s| \leqq k$ and $q > 1$ if $s < 0$. Let E be a C^∞ vector bundle over W. Then we have a unique extension of the section functor L_s^q to the vector bundles h^*E for all $h \in L_k^p(W)$, such that the union of the Banach* *spaces* $L^q_s(h^*E)$, $h \in L^p_k(W)$, *is a vector bundle of class* C^∞:

$$L^q_s(L^p_k(W)^*E) \to L^p_k(W)$$

provided $-n-k+n/p \leqq s-n/q \leqq k-n/p$.

In particular, if $E = T_2W$, $s = k$, $q = p$, then this is the tangent bundle of $L^p_k(W)$, where the tangent field of a differentiable curve $t \to f_t$ in $L^p_k(W)$ is given by the formula:

$$\left(\frac{d}{dt}f_t\right)(x) = \frac{d}{dt}f_t(x), \quad x \in N.$$

PROOF. (See also Theorem 6.1 and 5.1 in [4].) First there is an open neighborhood $\mathscr{D}$ of the diagonal in $W \oplus_N W$ and a bundle equivalence $\phi: p_1^*E \to p_2^*E$, where $p_i: \mathscr{D} \to W$ are the projections $p_i(a_1, a_2) = a_i$. The existence of $\mathscr{D}$, ϕ follows from the fact, that p_1 and p_2 coincide on the diagonal and we may even assume, that ϕ is the identity on the diagonal. We mention here two explicit constructions of ϕ. We choose some RMC structure for π and E and define $\mathscr{D}$ such that if $(a, b) \in \mathscr{D}$, then there is a unique minimal geodesic $c:[0, 1] \to W$, with $c(0) = a$ and $c(1) = b$. If $\zeta_0 \in E_a$, then we put $\phi(a, b)\cdot\xi_0 = \xi(1)$, where: (1) $\xi(t)$ is a parallel field, $\nabla\xi = 0$, along c and $\xi(0) = \xi_0$. (2) $\xi(t)$ is a Jacobi field, $\nabla^2\xi + R(\xi, \partial c)\partial c = 0$, along c and $\xi(0) = 0$, $\nabla\xi(0) = \xi_0$. Here we may have to choose $\mathscr{D}$ still smaller.

Now let $h, f \in C^0(W)$, such that $(h, f) \in C^0(\mathscr{D})$. Then

$$(h, f)^*\phi = \phi(h, f) \in C^0(L(h^*E, f^*E)).$$

If moreover $h \in C^\infty(W)$ and $f \in L^p_k(W)$ is given by $f = \exp \circ \xi$, $\xi \in L^p_k(V_h)$, then we define

$$L^q_s(f^*E) = \{\phi(h, f)\cdot\eta \mid \eta \in L^q_s(h^*E)\}.$$

If $h' \in C^\infty(W)$ and close to h, we have

$$\phi(h', f)^{-1} \circ \phi(h, f) = \gamma_{h,h'}, f = \exp \circ \xi,$$

where

$$\gamma_{h,h'}: V_h \to L(h^*E, h'^*E), \; V_h \subset T_hW,$$

is a C^∞ fiber map. We then have an induced C^∞ map

$$L^p_k(\gamma_{h,h'}): L^p_k(V_h) \to L^p_k(L(h^*E, h'^*E))$$

and this map followed by the continuous linear inclusion

$$L^p_k(L(h^*E, h'^*E)) \subset L(L^q_s(h^*E), L^q_s(h'^*E))$$

(see Theorem 4) is the coordinate transformation function for the bundle $L^q_s(L^p_k(W)^*E)$.

We have the tangent bundle, if $E = T_2W$, $k = s$ and $q = p$, the proof being analog to the proof of Theorem 5.1 [4]. If we define ϕ by

$$\phi(a, b)\cdot\eta = \frac{d}{dt}\exp(\xi + t\eta)|_{t=0}, \; \xi, \eta \in (T_2W)_a,$$

$\exp\xi = b$, then ϕ is the tangent trivialization of $TL_k^p(W) = L_k^p(L_k^p(W)^*T_2W)$ corresponding to the charts for $L_k^p(W)$ given by exp. This ϕ coincides with the ϕ given by the Jacobi field as in (2) above. In general, we will call a particular $\phi: p_1^*E \to p_2^*E$ a translation map for E.

THEOREM 11. *Let W have a RMC structure and let j be a positive integer with $j \leqq 2k$ and $j \leqq pk$, where $k > n/p$. Then we have a unique extension of the j-times covariant differentiation $f \in C^\infty(W) \mapsto \nabla^j f \in C^\infty(L^j(TN, T_fW))$ to a C^∞ section ∇^j in*

$$L_{k-j}^p(L^j(TN, L_k^p(W)^*T_2W)).$$

In a local trivialization over the chart $\exp_h: L_k^p(V_h) \to L_k^p(W)$, *the local representative*

$$\nabla_h^j: L_k^p(V_h) \to L_{k-j}^p(L^j(TN, T_hW))$$

of ∇^j, is given by

$$\nabla_h^j(\xi) = \nabla^j\xi + Q_j(\xi), \quad \text{with } Q_j \in PD_{j-1}^j(V_h, L^j(TN, T_h W)).$$

PROOF. We obtain the local formula for $\xi \in C^\infty(V_h)$ from Lemma 2 and by Corollary 7.1

$$Q_j: L_k^p(V_h) \to L_{k-j}^p(L^j(TN, T_hW)),$$

is of class C^∞. Now $L_{k-j}^p(L^j(TN, L_k^p(W)^*T_2W))$ is a C^∞ vector bundle by Theorem 10 and we are done, since $C^\infty(W)$ is dense in $L_k^p(W)$.

THEOREM 12. *Let E be a C^∞ vector bundle over W and let both π and E have a RMC structure. Let j, s, k be integers with $0 < j \leqq s \leqq k$ and let $k > n/p$, $s - n/q \leqq k - n/p$. Then j-times covariant differentiation of sections in $L_s^q(f^*E)$, $f \in C^\infty(W)$, has a unique extension to a C^∞ section ∇^j in the bundle*

$$L(L_s^q(L_k^p(W)^*E), L_{s-j}^q(L^j(TN, L_k^p(W)^*E))).$$

In a local trivialization given by $\exp_h$ *and some translation map ϕ for E, ∇^j is given by*

$$\nabla_h^j(\xi)\cdot\eta = \nabla^j\eta + \sum_{\nu=0}^{j-1} P_{j,\nu}(\xi)\cdot\nabla^\nu\eta$$

where $P_{j,\nu} \in PD_{j-\nu}^{j-\nu}(V_h, L(L^\nu(TN, h^*E), L^j(TN, h^*E)))$, $P_{j,\nu}(0) = 0$.

PROOF. Let $f = \exp \circ \xi$, $\xi \in C^\infty(V_h)$, and $\eta \in L_s^q(h^*E)$, then

$$\begin{aligned}\nabla_h(\xi)\cdot\eta &= \phi(h,f)^{-1}\cdot\nabla(\phi(h,f)\cdot\eta)\\ &= \nabla\eta + \phi(h,f)^{-1}\cdot\nabla(\phi(h,f))\cdot\eta,\end{aligned}$$

which proves the local formula for $j = 1$. The formula then follows for all j by induction. Now

$$\begin{aligned}P_{j,\nu}: L_k^p(V_h) &\to L_{k-j+\nu}^p(L(L^\nu(TN, h^*E), L^j(TN, h^*E)))\\ &\subset L(L_{s-\nu}^q(L^\nu(TN, h^*E), L_{s-j}^q(L^j(TN, h^*E)))\end{aligned}$$

is of class C^∞, since the first mapping by P is of class C^∞ by Corollary 7.1 and the second is a continuous linear inclusion by Theorem 3. Q.E.D.

THEOREM 13. *Let $\pi: W \to N$ have a RMC structure and let E be a C^∞ vector bundle over W with some RMC structure. Let $0 \leqq s \leqq k, k > n/p, s - n/q \leqq k - n/p$. Then*

$$\|\xi\|_{(q,s)} = \sum_{j=0}^{s} (\textstyle\int_N \|\nabla^j \xi\|^q)^{1/q}, \qquad \xi \in L_s^q(f^*E), f \in L_k^p(W),$$

defines a Finsler structure on the vector bundle $L_s^q(L_k^p(W)^*E)$, which is Lipschitz. Moreover, in the case $p = q = 2$, we obtain a C^∞ Riemannian metric for $H^s(H^k(W)^*E)$ by

$$\langle \xi, \eta \rangle_s = \sum_{j=0}^{s} \textstyle\int_N \langle \nabla^j \xi, \nabla^j \eta \rangle.$$

Taking $E = T_2 W$, $q = p$, $s = k$, we get a Lipschitz Finsler structure for the manifold $L_k^p(W)$ and a C^∞ Riemannian metric for $H^k(W)$.

PROOF. Locally, the norm of a vector $\eta \in L_s^q(h^*E)$ at the point $\xi \in L_k^p(V_h)$, $h \in C^\infty(W)$, is given by

$$\textstyle\sum_j \|\phi(h,f) \cdot \nabla_h^j(\xi) \cdot \eta\|_{(q,0)}, \qquad f = \exp \circ \xi,$$

where ϕ is the translation map for E used in the trivialization. This norm is equivalent to $\sum_j \|\nabla_h^j(\xi) \cdot \eta\|_{(q,0)}$ uniformly in ξ, for ξ in a neighborhood of 0, since $\phi(h,f)$ is an isomorphism on each fiber. Now

$$\left|\textstyle\sum_j \|\nabla_h^j(\xi_1) \cdot \eta\|_{(q,0)} - \sum_j \|\nabla_h^j(\xi_2) \cdot \eta\|_{(q,0)}\right| \leqq \sum_j \|(\nabla_h^j(\xi_1) - \nabla_h^j(\xi_2)) \cdot \eta\|_{(q,0)}$$

which shows that the norm is Lipschitz, as ∇_h^j is of class C^∞.

THEOREM 14. *Let $\pi: W \to N$ have a RMC structure, such that each fiber of W is a complete Riemannian manifold in the induced metric. Give $L_k^p(W)$, $k > n/p$, the Finsler structure $\| \ \|_{(p,k)}$ induced by the RMC structure of π as described in Theorem 13. Then $L_k^p(W)$ is a complete Finsler manifold and a complete Riemannian manifold if $p = 2$. Moreover, there is a constant c, such that*

$$|\xi|_0 \leqq c\|\xi\|_{(p,k)}, \quad \textit{for all } \xi \in L_k^p(L_k^p(W)^*T_2W).$$

PROOF. The existence of c is granted by Theorem 2. Let u, v be in the same component of $L_k^p(W)$ and let $\alpha: [0,1] \to L_k^p(W)$ be a C^1 curve connecting u, v. Then

$$d_0(u,v) \leqq \inf_\alpha \textstyle\int_0^1 \alpha'(t)_0 dt \leqq c \inf_\alpha \int_0^1 \|\alpha'(t)\|_{(p,k)} dt = c d_{(p,k)}(u,v),$$

where d_0 and $d_{(p,k)}$ denote the distance functions on $C^0(W)$ and $L_k^p(W)$ respectively. Therefore a Cauchy sequence in $L_k^p(W)$ is also a Cauchy sequence in $C^0(W)$, which is a complete manifold. Now if two points in $L_k^p(W)$ are in the same connected component of $L_k^p(W)$ and sufficiently close in $C^0(W)$, then they are in the domain of a single chart for $L_k^p(W)$. Therefore, all but finitely many points of a Cauchy sequence in a connected component of $L_k^p(W)$ must be in the domain of a single chart and converge to some point in the same component, because of the relative completeness of the domain.

2.3. *Variation integrals.* $\pi : W \to N$ will denote a C^∞ fiber bundle as before. A quasi-polynomial differential operator of order k and weight w from W to a vector bundle $E \to W$ is a section

$$P : C^\infty(W) \to C^\infty(C^\infty(W)^*E), \quad \text{i.e. } P(h) \in C^\infty(h^*E),$$

such that, choosing some RMC structure for W, P can be written in the form

$$P(h) = \textstyle\sum_\alpha (h^*A_\alpha)\cdot(\nabla^{\alpha_1}h, \ldots, \nabla^{\alpha_j}h)$$

where

$$A_\alpha \in C^\infty(L(L^{\alpha_1}(\pi^*TN, T_2W), \ldots, L^{\alpha_j}(\pi^*TN, T_2W); E))$$

and $1 \leqq \alpha_\nu \leqq k, |\alpha| \leqq w$. We denote this type of differential operators by $PD_k^w(W, E)$.

If $E = W \times \mathbf{R}$ then $P(h)$ is a function in N and $J(h) = \int_N P(h)^*1$ is called the variation integral associated to P (and the Riemannian metric on N used for the integration). $J = \int_N P$ is a well-defined function on $C^\infty(W)$ for any $P \in PD_k^w(W, \mathbf{R})$.

THEOREM 15. *Let E be a C^∞ vector bundle over W and $P \in PD_k^w(W, E)$. Let $1 \leqq p$, $q < \infty$, $r > n/p$, $1 \leqq k \leqq w$, $-r \leqq s \leqq r - k$, $-n - r + n/p \leqq s - n/q$, $q > 1$ if $s < 0$ and suppose one of the conditions* (1), (2) *or* (3) *in Theorem* 7 *is satisfied. Then we have a unique extension of P to a C^∞ section in $L_s^q(L_r^p(W)^*E)$. In particular, if $w = pk$, $k > n/p$, then P is a C^∞ section in $L_0^1(L_k^p(W)^*E)$ and if moreover $E = W \times \mathbf{R}$, then the corresponding variation integral $J = \int_N P$ is a C^∞ function on $L_k^p(W)$.*

PROOF. The theorem is an immediate consequence of Theorem 10 and Theorem 7, together with the fact that the integration, $\int_N : L_0^1(N, \mathbf{R}) \to \mathbf{R}$ is a continuous linear map.

We call $P \in PD_k^{2k}(W, \mathbf{R})$ an energy operator, iff P can be written in the form above, with

$$(h^*A_{(k,k)})(\nabla^k h, \nabla^k h) = \langle (h^*G)\cdot\nabla^k h, \nabla^k h\rangle,$$

where $G \in C^\infty(L(T_2W, T_2W))$ is positive definite on each fiber and $\langle\ ,\ \rangle$ denotes the Riemannian metric. This form of the principal part of P is in fact independent of the RMC structure used for the representation of P. We denote the set of energy operators of order k by $\mathscr{E}_k(W)$. If $P \in \mathscr{E}_k(W)$ we call $J(f) = \int_N P(f)$ the corresponding energy integral.

THEOREM 16. *Let N be a compact, connected C^∞ Riemannian manifold and let $\pi : W \to N$ be a C^∞ fiber bundle. Let $P \in \mathscr{E}_k(W)$, $k > n/2$, and let $J = \int_N P$ be the corresponding energy integral. Then*

(a) *J is a C^∞ function on $H^k(W)$ and has only C^∞ sections in W as critical points, provided N is without boundary.*

(b) *Suppose $A \subset H^k(W)$ is relatively compact in $C^0(W)$ and J is bounded on A. Then A is also relatively compact in $L_r^p(W)$, $r > n/p$, provided $r < k$ and $r - n/p < k - n/2$. If moreover $\|dJ\|$ is not bounded away from zero on some sequence f_j of points in A, then f_j has a converging subsequence, converging to a critical point of J.*

PROOF. J is of class C^∞ by Theorem 15. Suppose $f \in H^k(W)$ is a critical point of J, i.e. $dJ(f) = 0$. Locally in $H^k(V_h)$, J is of the form

$$J(\xi) = \langle (G \circ \xi)\cdot \nabla^k \xi, \nabla^k \xi \rangle_0 + \textstyle\int_N Q(\xi),$$

where $Q \in PD_k^{2k}$ and with zero principal part. Then

$$\begin{aligned} dJ(\xi)\cdot\eta &= 2\langle (G \circ \xi)\cdot \nabla^k \xi, \nabla^k \eta \rangle_0 + \langle (D_2 G \circ \xi)\cdot(\eta, \nabla^k \xi), \nabla^k \xi \rangle_0 + \textstyle\int_N DQ(\xi)\cdot\eta \\ &= \langle 2(G \circ \xi)\cdot(-\mathrm{Div})^k \nabla^k \xi + Q_0(\xi), \eta \rangle_0, \end{aligned}$$

with $Q_0 \in PD_{2k-1}^{2k}$, as is easily seen. Therefore, $dJ(\xi) = 0$ implies $\xi \in C^\infty(V_h)$ by Theorem 8.

We have $Q(\xi) = \sum_\alpha (A_\alpha \circ \xi)\cdot(\nabla^{\alpha_1}\xi, \ldots, \nabla^{\alpha_j}\xi)$ with $1 \leqq \alpha_\nu \leqq k$, $|\alpha| \leqq 2k$, $A_{(k,k)} = 0$ and

$$X_\alpha = \textstyle\int_N |(A_\alpha \circ \xi)\cdot(\nabla^{\alpha_1}\xi, \ldots, \nabla^{\alpha_j}\xi)| \leqq |A_\alpha \circ \xi|_0 \|\xi\|_{(p_1,\alpha_1)} \cdots \|\xi\|_{(p_j,\alpha_j)}$$

provided $\sum\limits_\nu 1/p_\nu \leqq 1$. Moreover, we have

$$\|\xi\|_{(p,\alpha)} \leqq \text{const.}\ (\|\xi\|_k)^{\alpha/k}(|\xi|_0)^{1-\alpha/k}$$

by a well-known interpolation inequality [**6**], provided $\alpha - n/p \leqq (\alpha/k)(k - n/2)$. With $\alpha_\nu p_\nu = 2k$ we have $\sum 1/p_\nu = |\alpha|/2k \leqq 1$ and $\alpha_\nu - n/p_\nu = (\alpha_\nu(k)(k - n/2)$. Then we obtain $X_\alpha \leqq \text{const.}\ |A_\alpha \circ \xi|_0(\|\xi\|_k)^{|\alpha|/k}(|\xi|_0)^{j-|\alpha|/k}$. If $j \geqq 2$ and $\alpha \neq (k, k)$, then $j - \alpha/k > 0$ and by the estimate above we can for any $\varepsilon > 0$ find $\delta > 0$ and C, such that

$$\textstyle\sum_\alpha X_\alpha \leqq \varepsilon \|\xi\|_k^2 + C, \quad \text{for } |\xi|_0 \leqq \delta,$$

since this can easily be done for $j = 0$ or 1. Now since $\langle (G \circ \xi)\cdot \nabla^k \xi, \nabla^k \xi \rangle_0 \geqq \lambda \|\nabla^k \xi\|_0^2$ for some $\lambda > 0$, there is for any $h \in C^\infty(W)$ a neighborhood Ω_h of h in $C^0(W)$, such that for $\xi \in H^k(V_h) \cap \Omega_h$ we have $\|\xi\|_k^2 \leqq \text{const.}\ J(\xi) + \text{const.}$ Then $\|\xi\|_k$ is bounded on $H^k(V_h) \cap \Omega_h \cap A$, which is therefore relatively compact in $L_r^p(V_h)$, since the inclusion $H^k(V_h) \subset L_r^p(V_h)$ is completely continuous. Since A is relatively compact in $C^0(W)$ we can choose finitely many Ω_h having the above property and covering A. It then follows, that A is relatively compact in $L_r^p(W)$.

Now let f_j, $j = 1, 2, \ldots$ be a sequence in A, such that $\|dJ(f_j)\|$ converges to zero. Assume that f_j converges in $C^0(W)$, else we would select a subsequence with this property. Then if $k = 1$ (and thus $n = 1$) f_j already converges in $H^k(W)$, as has been proved in [**5**]. Suppose $k \geqq 2$ and put $r = k - 1$ and $p = 2k/(k - 1)$, so that $r < k$ and $0 < r - n/p < k - n/2$. Then A is relatively compact in $L_r^p(W)$ and we may assume, that f_j converges in $L_r^p(W)$. We claim that f_j now converges in $H^k(W)$.

We choose an $h \in C^\infty(W)$ so close to the limit of f_j in $L_r^p(W)$, that f_j is in the domain of the chart $\exp_h : L_r^p(V_h) \to L_r^p(W)$ for all but finitely many j. In this chart we have

$$dJ(\xi)\cdot\eta = 2\langle (G \circ \xi)\cdot \nabla^k \xi, \nabla^k \eta \rangle_0 + \sum_{\nu=0}^{k} \textstyle\int_N Q_\nu(\xi)\cdot \nabla^\nu \eta,$$

with $Q_v \in PD_k^{2k-v}(V_h, L(L^v(TN, E_h), E_h))$ and $Q_k \in PD_{k-1}^k$. Now we have for ξ, $\eta \in H^k(V_h)$:

$$\begin{aligned} 2\lambda\|\nabla^k(\xi - \eta)\|_0^2 &\leqq \langle (G \circ \xi)\cdot\nabla^k(\xi - \eta), \nabla^k(\xi - \eta)\rangle_0 \\ &\quad + \langle (G \circ \eta)\cdot\nabla^k(\xi - \eta), \nabla^k(\xi - \eta)\rangle_0 \\ &= 2\langle (G \circ \xi)\cdot\nabla^k\xi, \nabla^k(\xi - \eta)\rangle_0 - 2\langle (G \circ \eta)\cdot\nabla^k\eta, \nabla^k(\xi - \eta)\rangle_0 \\ &\quad - \langle (G \circ \xi - G \circ \eta)\cdot\nabla^k(\xi + \eta), \nabla^k(\xi - \eta)\rangle_0 \\ &= dJ(\xi)\cdot(\xi - \eta) - dJ(\eta)\cdot(\xi - \eta) - \textstyle\sum_v \int_N (Q_v(\xi) - Q_v(\eta))\cdot\nabla^v(\xi - \eta) \\ &\quad - \langle (G \circ \xi - G \circ \eta)\cdot\nabla^k(\xi + \eta), \nabla^k(\xi - \eta)\rangle_0. \end{aligned}$$

Thus, if $\exp \circ \xi_j = f_j$, then

$$\begin{aligned} 2\lambda\|\nabla^k(\xi_j - \xi_i)\|_0^2 &\leqq \|dJ(\xi_j)\|\,\|\xi_j - \xi_i\|_k + \|dJ(\xi_i)\|\,\|\xi_j - \xi_i\|_k \\ &\quad + |G \circ \xi_j - G \circ \xi_i|_0\|\nabla^k(\xi_j + \xi_i)\|_0\|\nabla^k(\xi_j - \xi_i)\|_0 \\ &\quad + \textstyle\sum_v \|Q_v(\xi_j) - Q_v(\xi_i)\|_0\|\nabla^v(\xi_j - \xi_i)\|_0 \end{aligned}$$

and converges to zero, since $\|\xi_j\|_k$ is bounded for h sufficiently close to the limit of f_j and $Q_k : L_r^p(V_h) \to H^0(N, \boldsymbol{R})$ is of class C^∞ by Theorem 7. Q.E.D.

THEOREM 17. *Let $\pi : W \to N$ be a C^∞ fiber bundle with a RMC structure, such that each fiber of W is a complete Riemannian manifold. Suppose that either W is compact or N has a nonempty boundary. Let $k > n/p$ and let A be a subset of a connected component of $L_k^p(W)$, such that $\|\nabla f\|_{(p,k-1)}$ is bounded for $f \in A$. Then A is a relatively compact subset of $C^0(W)$. Moreover if $k > n/2$, then the energy function*

$$E_k : H^k(W) \to \boldsymbol{R},\ E_k(f) = \tfrac{1}{2}\|\nabla f\|_{k-1}^2$$

is a C^∞ function satisfying condition (C).

PROOF. Choose a $q < \infty$, such that $k - n/p \geqq 1 - n/q > 0$. Then by Theorem 2 we have a constant c, such that

$$\|\nabla f\|_{(q,0)} \leqq c\|\nabla f\|_{(p,k-1)}$$

and then by Lemma 3

$$d_2(f(x), f(y)) \leqq \text{const.}\, d_N(x, y)^{1-n/q} \text{ for } f \in A \text{ and } x, y \in N$$

or A is an equicontinuous subset of $C^0(W)$. Then A is a relatively compact subset of $C^0(W)$ by the Ascoli-Arzela Theorem, if $A(N)$ is bounded in W. If $x \in \partial N$, then all sections in A map x to the same point in W, since they belong to the same connected component of $L_k^p(W)$. Thus if ∂N is not empty, then the boundedness of $A(N)$ follows from the distance inequality above and else, if ∂N is empty, then from the compactness of W.

E_k is an energy integral of order k and since subsets of $H^k(W)$ on which E_k is bounded are relatively compact in $C^0(W)$, E_k satisfies condition (C) by Theorem 16.

COROLLARY. *Let W be as in Theorem* 17 *and let $J = \int_N P$ be an energy integral, $P \in \mathscr{E}_k(W)$. Then $J : H^k(W) \to \boldsymbol{R}$, $k > n/2$, satisfies condition* (C), *provided there is a*

q with $k - n/2 \geqq 1 - n/q > 0$, such that $\|\nabla f\|_{(q,0)}$ is bounded on any subset of $H^k(W)$ on which J is bounded.

REMARK. An interesting question is whether the corollary still holds with $q \leqq n$.

THEOREM 18. *Let $\pi: W \to N$ be a C^∞ fiber bundle with a RMC structure. Let $k > n/p$ and let Ω be a bounded subset of the connected component $L_k^p(W)_h$, $h \in C^\infty(W)$, of $L_k^p(W)$, in the intrinsic Finsler structure (Theorem 13). Then $\|\nabla f\|_{(q,0)}$ is bounded for $f \in \Omega$, for any q with $0 < 1 - n/q \leqq k - n/p$. In particular, Ω is relatively compact in $C^0(W)$, if W satisfies the conditions in Theorem 17.*

PROOF. Let $f \in \Omega$ and let $\alpha: [0,1] \to L_k^p(W)_h$ be a differentiable curve with $\alpha(0) = h$ and $\alpha(1) = f$. We have

$$\frac{d}{dt}\int_N \|\nabla\alpha(t)\|^q = q\int_N \|\nabla\alpha(t)\|^{q-2}\langle\nabla\alpha(t), \nabla_t(\nabla\alpha(t))\rangle$$

$$\leqq q\|\nabla\alpha(t)\|_{(q,0)}{}^{q-1}\|\nabla_t(\nabla\alpha(t))\|_{(q,0)},$$

by Holder inequality. Note that $t \to \nabla\alpha(t)(x)\cdot v$, $v \in T_xN$, is a vector field along the curve $t \to \alpha(t)(x)$ in W_x and $\nabla_t(\nabla\alpha(t))(x)\cdot v$ is defined to be the covariant derivative of this vector field. We see easily, by local computation, that

$$\|\nabla_t(\nabla\alpha(t))\| \leqq \|\nabla\alpha'(t)\| + a\|\alpha'(t)\| + b\|\nabla\alpha(t)\|$$

with some constants a, b, where $\alpha'(t) = d\alpha(t)/dt$. Then

$$\frac{d}{dt}\|\nabla\alpha(t)\|_{(q,0)} = \frac{1}{q}\|\nabla\alpha(t)\|_{(q,0)}{}^{1-q}\frac{d}{dt}\int_N \|\nabla\alpha(t)\|^q$$

$$\leqq (1 + a)\|\alpha'(t)\|_{(q,1)} + b\|\nabla\alpha(t)\|_{(q,0)}.$$

By the Comparison Theorem for ordinary differential equations $y' \leqq cg(t) + by$ implies

$$y(t) \leqq e^{bt}\left(\int_0^t cg(s)e^{-bs}ds + y(0)\right) \quad \text{for } t \geqq 0.$$

We therefore obtain

$$\|\nabla f\|_{(q,0)} \leqq e^b\|\nabla h\|_{(q,0)} + (1 + a)e^b\int_0^1 \|\alpha'(t)\|_{(q,1)}dt.$$

Now by Theorem 2 we have

$$\|\alpha'(t)\|_{(q,1)} \leqq \text{const.}\|\alpha'(t)\|_{(p,k)}.$$

Therefore, by choosing a sequence of curves α, such that the length of α converges to the distance between f and h, we obtain

$$\|\nabla f\|_{(q,0)} \leqq \text{const.}\|\nabla h\|_{(q,0)} + \text{const.}d_{(p,k)}(h, f),$$

which completes the proof.

COROLLARY. *Let W satisfy the conditions in Theorem 17 and let $J = \int_N P, P \in \mathscr{E}_k(W)$, be an energy integral on $H^k(W)$, $k > n/2$. If A is a closed bounded subset of $H^k(W)_h$, then J is bounded on A and the set of critical points of J contained in A is compact. In particular, the set of critical points of J is locally compact.*

PROOF. Suppose f_j is a sequence of points in A, such that $J(f_j)$ is not bounded. We can assume, that f_j converges in $C^0(W)$ by Theorem 18. But then we obtain a sequence ξ_j in a coordinate neighborhood with $\|\xi_j\|_{(p,k)}$ bounded, which contradicts that $J(f_j)$ is not bounded. Now the set of critical points in A is compact by Theorem 16(b).

3. **Applications.** Here N and M will be compact, connected C^∞ Riemannian manifolds without boundary. Moreover, we suppose N is of dimension $n = 2$ or $n = 3$ and that M has everywhere nonpositive sectional curvature. R_N, R_M will denote the curvature tensors of N, M and S the Ricci tensor of N of mixed variance, i.e.

$$S(x)\cdot v = \textstyle\sum_i R_N(x)\cdot(v, e_i, e_i), \quad v \in T_xN,$$

where $\{e_i\}$ denotes an orthonormal base of T_xN. Then $S \in C^\infty(L(TN, TN))$ and we choose a number λ, such that $\lambda > |S|_0$. We define a variation integral $J: H^2(N, M) \to \boldsymbol{R}$ by

$$J(f) = \frac{1}{2}\|\mathrm{Div}\partial f\|_0^2 + \frac{\lambda}{2}\|\partial f\|_0^2.$$

We have using Lemma 1

$$\begin{aligned} 2J(f) &= -\langle \nabla \mathrm{Div}\partial f, \partial f\rangle_0 + \lambda\|\partial f\|_0^2 \\ &= \|\nabla\partial f\|_0^2 - \langle \textstyle\sum_i (R_M \circ f)\cdot(\partial f, \partial_i f, \partial_i f), \partial f\rangle_0 \\ &\quad + \langle \partial f S, \partial f\rangle_0 + \lambda\|\partial f\|_0^2 \\ &\geqq \mathrm{const.}\|\partial f\|_1^2. \end{aligned}$$

We have used

$$\begin{aligned} \nabla\mathrm{Div}\partial f\cdot v &= \textstyle\sum_i \nabla^2\partial f\cdot(v, e_i, e_i) \\ &= \textstyle\sum_i (\nabla^2\partial f\cdot(e_i, v, e_i) \\ &\qquad + (R_M \circ f)\cdot(\partial f\cdot v, \partial_i f, \partial_i f) - \partial f\cdot R_N\cdot(v, e_i, e_i)) \\ &= \mathrm{Div}\nabla\partial f\cdot v + \textstyle\sum_i (R_M \circ f)\cdot(\partial f\cdot v, \partial_i f, \partial_i f) - \partial f S\cdot v, \end{aligned}$$

where $\partial_i f = \partial f\cdot e_i$, and that M has nonpositive sectional curvature. It follows that J dominates the energy function E_2 and therefore, it satisfies condition (C) by Theorem 17. Moreover, the critical points of J are maps $f: N \to M$ of class C^∞ by Theorem 16(a).

We have, with $f \in C^\infty(N, M)$,

$$\begin{aligned} dJ(f)\cdot\eta &= \langle \mathrm{Div}\partial f, \Delta\eta + (R_M \circ f)\cdot(\eta, \partial_i f, \partial_i f)\rangle_0 + \lambda\langle\partial f, \nabla\eta\rangle_0 \\ &= \langle \Delta\mathrm{Div}\partial f + (R_M \circ f)\cdot(\mathrm{Div}\partial f, \partial_i f, \partial_i f) - \lambda\mathrm{Div}\partial f, \eta\rangle_0. \end{aligned}$$

Then $dJ(f) = 0$, or f is a critical point of J, iff

$$\Delta\mathrm{Div}\partial f + \textstyle\sum_i (R_M \circ f)\cdot(\mathrm{Div}\partial f, \partial_i f) - \lambda\mathrm{Div}\partial f = 0.$$

By taking the inner product with $-\operatorname{Div}\partial f$, we obtain

$$\|\nabla \operatorname{Div}\partial f\|_0^2 + \lambda \|\operatorname{Div}\partial f\|_0^2 - \sum_i \langle (R_M \circ f)\cdot(\operatorname{Div}\partial f, \partial_i f, \partial_i f), \operatorname{Div}\partial f\rangle_0 = 0$$

which implies that all three terms must be zero or $\operatorname{Div}\partial f = 0$. We have proved

THEOREM 19. *J is a C^∞ function satisfying condition* (C) *and the critical points of J are exactly the harmonic maps of N into M, i.e. maps $h: N \to M$ of class C^∞ with* $\Delta h = \operatorname{Div}\partial h = 0$.

Now as a consequence J takes its minimum in every component of $H^2(N, M)$, which must be in a critical point of J. Therefore, every continuous mapping $N \to M$ is homotopic to a harmonic mapping, which is the result of Eells and Sampson in [3], with no restriction on the dimension of N. Of course Theorem 19 is a stronger result than in [3], in the case $n = 2, 3$. In particular, the Lusternik-Schnirelman category theory can be applied, so the number of harmonic maps in a component of $H^2(N, M)$ is at least as large as the category of the component.

It is not known to me, whether it is in general possible to construct a variation integral satisfying condition (C) and having exactly the harmonic maps as critical points.

REFERENCES

1. A. P. Calderon, *Lebesque spaces of differentiable functions and distributions,* Proc. Sympos. Pure Math., Vol. 4, Amer. Math. Soc., Providence, R.I, 1961, pp. 33–49.

2. J. Eells, Jr., *A setting for global analysis,* Bull. Amer. Math. Soc. **72** (1966), 739–807.

3. J. Eells and J. H. Sampson, *Harmonic mappings of Riemannian manifolds,* Amer. J. Math. **86** (1964), 109–160.

4. H. I. Eliasson, *On the geometry of manifolds of maps,* J. Differential Geometry, **1** (1967), 169–194.

5. ———, *Morse theory for closed curves,* Symposium for infinite dimensional Topology, Louisiana State University, 1967.

6. L. Nirenberg, *On elliptic partial differential equations,* Ann. Scuola. Norm. Sup. Pisa, **13** (1959), 123–131.

7. R. S. Palais, *Lectures on the differential topology of infinite dimensional manifolds,* notes by S. Greenfield, Brandeis University, 1964–1965.

8. ———, *Foundations of global non-linear analysis,* Benjamin, New York, 1968.

9. ———, *Seminar on the Atiyah-Singer index theorem,* Ann. of Math. Studies No. 57, Princeton Univ. Press, Princeton, N.J., 1965.

10. R. S. Palais and S. Smale, *A generalized Morse theory,* Bull. Amer. Math. Soc. **70** (1964), 165–172.

11. G. de Rham, *Variétés différentiables,* Hermann, Paris, 1960.

12. J. C. Saber, *Manifolds of maps,* Dissertation, Brandeis University, 1965.

13. K. K. Uhlenbeck, *The calculus of variations and global analysis,* Dissertation, Brandeis University, 1968.

UNIVERSITY OF ICELAND

NOTE ON THE DIFFERENTIABILITY OF NONLINEAR SEMIGROUPS

TOSIO KATO

Recently Kōmura [1] proved the following theorem.

THEOREM. *Any strongly continuous semigroup* $\{U_t\}_{0 \le t < \infty}$ *of nonexpansive (nonlinear) operators on a closed convex subset D of a Hilbert space H into itself has strong generator* $-A_0$ *with domain* $D(A_0)$ *dense in D.*

Here A_0 is defined by $A_0 x = \lim_{t \downarrow 0} t^{-1}(x - U_t x)$; $D(A_0)$ consists of all $x \in D$ for which the limit exists. For various consequences of the theorem, see [1] and [2].

The purpose of this paper is to give a new proof of the theorem, which is somewhat simpler and more direct than the original one.

For the proof we set

$$A_t = t^{-1}(1 - U_t), \quad t > 0, D(A_t) = D. \tag{1}$$

We note that A_t is monotone. Thus $(1 + \lambda A_t)^{-1}$ exists for $\lambda > 0$.

LEMMA 1. *$D((1 + \lambda A_t)^{-1}) \supset D$, and $(1 + \lambda A_t)^{-1}x$ is continuous jointly in $\lambda > 0$ and $t > 0$ for each $x \in D$.*

PROOF. To see that $(1 + \lambda A_t)^{-1}x = y$ is defined for $x \in D$, we solve $y + \lambda A_t y = x$ for $y \in D$. This equation is equivalent to

$$y = Gy \equiv \frac{t}{\lambda + t}x + \frac{\lambda}{\lambda + t}U_t y.$$

But G maps D into itself because U_t has the same property and D is convex. Since G is a contraction with $|G|_{\text{Lip}} \le \lambda(\lambda + t)^{-1} < 1$ and since D is closed, G has a unique fixed point $y \in D$, which is the desired solution. The continuous dependence of y on λ, t follows from the fact that G is strongly continuous in λ, t and $|G|_{\text{Lip}} < 1$.

In what follows we fix $x \in D$ and set

$$y_{\lambda,t} = (1 + \lambda A_t)^{-1} x \in D, \quad \lambda > 0, t > 0. \tag{2}$$

We note the following relations, which are direct consequences of (2).

$$U_t y_{\lambda,t} - x = (1 + t/\lambda)(y_{\lambda,t} - x), \tag{3}$$

$$A_t y_{\lambda,t} = \lambda^{-1}(x - y_{\lambda,t}), \tag{4}$$

$$|y_{\lambda,t} - x| \le (\lambda/t)|x - U_t x|. \tag{5}$$

To prove (5), it suffices to note that (3) gives, with $y = y_{\lambda,t}$,

$$|y - x| = \lambda(\lambda + t)^{-1}|U_t y - x|$$
$$= \lambda(\lambda + t)^{-1}|U_t y - U_t x + U_t x - x| \le \lambda(\lambda + t)^{-1}(|y - x| + |x - U_t x|).$$

LEMMA 2. *Let $\varepsilon > 0$, $0 < \delta \le 1$ be such that $t \in (0, \delta]$ implies $|x - U_t x| \le \varepsilon$. If $nt = s \in (0, \delta]$ with n a positive integer, then*

$$(6) \qquad |y_{\lambda,t} - y_{\lambda,s}|^2 \le 2\varepsilon|y_{\lambda,t} - x|.$$

PROOF. We may assume $x = 0$ without loss of generality, for this simply amounts to shifting the origin of H. Also we write y_t for $y_{\lambda,t}$ for simplicity.

For $k = 1, 2, \ldots, n$ we have

$$|y_t - U_{(k-1)t}y_s|^2 \ge |U_t y_t - U_{kt}y_s|^2$$
$$= |(1 + t/\lambda)y_t - U_{kt}y_s|^2 \qquad \text{(by (3))}$$
$$\ge |y_t - U_{kt}y_s|^2 + (2t/\lambda)\operatorname{Re}(y_t, y_t - U_{kt}y_s).$$

Also

$$|y_t - U_s y_s|^2 = |y_t - (1 + s/\lambda)y_s|^2$$
$$\ge |y_t - y_s|^2 + (2s/\lambda)\operatorname{Re}(y_s, y_s - y_t).$$

If we add these $n + 1$ inequalities and notice that $U_s = U_{nt}$, all the squared terms cancel out. What remains gives, after canceling the factor $2s/\lambda$ (note $s = nt$ again)

$$|y_t|^2 + |y_s|^2 \le \frac{1}{n}\sum_{k=1}^{n}\operatorname{Re}(y_t, U_{kt}y_s) + \operatorname{Re}(y_s, y_t).$$

But since $|U_{kt}y_s - U_{kt}0| \le |y_s|$, we have $|U_{kt}y_s| \le |y_s| + |U_{kt}0| \le |y_s| + \varepsilon$ because $kt \in (0, \delta]$. Thus

$$|y_t|^2 + |y_s|^2 \le |y_t|(|y_s| + \varepsilon) + \operatorname{Re}(y_s, y_t)$$
$$\le \tfrac{1}{2}(|y_t|^2 + |y_s|^2) + \varepsilon|y_t| + \operatorname{Re}(y_s, y_t).$$

Hence follows (6) immediately, with $x = 0$.

LEMMA 3. *With ε, δ as in Lemma 2, we have*

$$(7) \qquad |y_{\lambda,t} - x| \le 2\varepsilon(1 + 4\lambda/\delta) \quad \text{for } \lambda > 0,\ t \in (0, \delta].$$

In particular $y_{\lambda,t}$ is bounded as $t\downarrow 0$ for each fixed λ.

PROOF. Again we may assume $x = 0$ and write y_t for $y_{\lambda,t}$. If $s \in [\delta/2, \delta]$, we have by (5)

$$(8) \qquad |y_s| \le \frac{2\lambda}{\delta}|U_s 0| \le \frac{2\lambda\varepsilon}{\delta}.$$

This proves (7) for $t \in [\delta/2, \delta]$. If $t \in (0, \delta/2)$, there is a positive integer n such that $nt = s \in [\delta/2, \delta]$. Thus we can use (6) to obtain

$$|y_t - y_s|^2 \le 2\varepsilon|y_t| \le 2\varepsilon|y_t - y_s| + 2\varepsilon|y_s|.$$

Using (8) to estimate the last term $|y_s|$, we obtain $|y_t - y_s| \le \varepsilon[1 + (1 + 4\lambda/\delta)^{1/2}]$ and, using (8) again, arrive at the inequality (7).

LEMMA 4. *$y_\lambda = \lim_{t\downarrow 0} y_{\lambda,t} \in D$ exists for each $\lambda > 0$.*

PROOF. Since $y_t = y_{\lambda,t}$ is bounded as $t\downarrow 0$ by Lemma 3, we have $|y_t - x| \le M$ = const. for $t \in (0, 1]$. Hence (6) gives

$$|y_t - y_s| \le (2M\varepsilon)^{1/2}, \quad nt = s \in (0, \delta]. \tag{9}$$

Now let $s, s' \in (0, \delta]$ and suppose that s/s' is rational. Then there is $t > 0$ such that $s = nt$, $s' = n't$ with positive integers n, n'. Then (9) is true for s and s', so that

$$|y_s - y_{s'}| \le 2(2M\varepsilon)^{1/2}. \tag{10}$$

Since y_s is continuous in s by Lemma 1, (10) is extended by continuity to any $s, s' \in (0, \delta]$. Since $\varepsilon > 0$ can be made arbitrarily small by a suitable choice of δ, $\lim y_t = y_\lambda$ exists. $y_\lambda \in D$ since D is closed and $y_{\lambda,t} \in D$.

LEMMA 5. *$y_\lambda \to x$ as $\lambda \downarrow 0$.*

PROOF. Letting $t \downarrow 0$ in (7), we obtain $|y_\lambda - x| \le 2\varepsilon(1 + 4\lambda/\delta)$. This is true for any $\varepsilon > 0$ if δ is chosen properly. If $4\lambda \le \delta$, we have $|y_\lambda - x| \le 4\varepsilon$. This proves the lemma.

LEMMA 6. *For any $s, \lambda > 0$, we have*

$$|A_s y_\lambda| \le \lambda^{-1}|y_\lambda - x|. \tag{11}$$

PROOF. For $t = s/n$, where n is a positive integer,

$$\begin{aligned} s|A_s y_{\lambda,t}| &= |(1 - U_s)y_{\lambda,t}| \\ &\le \sum_{k=1}^{n} |U_{(k-1)t}y_{\lambda,t} - U_{kt}y_{\lambda,t}| \\ &\le n|y_{\lambda,t} - U_t y_{\lambda,t}| = s|A_t y_{\lambda,t}|. \end{aligned}$$

If we fix s and let $n \to \infty$, then $t \downarrow 0$, $y_{\lambda,t} \to y_\lambda$, and $A_t y_{\lambda,t} \to \lambda^{-1}(x - y_\lambda)$ by (4). Since A_s is a continuous operator, we obtain (11).

Now we have all the material to complete the proof of the theorem. (11) implies that $U_t y_\lambda$ is Lipschitz continuous in t, for $|U_{t+h}y_\lambda - U_t y_\lambda| \le |U_h y_\lambda - y_\lambda| = h|A_h y_\lambda|$. It follows that $U_t y_\lambda$ is differentiable, and hence belongs to $D(A_0)$, for almost all $t > 0$. Since $\lim_{t\downarrow 0} U_t y_\lambda = y_\lambda$, y_λ is in the closure of $D(A_0)$. Since $x = \lim_{\lambda\downarrow 0} y_\lambda$, the same is true of x. Since $x \in D$ was arbitrary, $D(A_0)$ is dense in D. (Actually it is known that $y_\lambda \in D(A_0)$, see **[2]**.)

References

1. Y. Kōmura, *Differentiability of nonlinear semi-groups* (to appear).

2. M. G. Crandall and A. Pazy, *Nonlinear semi-groups of contractions and dissipative sets*, J. Functional Anal. **3** (1969), 376–418.

University of California, Berkeley

SCATTERING THEORY AS THE ANALYSIS OF ELLIPTIC FIXED POINTS

I. E. SEGAL

Elliptic fixed points. Let M be a smooth (i.e. C^∞) Banach manifold, let G be a given topological group, and let T be a continuous isomorphism of G into the group of C^∞ homeomorphisms of M; for $g \in G$, let $T_g(x)$ denote the transform of $x \in M$ by g. If x_0 is any fixed point under T, there is a naturally induced action of G in the tangent plane M_0 to M at x_0, and the mapping $g \to \partial_{x_0} T_g$ from g to the Frechet differential at x_0 of T is a linear representation of G on M_0. An *elliptic* fixed point may be defined as one such that this representation is uniformly bounded, relative to the Banach norm in M_0. This implies in particular that the spectrum of $T_g^{(0)} = \partial_x T_g$ lies on the unit circle. For an abelian group the converse holds; and if in addition, M is a Hilbert manifold, an elliptic fixed point is one such that induced linear representation of G is similar to an orthogonal representation.

While there has been intensive investigation of the theory of hyperbolic fixed points of flows, elliptic fixed points have apparently been considered too complicated to be appropriate subjects for investigation at this time. My main theme is that this is not actually the case; particularly in the case of certain infinite-dimensional manifolds, the asymptotic structure of the flow near an elliptic fixed point has a conceptually simple structure; and that nonlinear scattering theory, say for relativistic partial differential equations, consists essentially in the analysis of this structure.

One of the simplest relevant examples is associated with an equation of the form

$$(*) \qquad x'' + m^2 x' + f(x) = 0,$$

where x is an unknown real function of a variable $t \in R^1$, and f is a given function such that $f(0) = f'(0) = 0$. If for example $f(x) = gx^p$, where $g > 0$ and p is an odd integer > 1, this equation has a unique regular solution, global in t, for arbitrarily prescribed $x(0)$ and $x'(0)$; defining T_t as the map $(x(0), x'(0)) \to (x(t), x'(t))$, a one-parameter flow in the plane is obtained. The origin is evidently a fixed point, and the induced linear representation in the tangent plane at the origin is readily determined; it is worth noting that, in a more-or-less canonical way, it is the same as the one-parameter flow associated with the first-order variational equation to equation (*) at the solution $x(t) = x'(t) = 0$ for all t, i.e. the equation

$$(**) \qquad x''(t) + m^2 t = 0,$$

described in the applied literature as "simple harmonic motion". This flow may be described as rotation around the origin at constant angular velocity, showing that the origin is an elliptic fixed point.

Continuing with the equation in question, it is easily verified that $(x'(t))^2 + m^2(x(t))^2 + (2/(p+1))g(x(t))^{p+1} = \text{const.}$, from which it is not difficult to deduce that the flow beginning at any point other than the origin remains bounded away from the origin and ∞, and is periodic as a function of a modified time t', the period, etc. depending on the const. in question. As const. $\to 0$, the flow approaches that associated with simple harmonic motion; but the constant remains actually constant in time. Thus the nonlinear flow is no better approximated by the linear flow after the passage of a long time than it was to begin with.

This behavior, relatively intricate for such a simple equation, is difficult to abstract and generalize. And in the case of an analogous equation in two or more finite dimensions, consisting in the perturbation of N-dimensional simple harmonic motion by a nonlinear term, the situation is much more complicated, even for relatively simple nonlinear perturbations. In the infinite-dimensional case, the corresponding situation might appear from this to be hopelessly complicated, but the fact is that it is tractable, in typically relevant cases, as of certain relativistic nonlinear partial differential equations, by virtue of the circumstances that as $|t| \to \infty$, $x(t) \to 0$ in a sense which causes the nonlinear term to tend to zero, although $x(t)$ remains bounded away from zero in a certain Hilbert norm (the "energy" norm, which I shall define later). This is possible only in a function space; in a finite-dimensional linear space, all Banach norms are equivalent, but in a function space there are simple inequivalent norms; and concrete instances of this inequivalence, far from being pathological, are generic in connection with the asymptotics of many nonlinear evolutionary equations which have been investigated.

The abstract scattering problem. Let me first describe the problem in intuitive terms. An almost universal feature of nonlinear relativistic theoretical physics, whether classical or quantum, is the tendency to describe the states of the "true", nonlinear, physical system, in terms of the states of an associated hypothetical, linear, idealized if not fictitious physical system. The fact is that, empirically, the effect of the nonlinear system is often observed as transitions between states of the associated linear system. The mathematical transformation which described these transitions is the "scattering", or "collision", or "dispersion" operator, in its most common form.

Now to make this precise, assume given a Banach manifold M, together with a one-parameter flow T_t with elliptic fixed point 0; let M_0 denote the tangent space at 0, and $T_t^{(0)}$ the corresponding linear flow. The typical problem is that one is given a vector x_0 in M_0, which is to be thought of physically as the mathematical representation of the system at an infinitely early time, before any interaction (including self-interaction) took place; at this infinitely early time, x_0 develops temporally in a linear fashion, in accordance with the representation $T_t^{(0)}$, which like uniform motion in Newtonian theory, does not derive from the action of any forces. One then wants first of all to find an element x of M, whose temporal development $T_t x$ is in some sense asymptotic to that of x_0 as $t \to -\infty$; next, having done this, one wants to do the reverse as $t \to +\infty$, i.e. find an element $x_1 \in M_0$ such that the temporal development of x is asymptotic to the (linear) temporal development of x_1 as $t \to +\infty$. The mapping $x_0 \to x_1$ is then the S- (for "scattering") operator, which is what one actually sees; the total motion of the system, relative to what the motion would have been in the absence of interactions, from the beginning of the effect of the interaction to its end. The interim operator $x_0 \to x$ is of theoretical interest, and is called the "wave" operator.

In order to compare a vector in M_0 with a vector in M, more structure is required. To this end, assume the existence of a C^∞ homeomorphism P_t from M_0 to M; in practice, this comes from the circumstance that M and M_0 are solution manifolds of evolutionary partial differential equations, and P_t is defined as the mapping which carries any element of M_0 into the element of M which has the same Cauchy data at time t, as will be illustrated shortly. In terms of this one-parameter family of homeomorphisms, a given $x \in M$ is asymptotic to an element $x_0 \in M_0$ as $t \to -\infty$ in case $T_{(-t)}P_t x_0 \to u$; and similarly for $t \to +\infty$.

Consider for example the case of the differential equation

$$(\asymp) \qquad x''(t) + M^2 x(t) + F(x(t)) = 0,$$

where $x(t)$ has values in the Hilbert space $\mathscr{H}$, on which M is a strictly positive selfadjoint operator, and F is a given nonlinear operator, which, together with its Frechet differential, vanishes at the origin. Actually, since M may be unbounded, it is better to consider the corresponding integral equation

$$(\asymp') \qquad x(t) = x_0(t) + \int_{t_0}^{t} \frac{\sin[(t-s)M]}{M} F(x(s))ds,$$

where t_0 is a given time and $x_0(\cdot)$ is a given solution of the first-order variational equation at the origin,

$$(\asymp_0) \qquad x''(t) + M^2 x(t) = 0,$$

which is also better taken in its integrated form

$$(\asymp'_0) \qquad (x(t), x'(t)) = \begin{pmatrix} \cos Mt & \dfrac{\sin Mt}{M} \\ -M \sin Mt & \cos Mt \end{pmatrix} (x(0), x'(0)).$$

Imposing a manifold structure by choosing the "energy" norm in the Cauchy data space, $\|(f, g)\|^2 = \|Mf\|_H^2 + \|g\|_H^2$, then under appropriate assumptions on F, the solution manifold M of equation ($\asymp'$), where the given $x_0(\cdot)$ is required to have finite energy norm, is a C^∞ manifold; the zero solution is an elliptic fixed point for the flow T_t defined by the given equation; and the mappings P_t exist and are unique. It suffices for example if the mapping F is Lipschitzian from the domain $\mathscr{D}_M$ of M, in its natural Hilbert metric, to $\mathscr{H}$; or if it is Lipschitzian on each bounded subset of $\mathscr{H}$, and if in addition there exists a nonpositive real function $E(\cdot)$ on $\mathscr{H}$ such that $\partial_x E(x)y = \langle F(x), y\rangle$ for $x, y \in \mathscr{H}$. Either of these assumptions is satisfied by certain classes of nonlinear relativistic partial differential equations. The tangent space M_0 is naturally isomorphic to the solution manifold of equation $(\asymp'_0)$, with linear flow $T_t^{(0)}$ defined by that equation.

What does the problem of the existence of the wave operator mean in concrete analytical terms? Suppose $x_0(\cdot)$ is a given tangent vector in M_0, identified with a solution of equation $(\asymp'_0)$; let $x(\cdot)$ denote the hypothesized solution of equation $(\asymp')$, which is asymptotic to $x_0(\cdot)$ as the time $t \to -\infty$. It is not difficult to verify that the actual equation determining $x(\cdot)$ is equation $(\asymp')$, with $t_0 = -\infty$. There is of course no apparent reason why the infinite integral

$$\int_{-\infty}^{t} \frac{\sin[(t-s)M]}{M} F(x(s))ds$$

should converge, and some reason to think that it will not, since the integrand as a function of the time variable s looks about the same size for all s. The one-dimensional analog to this integral would be, to take as an example the case $F(x) = x^3$, the integral $\int_{-\infty}^{t} m^{-1} \sin[(t-s)m]x(s)^3 ds$; if $x(\cdot)$ represents a perturbation of simple harmonic motion, one should expect the integral to diverge, at least absolutely. An important part of the problem is to find a space, in which to look for a solution, on which the integral in question is convergent.

Having solved the problem of the existence of the wave operator (cf. [**1**] for results on relativistic equations), there is even less reason why the integral involved in determining the tangent vector $x_1(\cdot)$, to which $x(t)$ is hopefully asymptotic as $t \to +\infty$, should converge. This vector is defined by the equation

$$x(t) = x_1(t) - \int_{t}^{\infty} \frac{\sin[(t-s)M]}{M} F(x(s))ds.$$

While the behavior of $F(x(s))$ for $s \to -\infty$ might be expected to resemble somewhat that of $F(x_0(s))$ as $s \to -\infty$, and be thereby somewhat accessible, this is not at all true as $s \to +\infty$. The proof of the convergence of the infinite integral

$$\int_{-\infty}^{\infty} \frac{\sin[(t-s)M]}{M} F(x(s)),$$

which is the key analytical problem in showing that $x(\cdot)$ is asymptotic to some tangent vector $x_1(\cdot)$ as the time approaches $+\infty$, requires further assumptions and is of a more complex nature than the proof of the existence of the wave operator (cf. [**2**], [**3**]).

The canonical symplectic structure and the energy. An important feature of equation $(\asymp')$ or of the corresponding manifold M, is the existence of a natural symplectic structure in M making possible a full analogy between M and finite-dimensional Hamiltonian systems, and also a purely mathematical definition of the "energy". This subject can be treated only tangentially here, but it serves to remove the vestige of physical allusion involved in the use of the "energy" norm, and to strengthen the analogy with the case of an ordinary differential equation; and the relevant points will be briefly indicated.

There is a natural correspondence between tangent vectors to the manifold M and solutions to the first-order variational equation

$$y(t) = x_0(t) + \int_{t_0}^{t} \frac{\sin[(t-s)M]}{M} \partial_{x(s)} F(x(s)) y(s) ds$$

(cf. [**4**]). If y and z are any two such tangent functions at the point $x(\cdot) \in M$, a symplectic form Ω is defined on M by the equation

$$\Omega_{x(\cdot)}(y, z) = \langle y(t), z'(t) \rangle - \langle y'(t), z(t) \rangle;$$

the right-hand side is t-independent, corresponding to the invariance of Ω under the group T_t, and indeed in the case of a relativistic equation, this form is Lorentz invariant.

The form Ω in particular determines, as on any symplectic manifold, a canonical Poisson bracket $\{\cdot,\cdot\}$ for functions on M, in terms of which the "energy" E on M is determined uniquely, if it exists, by the equation $\{E, K\} = (\partial/\partial t)K$, for an arbitrary smooth function K on M. It suffices to take the special cases in which K has the form $\langle x(t), v \rangle$ for some fixed t and v. In the infinite-dimensional case the global existence question for Hamiltonians (i.e. the energy function E), for given symplectic vector fields on the solution manifold M, such as are determined by the differential equation, have yet to be explored; but in practice, in the case of second-order hyperbolic equations which are temporally invariant, it is generally not difficult to solve the foregoing equation explicitly for E. This is, however, not the case for equations as complicated as those of general relativity, where the concept and form of the energy are still under investigation.

Relativistic partial differential equations or infinite-dimensional manifolds. From the standpoint of the theory of differentiable flows, the assumption that the flow comes from a partial differential equation may seem to involve an *ad hoc* element, and the question may arise of whether the relevant structural elements may not be directly postulated, within the framework of differentiable flow theory. In this connection it seems appropriate to attempt to isolate the key physical assumptions involved in the relativistic partial differential equations which "arise". These seem to be notably the following.

First, *relativistic invariance*, in the specific form that the "phase space" M is to be a Lorentz-invariant subspace of a relativistically invariant finite-dimensional vector bundle over space-time.

Next, *locality*, in the sense that the equations which pick out the submanifold M from the bundle in question depend only on the values of the cross-sections involved at one and the same point x of space-time. Thus, if $\phi(\cdot)$ and $\psi(\cdot)$ are cross-sections of the scalar bundle, terms of the form $A(\phi(x), \psi(x))$, where A is any given function of two scalar variables, are admissible; but not terms of the form $\phi(x)\psi(x')$, with $x \neq x'$.

Third, *positivity of the energy*, at least for all classical (unquantized) equations of second order. This formulates a kind of physical stability, and is quite well founded. On the other hand, in the case of equations which have no classical significance, but are physically applicable when "quantized", such as the Dirac equation, the energy may be indefinite in sign in the case of the classical equation, since the "quantization" process produces a positive energy (operator, not function).

Fourth, *quasi-elementarity*: the cross section whose values are the origin should be a fixed point, under the action of the Poincaré group, and the induced linear representation of the Poincaré group in the tangent space M_0 should be a finite direct sum of irreducible representations. This means roughly that only a finite number of different particles or waves are involved in the interaction under consideration.

While it might be of interest to examine these assumptions from the standpoint of differentiable flows on symplectic manifolds, the fact is that the only known nontrivial examples of such systems are those defined by relativistic nonlinear partial differential equations (i.e., the solution manifolds thereof). When these equations have been studied as long as algebraic manifolds, we may have an effective intrinsic characterization for them, but in the meantime it would seem best not to get too far away from specific equations, or the ideas that emerge from their treatment.

Concluding remarks. This has been largely a talk about software, to use the language of the computer industry; there is no time to go into the hardware, i.e. the existence theorems for the wave and scattering operators (cf. loc. cit.). Let me only mention two aspects. First, adequate hardware has been manufactured for dealing with fairly broad classes of scalar relativistic equations, in particular, for the equations which "arise": $\square \phi = m^2\phi + g\phi^p$ in four-dimensional space-time, when $m > 0$, $g > 0$, and p is an odd integer > 1 (the cases in which $g < 0$ or p is even are excluded by the indefiniteness of the associated energy). Many interesting questions remain, such as that of the asymptotic behavior outside a neighborhood of the fixed point $\phi = 0$; or naturally emerge, such as the inverse scattering question (can the nonlinear interaction function, here $g\phi^p$, be recovered from the S-operator?). Second, it would seem perhaps possible and useful to abstract these results, and obtain conditions applicable to a general class of elliptic fixed points.

References

1. I. Segal, *Quantization and dispersion for non-linear relativistic equations*, Proc. Conf. Math. Th. El. Parts., M.I.T. Press, Cambridge, Mass. (1966), 79–107.

2. ———, *Dispersion for non-linear relativistic equations*. II, Ann. École Norm. Sup. (4) **1** (1968), 459–497.

3. W. Strauss, *Scattering for the equation* $\square u = f(u)$, J. Functional Anal. (to appear).

4. I. Segal, *Differential operators in the manifold of solutions of a non-linear differential equation*, J. Math. Pure Appl. **44** (1965), 71–132.

Massachusetts Institute of Technology

TOPOLOGY OF ELLIPTIC OPERATORS

MICHAEL F. ATIYAH [1]

In these notes I will deal with topological questions concerning elliptic operators, and more particularly with the index theorem. The analytical part of the theory will be reviewed only briefly and we refer to L. Nirenberg's lectures [2] for more details.

I. **The symbol and the index of an elliptic operator.** Throughout this section let X be a closed smooth (i.e. C^∞) manifold of dimension n, and let E, F be smooth complex vector bundles over X of fiber dimension N. The linear space of smooth cross-sections of E (resp. F) is denoted by $C^\infty(E)$ (resp. $C^\infty(F)$). Furthermore let k be a nonnegative integer.

DEFINITION. A *differential operator* P (*of order* k) *from* E *to* F is a linear map

$$P: C^\infty(E) \to C^\infty(F)$$

which locally, in terms of coordinates, can be expressed as a matrix of polynomials in partial derivatives of order $\leq k$. More precisely, if $x_1, \ldots, x_n$ are local coordinates of X with domain U, and if $\tau_1 : E|U \cong U \times C^N$, $\tau_2 : F|U \cong U \times C^N$ are smooth local trivializations of E, resp. F, then the linear operator $C^\infty(E) \to (C^\infty(U))^N = C^\infty(F|U)$ induced by P has the form $\sum_{|\alpha| \leq k} a_\alpha(x) D_x^\alpha$, where for each multi-index $\alpha = (\alpha_1, \ldots, \alpha_n)$ with $|\alpha| = \sum \alpha_i$, a_α is an $N \times N$-matrix of elements of $C^\infty(U)$ (i.e. smooth complex valued functions), and D_x^α denotes the partial derivative $\partial^{\alpha_1}/\partial x_1^{\alpha_1} \ldots \partial^{\alpha_n}/\partial x_n^{\alpha_n}$.

P is *elliptic* (*of order* k) if for each local representation as above, the highest order term

$$p_k(x, \xi) = \sum_{|\alpha| = k} a_\alpha(x) \xi^\alpha$$

is a nonsingular matrix for $x \in U$ and $\xi \in \boldsymbol{R}^n - \{0\}$.

The system of functions $p_k(x, \xi)$, given for every local representation of P as above, is called the *leading symbol of* P.

There is an intrinsic geometrical interpretation of the leading symbol of an elliptic differential operator P. We fix a Riemannian metric on X. Thus the tangent bundle $T(X)$ and the cotangent bundle $T^*(X)$ are canonically isomorphic, and we denote by $B(X)$ (resp. $S(X)$) the corresponding unit ball bundle (resp. sphere bundle). If $\pi^*(E)$, $\pi^*(F)$ are the pullbacks of E, F under the projection $\pi: S(X) \to X$, then the leading symbol of P determines a well-defined vector bundle isomorphism over $S(X)$:

$$\sigma(P): \pi^*(E) \xrightarrow{\cong} \pi^*(F),$$

---

[1] Notes by U. Koschorke.

given by $p_k(x, \xi): E_x \cong F_x$. This isomorphism $\sigma(P)$ gives a geometrical way of looking at the leading symbol of P.

It is a classical result that the dimension of both the kernel and the cokernel of an elliptic differential operator P are finite. The difference of the dimensions,

$$\text{index } P = \dim \ker P - \dim \operatorname{coker} P,$$

is known to depend only on the highest order term, i.e. $\sigma(P)$. Thus it is natural to try to connect the global data given by the symbol to other global data, such as the index. In fact our main goal is to give an explicit formula expressing the index of an elliptic operator P in terms of $\sigma(P)$.

For this purpose it is convenient to enlarge our class of operators by introducing pseudo-differential operators (cf. [2]). Essentially all smooth isomorphisms from $\pi^*(E)$ to $\pi^*(F)$ can occur as symbols of such operators. On the other hand, an elliptic pseudo-differential operator still has a well-defined index which depends only on the homotopy class (through smooth isomorphisms) of its symbol. Therefore, since smooth maps on one side and continuous maps on the other side have a similar homotopy behavior, the index gives rise to an integer valued function on the set of homotopy classes of continuous isomorphisms between $\pi^*(E)$ and $\pi^*(F)$. Thus the use of pseudo-differential operators enables us to replace the set of polynomial symbols by simpler geometrical objects, and to perform more drastic deformations.

We now indicate briefly an alternative topological approach which would allow us to avoid the analytical concept of pseudo-differential operators. Let $A(N, n, k)$ be the set of $N \times N$-matrices $p(\xi)$ whose entries are homogeneous polynomials of order k in $\xi_1, \ldots, \xi_n$, such that $\det p(\xi) \neq 0$ for $\xi \in \boldsymbol{R}^n - \{0\}$. In an obvious way this set can be imbedded in some euclidean space and hence has a natural topology. $A(N, n, k)$ is the typical fiber of a fiber bundle over X whose sections correspond to the symbols of elliptic differential operators of order k from E to F. Thus, if we deal with deformations of symbols in the framework of elliptic differential operators we need some information on the homotopy properties of $A(N, n, k)$. However, not much is known on $A(N, n, k)$, and even the question whether it is empty or not is rather hard to decide. For example $A(N, n, 1)$ is not empty if and only if the $(N-1)$-sphere S^{N-1} admits $(n-1)$ linearly independent tangent vector fields. This last problem, which has been solved by J. F. Adams, is highly nontrivial.

Nevertheless the situation is not quite that hopeless. If we imbed $A(N, n, 2j)$ into $A(N, n, 2(j+1))$ by means of the map $p(\xi) \to (\sum \xi_i^2) p(\xi)$ (which corresponds by the Fourier transform to composition with the Laplace operator), then the direct limit $\lim_{j \to \infty} A(N, n, 2j)$ can be interpreted as the set of even polynomial maps from the sphere S^{n-1} into $\mathrm{GL}(N, \boldsymbol{C})$ (or, equivalently, as the set of polynomial maps from the projective space P^{n-1} into $\mathrm{GL}(N, \boldsymbol{C})$). This set is dense in the space of all even continuous maps from S^{n-1} into $\mathrm{GL}(N, \boldsymbol{C})$. We could use this observation in order to deduce an index function defined on the set of homotopy classes of even continuous isomorphisms between $\pi^*(E)$ and $\pi^*(F)$. Without resorting to pseudo-differential operators we thus would simplify the computation of the index of an elliptic differential operator. However, we shall not pursue this line.

We now resume our previous study of the index of elliptic (pseudo-) differential operators and we shall try to understand the topological significance of the symbol.

First we consider the case when E, F are trivial line bundles. The isomorphisms between $\pi^*(E)$ and $\pi^*(F)$ are simply maps from $S(X)$ into $\mathrm{GL}(1, \boldsymbol{C}) = \boldsymbol{C}^*$ (which retracts onto the unit circle S^1), and the index is defined on the set of homotopy classes $[S(X), S^1] \cong H^1(S(X))$ (cohomology with integer coefficients). It follows from [2] that the index of an elliptic operator P is zero if for each $x \in X$ $\sigma(P)$ is constant on the sphere $S(X)_x$ over x. Hence the index vanishes on the image of τ^* in the following exact cohomology sequence

$$\begin{array}{ccccccc} \to H^1(B(X)) \to & H^1(S(X)) & \xrightarrow{\delta} & H^2(B(X), S(X)) & \xrightarrow{j^*} & H^2(B(X)) \to \dots \\ \cong \quad \nearrow_{\pi^*} & \downarrow \text{index} & & & & \\ H^1(X) & \boldsymbol{Z} & & & & \end{array}$$

If $\dim X = n > 2$, then, by a classical result of Thom, $H^2(B(X), S(X)) = 0$, and π^* is surjective. Hence the index of every elliptic operator from the trivial line bundle into itself is zero (as proved also in [2]). If X is a Riemann surface of genus g, we have a commutative diagram of canonical homomorphisms

$$\begin{array}{ccc} H^2(B(X), S(X)) & \xrightarrow{j^*} & H^2(B(X)), \\ \cong & & \cong \\ H^0(X) & \xrightarrow{\cdot \chi(X)} & H^2(X), \\ \cong & & \cong \\ \boldsymbol{Z} & \longrightarrow & \boldsymbol{Z}, \end{array}$$

where the bottom map is given by multiplication with the Euler characteristic $2 - 2g$ of X. Thus if X is not the torus, j^* is injective and π^* is surjective, and again index $\equiv 0$. In the case of the torus however there might exist elliptic operators from the trivial line bundle into itself with nonvanishing index.

Next we allow E, F to be arbitrary line bundles. We denote by $S'(X) = B^+(X) \cup_{S(X)} B^-(X)$ the n-sphere bundle over X obtained by gluing together two copies $B^+(X)$ and $B^-(X)$ of the unit ball bundle $B(X)$. Then any isomorphism σ between $\pi^*(E)$ and $\pi^*(F)$ gives rise to a line bundle over $S'(X)$: we lift E to $B^+(X)$ and F to $B^-(X)$ and identify them along $S(X)$ by means of σ. This gives a perfectly good alternative description of σ. On the other hand each line bundle over $S(X)$ can be constructed starting from such a σ. Hence we can define the index on $H^2(S'(X), \boldsymbol{Z}) \cong \mathrm{Vect}^1(S'(X))$. (Here $\mathrm{Vect}^1(S'(X))$ denotes the set of isomorphism classes of complex line bundles over $S'(X)$, made into a group by means of the tensor product. The group isomorphism $\mathrm{Vect}^1(S'(X)) \cong H^2(S'(X))$ is given by the first Chern class.) Furthermore it follows from the construction that our index function vanishes on $\pi'^*(H^2(X))$ where $\pi' : S'(X) \to X$ is the bundle projection. As π' admits a section, $H^2(S'(X))$ is the direct sum of $\pi'^*(H^2(X))$ and $H^2(S'(X), B^-(X))$, and it is clear that we lose no information when we restrict the index function to $H^2(S'(X), B^-(X)) = H^2(B(X), S(X))$.

If $\dim X > 2$, again index $\equiv 0$ in this setting. However, if X is a Riemann surface other than the torus, $H^2(B(X), S(X))$ is isomorphic to Z, and for arbitrary line bundles E, F there is no longer any reason to think that the index of all elliptic operators from E to F is zero. In fact we can give a simple example of an elliptic differential operator with nonvanishing index: the operator

$$\bar{\partial} : f \to \frac{\partial f}{\partial \bar{z}} d\bar{z}$$

is elliptic, $\ker \bar{\partial}$ consists of the constants, $\operatorname{coker} \bar{\partial} \cong \ker \bar{\partial}^*$ consists of the anti-holomorphic differentials and hence (from the classical theory of Riemann surfaces) $\operatorname{index} \bar{\partial} = 1 - g \neq 0$ (since the genus is $g \neq 1$).

Now let X again be a compact manifold of arbitrary dimension, and let E, F be N-dimensional trivial bundles over X. Then the index function takes the form

$$\text{index} : [S(X), \mathrm{GL}(N, \boldsymbol{C})] \to \boldsymbol{Z}.$$

As we saw earlier, $[S(X), \mathrm{GL}(N, \boldsymbol{C})]$ has a nice cohomological interpretation when $N = 1$. For $N > 1$ however, this group is not even always commutative. It is therefore convenient to stabilize it.

Thus let

$$C^* = \mathrm{GL}(1) \subset \mathrm{GL}(2) \subset \ldots \subset \mathrm{GL}(N) \subset \ldots \subset \mathrm{GL}(2N) \subset \ldots$$

be the obvious sequence of imbeddings, and denote by $\mathrm{GL}(\infty)$ the topological group $\lim_{N\to\infty} \mathrm{GL}(N)$. If Y is a compact space and $A, B : Y \to \mathrm{GL}(N)$ are continuous maps, then by a rotation argument one can show that the product map $A \cdot B$ is homotopic to the direct sum map $A \oplus B$ as maps from Y into $\mathrm{GL}(2N)$. Hence the group

$$K^1(Y) := [Y, \mathrm{GL}(\infty)] = \lim_{N\to\infty} [Y, \mathrm{GL}(N)]$$

is abelian.

It is now easy to see the above form of the index function gives rise to a group homomorphism index : $K^1(S(X)) \to \boldsymbol{Z}$. We merely have to note that if P and Q are elliptic operators of order 0, then index $(P \oplus Q) =$ index $P +$ index Q, and $\sigma(P \oplus Q) = \sigma(P) \oplus \sigma(Q)$; for the identical operator on a trivial bundle of dimension N the index vanishes and the symbol is a constant function with value $\mathrm{Id} \in \mathrm{GL}(N)$. Furthermore, in analogy with the case $N = 1$, this homomorphism vanishes on the image of $\pi^* : K^1(X) \to K^1(S(X))$.

Finally let P be an elliptic operator from an arbitrary smooth vector bundle over X into another. As in the case of line bundles, the symbol of P determines a vector bundle $V_{\sigma(P)}$ over $S'(X) = B^+(X) \cup_{S(X)} B^-(X)$, and we obtain a homomorphism of semigroups index: $\mathrm{Vect}\,(S'(X)) \to \mathbf{Z}$, defined by $V_{\sigma(P)} \mapsto \operatorname{index} P$. Here $\mathrm{Vect}\,(S'(X))$ denotes the semigroup of isomorphism classes of vector bundles over $S'(X)$, with addition given by the direct sum. By the following method, due to Grothendieck, we can make $\mathrm{Vect}\,(S'(X))$ into a group.

Quite generally let A be an abelian semigroup. Define B to be the quotient of $A \times A$ under the following equivalence relation: $(a_1, a_2) \sim (a_1', a_2')$ if there exist $a, a' \in A$ such that $(a_1 + a, a_2 + a) = (a_1' + a', a_2' + a')$. Then B is an abelian group; the map $j: A \to B$ which maps $a \in A$ into the equivalence class of $(a, 0)$ is additive; and each additive map from A into a group factors uniquely through j.

In particular if $A = \mathrm{Vect}\,(Y)$ is given by the (complex) vector bundles over a compact space Y, one denotes by $K(Y)$ the corresponding group. Furthermore, if Z is a closed subspace of Y, we define $K(Y, Z)$ to be the kernel of the obvious restriction $K(Y/Z) \to K(Z/Z)$. Here Y/Z is the space obtained from Y by collapsing Z to a point, and Z/Z is this resulting point.

We can now interpret our index function as a group homomorphism from $K(S'(X))$ into $\mathbf{Z}$. Again, as in the case of line bundles, we lose no information when we restrict it to $K(S'(X), B^-(X)) = K(B(X), S(X))$. Thus we will consider it from now on as a group homomorphism index: $K(B(X), S(X)) \to \mathbf{Z}$.

In examining the topological role of the symbol of elliptic operators we have been led to introduce the functors K and K^1. In fact they fit into an extraordinary cohomology theory—K-theory—which is the appropriate setting for the study of the index. In particular we have coboundary operators, and our two last interpretations of the index are related to one another by the following commutative diagram

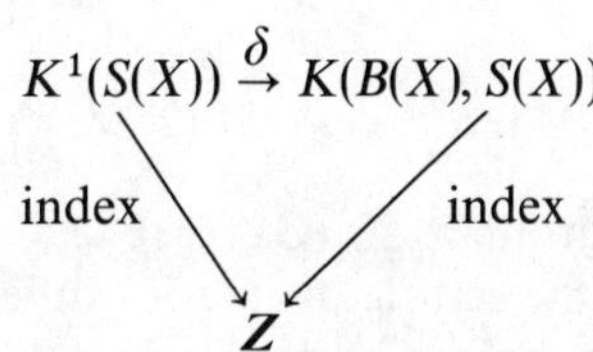

II. **Classical Examples.** We will now give some concrete examples of elliptic differential operators. This will not only motivate our investigation of the index problem, but also give us some idea what the index formula should look like.

First we study purely locally homogeneous elliptic differential operators with constant coefficients. Such an operator can be given by its symbol over one base point, i.e. by an $N \times N$ matrix $p(\xi)$ (whose entries are homogeneous polynomials of a fixed degree in $\xi = (\xi_1, \ldots, \xi_n)$) such that $\det(p(\xi)) \neq 0$ for $\xi \neq 0$. In this terminology e.g. the Laplace operator Δ corresponds to $\sum \xi_i^2 \cdot \mathrm{Id}$.

Since many important elliptic operators are of order 1 or can be decomposed into first order operators, it is reasonable to ask whether $-\Delta$ admits a square root. Thus we have to look for $N \times N$-matrices $A_1, \ldots, A_n$ such that

$$\left(\sum_{i=1}^{n} A_i\xi_i\right)^2 = -(\sum \xi_i^2)\,\mathrm{Id},$$

or, equivalently, such that

$$(*) \qquad A_i^2 = -\mathrm{Id} \quad \text{for } i = 1, \ldots, n; \qquad A_iA_j = -A_jA_i \quad \text{for } i \neq j.$$

These relations generate abstractly the complexified Clifford algebra $C_n = \mathrm{Cliff}_{\boldsymbol{C}}(\boldsymbol{R}^n)$ of $\boldsymbol{R}^n$, and the solutions of (*) correspond to the representations of C_n on $\boldsymbol{C}^N$. If n is even, C_n is simple and hence a matrix algebra. Therefore the equations (*) have a solution with $N^2 = 2^n = \dim C_n$, i.e. $N = 2^{n/2}$. The corresponding first order differential operator is called the *Dirac operator*.

Since C_n is simple for even n, there is no square root of $-\Delta$ when $0 < N^2 < 2^n$. However, it follows from the general theory of Clifford algebras that the Dirac operator can be written in the form

$$\begin{pmatrix} 0 & D^+ \\ D^- & 0 \end{pmatrix}$$

where D^+ and D^- are adjoint to one another. Hence D^+ is a solution to the modified condition $D^+ \cdot D^{+*} = -\Delta$.

We will later give an explicit description of the Dirac operator and of D^+, and we will see that e.g. for $n = 2$, D^+ is just the Cauchy-Riemann operator $\bar{\partial}$. Now we only state two important properties of D^+. First, D^+ has minimal N, i.e. D^+ is defined on systems of $N = 2^{n/2-1}$ complex valued functions, and there is no first-order elliptic differential operator over an open subset of $\boldsymbol{R}^n$ defined on a small number of functions. This follows from the results of J. F. Adams, and from the remarks in §I concerning the space $A(N, n, 1)$. Second, we have the "*local Bott theorem*": the symbol of D^+

$$\sigma(D^+): S^{2n-1} \to \mathrm{GL}(2^{n-1}, \boldsymbol{C})$$

generates the infinite cyclic group $\pi_{2n-1}(\mathrm{GL}(\infty))$.

Next we indicate how, in the setting of exterior differential forms, one can find another square root of the Laplace operator. For $0 \leq p \leq n$ let Ω^p denote the space of differentiable p-forms over an open set $U \subset \boldsymbol{R}^n$, and let $d: \Omega^p \to \Omega^{p+1}$ be the exterior derivation. We are going to make d into a square system. Thus, for p-forms α, β with compact support, define an inner product by

$$\langle \alpha, \beta \rangle = \quad \alpha \wedge * \beta,$$

where $*: \Omega^p \to \Omega^{n-p}$ is the $C^\infty(U)$-linear operator with $*(dx^1 \wedge \ldots \wedge dx^p) = dx^{p+1} \wedge \ldots \wedge dx^n$ etc. We can then define the formal adjoint $d^*: \Omega^{p+1} \to \Omega^p$ of d by the following condition: $\langle d\alpha, \beta \rangle = \langle \alpha, d^*\beta \rangle$ for $\alpha \in \Omega^p$, $\beta \in \Omega^{p+1}$ with compact support. Clearly

$$D = d + d^*: \Omega = \sum_{p=0}^{n} \Omega^p = C^\infty(U \times \Lambda^* C^n) \to \Omega,$$

is a first order differential operator, and we have

$$D^2 = DD^* = dd^* + d^*d = -\Delta.$$

Thus, being a square root of the Laplacian, D is elliptic. Since $\dim(\Lambda^* C^n) = 2^n$, the symbol of D corresponds to the regular representation of $\mathrm{Cliff}_C(R^n)$ on itself.

Similarly if U is an open subset of $R^{2n} = C^n$ let $\Omega^{0,p}$ denote the space of exterior differentials of the form $a_{i_1,\ldots,i_p}\, d\bar{z}_{i_1} \wedge \ldots \wedge d\bar{z}_{i_p}$. Then the exterior derivation $\bar{\partial}$ gives rise to a first order elliptic operator $\bar{\partial} + \bar{\partial}^*$ whose square is the Laplace operator. Hence the symbol of $\bar{\partial} + \bar{\partial}^*$ defines a (nonzero) homomorphism from $C_{2n} = \mathrm{Cliff}_C(R^{2n})$ into the matrix algebra $\mathrm{End}_C(\Lambda^*(C^n))$. As both algebras have dimension 2^{2n} and as a full matrix algebra is simple, it follows that this homomorphism is in fact an isomorphism

$$\mathrm{Cliff}_C(R^{2n}) \cong \mathrm{End}_C(\Lambda^*(C^n)).$$

This gives an explicit description of the minimal representation of C_{2n} which we used when defining the Dirac operator. Thus $\bar{\partial} + \bar{\partial}^*$ is just the Dirac operator, and its splitting into

$$\begin{pmatrix} 0 & \bar{D}^+ \\ \bar{D}^- & 0 \end{pmatrix}$$

can be given by

$$\bar{D}^+ = (\bar{\partial} + \bar{\partial}^*)|\sum_{p=0}^{n} \Omega^{0,2p}, \qquad \bar{D}^- = (\bar{\partial} + \bar{\partial}^*)|\sum_{p=0}^{n} \Omega^{0,2p+1}.$$

Next we study global versions of some of the local elliptic operators investigated so far. Thus let X be a closed Riemannian manifold. Then we have the exterior differentiation d operating on the space Ω of (complexified) global exterior differential forms, and the Riemannian structure defines an adjoint d^*. Also $D = d + d^*$ is an elliptic differential operator over X and hence has an index. Since D is selfadjoint, this index vanishes.

D does not preserve the degree of a differential form; however, it preserves the parity. Thus, if we define $\Omega^+ = \sum_{p\,\mathrm{even}} \Omega^p$, $\Omega^- = \sum_{p\,\mathrm{odd}} \Omega^p$, we can split D into two operators

$$D^+: \Omega^+ \to \Omega^-, \qquad D^-: \Omega^- \to \Omega^+$$

which are adjoint to one another. D^+ is elliptic (since D is), and its index is given by

$$\mathrm{index}\, D^+ = \dim\ker D^+ - \dim\ker D^-.$$

Next we show that $\ker D = \ker D^2$. We have only to see that for $u \in \ker D^2$ Du vanishes. But this is clear because $0 = \langle D^2u, u\rangle = \langle D^*Du, u\rangle = \langle Du, Du\rangle$.

Now $D^2 = -\Delta$ is the Laplace operator. By definition its kernel is formed by the harmonic forms on X. Thus, if H^p denotes the space of harmonic p-forms,

$$\ker D^+ = \sum_{p\,\text{even}} H^p, \qquad \ker D^- = \sum_{p\,\text{odd}} H^p,$$

and we have

$$\operatorname{index} D^+ = \sum (-1)^p \dim H^p.$$

According to the main theorem of Hodge theory, H^p is isomorphic to the cohomology group $H^p(X, \mathbf{C})$. Thus $B^p = \dim H^p$ is the pth Betti number of X, and the index of D^+ is just the Euler characteristic $\chi(X) = \sum(-1)^p B^p$ of X.

It is a well-established result in algebraic topology that $\chi(X)$ vanishes if X admits a tangent vector field without zeros. Now we can use the above index interpretation of $\chi(X)$ to give a proof of this fact which stresses more the operator point of view.[2] We first observe that the symbol $\sigma(D)$ is given by exterior and interior multiplication on the left in $\Lambda^*(T^*(X))$. A nonvanishing vector field on X gives rise, by exterior and interior multiplication on the right, to a bundle isomorphism which commutes with $\sigma(D)$, and interchanges $\sigma(D^+)$ with $\sigma(D^-)$. Therefore index $D^- =$ index D^+. On the other hand, since $D^- = D^{+*}$, index $D^- = -$index D^+. We thus obtain finally that $\chi(X) =$ index $D^+ = 0$.

Now assume that X is an orientable compact Riemannian manifold with $\dim X = 4k$. Then there is a canonical isomorphism $*: \Omega^p \to \Omega^{4k-p}$ such that on $\Omega^p \quad (*)^2 = (-1)^p$. If we define

$$\tau : \Omega \to \Omega$$

by $\tau|\Omega^p = (-1)^{(p(p-1)+2k)/2} \cdot *$, τ is an involution. Hence Ω splits into the eigenspace Ω_+ of $+1$ and the eigenspace Ω_- of -1. Since $D\tau = -\tau D$, D also can be decomposed into the direct sum of the elliptic operators $D_+ : \Omega_+ \to \Omega_-$, $D_- : \Omega_- \to \Omega_+$. D_+ and D_- are adjoint to one another, and therefore

$$\operatorname{index} D_+ = \dim \ker D_+ - \dim \ker D_-.$$

Clearly $\ker D_\pm = \{\alpha \text{ harmonic form} | \tau = \pm\tau\alpha\}$. Moreover we may restrict our attention to real harmonic $2k$-forms, because the contributions to the index from the other dimensions cancel out, and on H^{2k} $\tau = *$ is real. Now if $\alpha \neq 0$ is a real harmonic $2k$-form with $\alpha = \pm * \alpha$, then $\pm \int_X \alpha \wedge \alpha = \int_X \alpha \wedge * \alpha > 0$, and hence the corresponding expression in real cohomology $\pm(\alpha \cup \alpha)[X]$ is positive. Thus index D_+ is the difference of the dimensions of maximal subspaces of $H^{2p}(X, \mathbf{R})$ on which the nondegenerate quadratic form $\alpha \mapsto (\alpha \cup \alpha)[X]$ is positive, resp. negative, definite. In other words the index of D_+ equals the signature (or index) of the manifold X.

Let us remark here that one can also interpret sign (X) as the index of the elliptic operator $d^*d - dd^* = (d + d^*)(d - d^*): \Omega_+^{2k} \to \Omega_-^{2k}$.

---

[2] For fuller details and generalizations of this argument see: M. F. Atiyah, *Vector fields on manifolds*, Arbeitsgemeinschaft für Forschung des Landes Nordrhein-Westfalen (1969).

The signature $\operatorname{sign}(X)$ of X is related to the Euler characteristic by $\operatorname{sign}(X) \equiv \chi(X) \pmod 2$. However, $\operatorname{sign}(X)$ is a much more sophisticated invariant. For example if X has a nonvanishing vector field, we can still have $\operatorname{sign}(X) \neq 0$. The argument which shows that in this case $\chi(X)$ vanishes, does not necessarily extend to the signature, because the definition of D_+ involves the Riemannian metric on X in a much more essential way than the definition of D^+ does. However, if the vector field preserves the Riemannian metric, i.e. if it is an infinitesimal isometry, then $\operatorname{sign}(X) = 0$.

As an example we determine $\chi(X)$ and $\operatorname{sign}(X)$ when X is the projective space $P^2(\boldsymbol{C})$. One knows that there is a class $c_1 \in H^2(P^2(\boldsymbol{C}), \boldsymbol{R})$ such that $H^*(P^2(\boldsymbol{C}), \boldsymbol{R})$ is the truncated polynomial algebra $\boldsymbol{R}[c_1]/(c_1^3)$. Hence $\chi(P^2(\boldsymbol{C})) = 3$ and $\operatorname{sign}(P^2(\boldsymbol{C})) = 1$.

In the last examples we assumed the main theorem of Hodge theory without proof. However, we can deduce it, and even a more general theorem on elliptic complexes, from [2].

First we digress to show how the regularity results of [2], expressed in terms of the Sobolev spaces H^s, lead quickly to statements in C^∞. Thus let

$$P: C^\infty(E) \to C^\infty(F)$$

be an elliptic pseudo-differential operator of order m and let

$$P_s: H^s(E) \to H^{s-m}(F)$$

be the induced operator on the Sobolev spaces. Let P^t, P_s^t denote the corresponding transposed operators. Then the regularity results in [2] assert that

(i) $\operatorname{Ker} P_s^t = \operatorname{Ker} P^t$ for all s.

(ii) P_s has closed range.

Thus if $f \in C^\infty(F) \subset H^{s-m}(F)$ we can write $f = P_s(e) + h$, where $e \in H^s(E)$ and $h \in \operatorname{Ker} P^t \subset C^\infty(F)$. By the regularity (hypoellipticity) of P this equation implies that $e \in C^\infty(E)$. Thus we deduce that $C^\infty(F)$ is the direct sum of $P(C^\infty(F))$ and $\operatorname{Ker} P^t$.

Now let

$$0 \to C^\infty(E^0) \xrightarrow{d_0} C^\infty(E^1) \xrightarrow{d_1} C^\infty(E^2) \xrightarrow{d_2} \cdots \to C^\infty(E^m) \to 0$$

be a complex of differential operators over the closed manifold X, i.e. $E^0, E^1, \ldots, E^m$ are smooth complex vector bundles, and $d_0, d_1, \ldots$ are differential operators of a fixed order such that $d \cdot d = 0$. If for each $x \in X$ and $\xi \in T_x^*(X) - \{0\}$ the leading symbol sequence

$$0 \to E_x^0 \xrightarrow{\sigma(d_0)_\xi} E_x^1 \xrightarrow{\sigma(d_1)_\xi} E_x^2 \to \cdots \to E_x^m \to 0$$

is exact, then the complex is called *elliptic*. This definition extends in a natural way the notion of elliptic operators ($m = 1$). On the other hand an elliptic complex can easily be transformed into an elliptic operator. If an elliptic complex as above is given, we can choose hermitian inner products on $E^0, \ldots, E^m$ and a measure on

X and thus introduce adjoint operators d_i^*. Clearly the exactness of the symbol sequence implies the ellipticity of the operator

$$D = d + d^* : C^\infty(\bigoplus_0^m E^i) \to C^\infty(\bigoplus_0^m E^i).$$

Now define the ith cohomology H^i of our complex by

$$H^i\,(\text{complex}) = \text{Ker}\, d_i/\text{Image}\, d_{i-1}.$$

We claim that this vector space is of finite dimension, or more precisely that the obvious map

$$\text{Ker}\,(D^2|C^\infty(E^i)) \subset \text{Ker}\, d_i \to H^i$$

is an isomorphism. In fact as we have already explained the regularity theorems in [2] imply that each $u \in C^\infty(E^i)$ can be written in the form $u = d_{i-1}v + d_i^* w + h$, where h is harmonic, i.e. $D^2(h) = 0$. In particular if $u \in \text{Ker}\, d_i$, then $d_i d_i^* w = 0$ and hence is exact off the zero section of $T^*(X)$. A Riemannian metric on X boundary and a harmonic element. In addition this decomposition is unique because the image of d is orthogonal to the kernel of d^* and thus to the kernel of D^2. This proves the assertion above. Thus we can define the Euler characteristic of the elliptic complex by

$$\chi = \sum(-1)^i \dim H^i\,(\text{complex}).$$

This number generalizes naturally the notion of the index of an elliptic operator. On the other hand it is the index of the elliptic operator

$$D^+ : C^\infty(\bigoplus_{i\,\text{even}} E^i) \to C^\infty(\bigoplus_{i\,\text{odd}} E^i)$$

associated to the complex.

A classical example of an elliptic complex is the complex Ω of exterior forms over a closed manifold X

$$0 \to \Omega^0(X) \xrightarrow{d} \Omega^1(X) \xrightarrow{d} \Omega^2(X) \to \ldots \to \Omega^n(X) \to 0.$$

Its symbol sequence is given by exterior multiplication with cotangent vectors, and hence is exact off the zero section of $T^*(X)$. A riemannian metric on X determines a choice of the structures needed in the construction above, and we have a canonical isomorphism between the space of harmonic p-forms on X and $H^p(\Omega)$. Together with de Rham's theorem, which gives an isomorphism between $H^p(\Omega)$ and $H^p(X, C)$, this yields the main result of Hodge's theory.

If X is a compact, complex analytic manifold of complex dimension n, the Dolbeault complex

$$0 \to \Omega^0 \xrightarrow{\bar\partial} \Omega^{0,1} \xrightarrow{\bar\partial} \Omega^{0,2} \to \ldots \to \Omega^{0,n} \to 0$$

provides an even more interesting and delicate example of an elliptic complex. Here the symbol sequence is given by exterior multiplication in the complex exterior algebra of $\overline{T}^*(X)$. If we introduce a metric, we again get an isomorphism between the space of harmonic p-forms and $H^p(\text{complex}) = \operatorname{Ker} \bar{\partial}^{0,p}/\operatorname{Im} \bar{\partial}^{0,p-1} \cong H^p(X, \mathcal{O})$, where $\mathcal{O}$ is the sheaf of holomorphic functions. Thus the Euler characteristic of X with coefficients in $\mathcal{O}$

$$\chi(X, \mathcal{O}) = \sum(-1)^p \dim H^p(X, \mathcal{O})$$

is the index of the elliptic operator associated to the Dolbeault complex.

In the more special case when X is algebraic, i.e. a complex analytic submanifold of a projective space, the number, $g_p = \dim H^p(X, \mathcal{O})$ = dimension of the space of holomorphic p-forms, can be interpreted as a higher dimensional genus (for Riemann surfaces g_1 is just the usual genus). $\chi(X, \mathcal{O}) = 1 - g_1 + g_2 - g_3 + \dots$ is also called the arithmetic genus of X.

A holomorphic vector bundle over a complex manifold X gives rise to a generalized Dolbeault complex,

$$0 \to \Omega^0(V) \xrightarrow{\bar{\partial}_V} \Omega^{0,1}(V) \xrightarrow{\bar{\partial}_V} \Omega^{0,2}(V) \to \dots.$$

Here $\Omega^{0,p}(V)$ is defined by the use of the tensor product; for instance $\Omega^0(V) = C^\infty(V)$. Locally the definition of $\bar{\partial}_V$ is clear. Since it annihilates the holomorphic transition matrices of V, the extension to a global operator $\bar{\partial}_V$ poses no problem.

The index of the elliptic operator $\bar{D}_V^+$ associated to the complex above has the form

$$\operatorname{index} \bar{D}_V^+ = \chi(X, V) = \sum(-1)^p \dim H^p(X, \mathcal{O}(V)).$$

In favorable situations $H^p(X, \mathcal{O}(V)) = 0$ for $p > 0$, and $\chi(X, V) = \dim H^0(X, \mathcal{O}(V))$. Now $H^0(X, \mathcal{O}(V))$ is the space of holomorphic sections of V, and it is important to determine its dimension. If V is a line bundle with a holomorphic section $\phi \not\equiv 0$, this question can be reduced to another classical problem: the holomorphic sections ψ of V are in one-to-one correspondence with the meromorphic functions ψ/ϕ on X with a prescribed set of poles.

These few remarks indicate that a formula expressing $\operatorname{index} \bar{D}_V^+ = \chi(X, V)$ in terms of topological invariants of X and V is of fundamental interest. The solution, which includes the Riemann-Roch Theorem, has been given by F. Hirzebruch [**3**]. For example if V is a trivial line bundle we get

$$\chi(X, V) = \tfrac{1}{2}(c_1(X))[X] \quad \text{for } \dim X = 1,$$
$$\chi(X, V) = \tfrac{1}{12}(c_1^2(X) + c_2(X))[X] \quad \text{for } \dim X = 2 \text{ etc.},$$

where $c_i(X) \in H^{2i}(X)$ is the ith Chern class of TX. Hirzebruch proved his formulas only for algebraic manifolds; however the general index theorem implies that they are still valid in the general case of a complex manifold X.

Hirzebruch gave also a formula for the index of the operator D_+ canonically associated to a real oriented Riemannian manifold X of dimension $4k$. He showed that the signature of X equals the L-genus. Thus the index of D_+ has the form $\frac{1}{3}p_i[X]$ for $k = 1$, $(1/45)(7p_2 - p_1^2)[X]$ for $k = 2$ etc., where p_i is the ith Pontrjagin class of X. The polynomials used here are related to, but different from, the polynomials involved in the formulas for index $D^+ = \chi(X, V)$ above.

III. **The index theorem.** In this section we will state the index theorem for elliptic operators, and outline its proof. In other words, we will give an explicit description of the homomorphism index: $K(B(X), S(X)) \to \boldsymbol{Z}$ (defined in §I) in terms of the topological invariants of the compact manifold X. For this purpose we will use the language of K-theory which seems to be most appropriate to the problem. However, it is not difficult to transform the result into a formula involving Chern character, Todd class, and other objects of algebraic topology which might be more familiar.

First we recall some additional notions and results of K-theory. In §I we introduced, for compact Y, the group $K(Y)$ of virtual complex vector bundles over Y. For a locally compact space Y we now define $K(Y)$ to be the kernel of the restriction map $K(Y^+) \to K(+) = \boldsymbol{Z}$, where Y^+ denotes the one-point compactification of Y (with the obvious base point $+$). With this notation if Z is a closed subset of a compact space Y, then $K(Y, Z) = K(Y - Z)$. We can interpret elements of this group as equivalence classes of isomorphisms σ of the form $\sigma: E|Z \cong F|Z$, where E, F are complex vector bundles over Y.

In the special case when X is a compact manifold (with a Riemannian metric), the tangent bundle TX is diffeomorphic to its open ball bundle, and hence $K(TX) = K(B(X), S(X))$. The description of this group by means of isomorphisms σ is particularly meaningful, because here such isomorphisms correspond to symbols of elliptic operators. The index homomorphism thus takes the following form:

$$\begin{aligned} \text{index}: K(T(X)) &\to \boldsymbol{Z} \\ [\sigma(P)] &\mapsto \text{index } P. \end{aligned}$$

We shall refer to this as the *analytical index*.

The tensor product of vector bundles defines a ring structure on $K(X)$, and via the projection, $K(T(X))$ is a module over $K(X)$. A typical example of this module multiplication is the following: if X is a complex manifold, let $[V] \in K(X)$ be represented by a holomorphic vector bundle V, and let $[\sigma(\bar{\partial})]$ (resp. $[\sigma(\bar{\partial}_V)]$) $\in K(T(X))$ be given by the symbol of the operator $\bar{D}^+$ (resp. $\bar{D}_V^+$) corresponding to the Dolbeault complex (of V). Then

$$[\sigma(\bar{\partial}_V)] = [\sigma(\bar{\partial})] \cdot [V].$$

Moreover we can assert the following "*global Bott theorem*": For a compact complex manifold X, $K(T(X))$ is a free module over $K(X)$ with generator $[\sigma(\bar{\partial})]$, i.e. we have an isomorphism of the form

$$K(X) \to K(T(X)) \qquad [V] \mapsto [\sigma(\bar{\partial})] \cdot [V].$$

These results which can be extended to quasi-complex manifolds, show that the situation in the Riemann-Roch theorem is quite typical for the index question. In fact the partial answers contained in the classical Riemann-Roch-Hirzebruch theorem provide the main guide to the general index formula.

The global Bott theorem above holds in greater generality: if E is a complex vector bundle over any (compact) space X, then there is an isomorphism $K(X) \cong K(E)$ given by multiplication with the fundamental element $\lambda_E \in K(E)$ which is derived from the exterior algebra of E. For X = point this assertion is equivalent to the "local Bott theorem" stated in connection with the Dirac operator. The isomorphism $K(X) \cong K(E)$ also holds for locally compact X.

Now we are well prepared to define, for a closed smooth manifold X, a homomorphism i_t from $K(T(X))$ into $\boldsymbol{Z}$, called the *topological index*, which will turn out to coincide with the analytical index. Embed X in a high dimensional euclidean space E and choose a tubular neighborhood N. Thus N is open in E, but can be thought of as a (real) vector bundle over X, with projection p. Then i_t is defined to be the composition of three homomorphisms:

$$i_t : K(T(X)) \xrightarrow{i_3} K(T(N)) \xrightarrow{i_1} K(TE) \xrightarrow{i_2} \boldsymbol{Z}.$$

Here i_1 is induced by the map $(TE)^+ \to (TN)^+$ which collapses the complement of the open set TN in TE into the compactifying point $+ \in (TN)^+$. The tangent bundle of the euclidean space E can be identified with $E \otimes_{\boldsymbol{R}} \boldsymbol{C}$, and i_2 is defined to be the canonical isomorphism from $K(E \otimes_{\boldsymbol{R}} \boldsymbol{C})$ to $K(\text{point}) = \boldsymbol{Z}$, given by the (local) Bott theorem. A similar (but global) argument leads to the definition of i_3: since $p : N \to X$ is a real vector bundle over X, $Tp : TN \to TX$ can be considered as a complex vector bundle (for $x \in X$ and $\xi \in T_x(X)$ the fiber $(Tp)^{-1}(\xi)$ is $N_x \oplus T(N_x) \cong N_x \otimes_{\boldsymbol{R}} \boldsymbol{C}$, where $N_x = p^{-1}(x)$). Then the (generalized) global Bott theorem above yields the isomorphism $i_3 : K(TX) \to K(TN)$.

The homomorphism i_t is well defined and in particular independent of the choice of the embedding. In fact let $f : X \to E$, $g : X \to E'$ be two embeddings of X into euclidean spaces. Clearly the product embedding $f \times g : X \to E \times E'$ is homotopic through embeddings to both $f \times 0$ and $0 \times g$. Thus, since the construction above is stable and homotopy invariant, we obtain the same topological index, whether we use f or g.

Now we can state the central theorem of these notes.

INDEX THEOREM. *The analytical index equals the topological index.*

We now proceed to outline the proof of the index theorem. The idea is to interpret the transformations i_1, i_2 and i_3 (introduced in the construction of the topological index merely at the symbol level) as operations on operators themselves. At the same time we have to show that the analytical index is not changed by these operations. Thus conceptually the proof of the index theorem is rather simple: given an elliptic operator on X one has to construct an operator of the same index first on N, and finally on the sphere E^+, where the computation of the index is much easier. This method has much in common with A. Grothendieck's approach to the Riemann-Roch problem.

As a preliminary step in the proof, we have to introduce, on noncompact manifolds, an appropriate class of elliptic operators which still have finite dimensional solution spaces. Thus, if U is a possibly noncompact manifold, let P be an elliptic pseudo-differential operator of order zero on U which equals the identity at ∞, i.e. outside a compact subset K of U. This means $P\phi = \phi$ for all smooth sections ϕ with $\operatorname{supp}\phi \cap K = \emptyset$. We furthermore put the same condition on

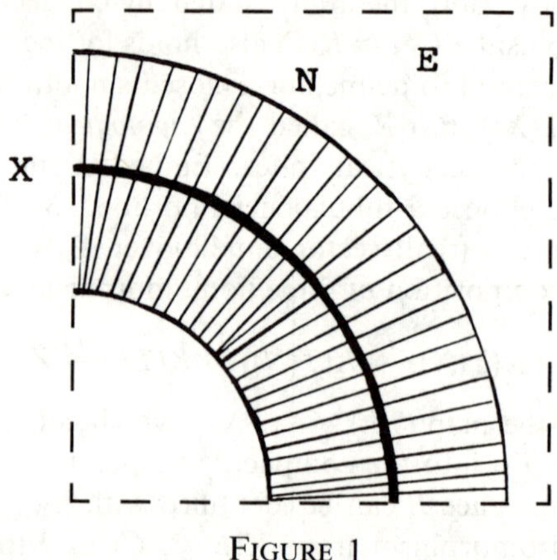

FIGURE 1

the transpose P^t. An equivalent way of describing these conditions is to say that P is represented by a distribution $K(x, y)$ on $U \times U$ which outside $K \times K$ is a delta function on the diagonal.

These restrictions for P guarantee that the main results of the theory of elliptic operators are still valid. In particular the kernel and the cokernel of P are finite dimensional. Furthermore each element of $K(TU)$ can be represented as the symbol class $[\sigma(P)]$ of such an operator P, and the index of P gives rise to a well-defined homomorphism index: $TU \to \mathbf{Z}$.

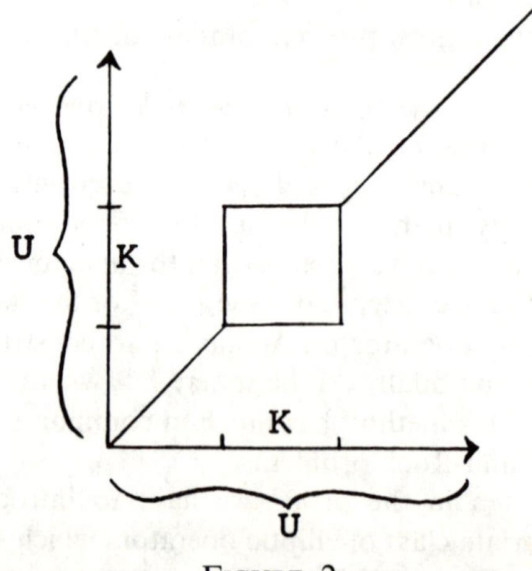

FIGURE 2

If $U \subset V$ is open in a manifold V, we can extend an operator P as above to an operator P_V on V by taking the identity outside U. If $P_V\phi = 0$ on V, then $\operatorname{supp}\phi \subset K \subset U$, since $P\phi = \phi$ outside K. Hence $\operatorname{Ker} P_V$ is isomorphic to $\operatorname{Ker} P$ by the obvious restriction map. The same applies to the transposed P^t, and hence $\operatorname{index} P = \operatorname{index} P_V$.

Now we come back to the situation in the construction of the topological index. Our extension of the analytical index function to noncompact manifolds now yields the following diagram:

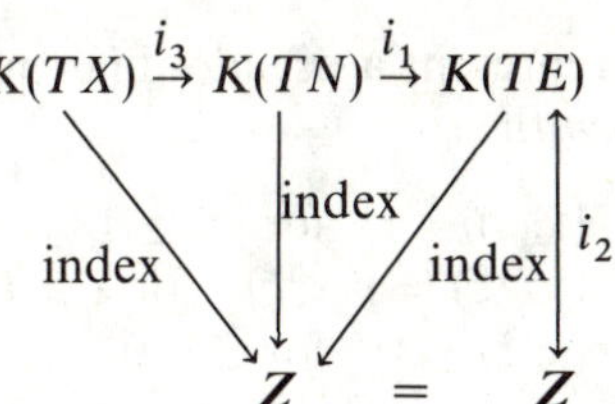

The index theorem is proved once we have shown that all three triangles in this diagram are commutative. Hence we proceed in three steps.

(1) Since N is an open submanifold of E, we can apply the above extension argument, and it is clear that i_1 commutes with the analytical index.

(2) In order to prove the commutativity of the triangle involving i_2 we have to represent the fundamental generator $[\lambda]$ of $K(TE)$ by an elliptic operator and to show that its index is equal to 1. Here λ is derived from the exterior algebra of $TE = \boldsymbol{C}^m$ ($m = \dim_{\boldsymbol{R}} E$) in the usual way: by means of the hermitian metric in $\boldsymbol{C}^m$ we make the sequence (Λ):

$$0 \to \Lambda^0(\boldsymbol{C}^m) \xrightarrow{\wedge u} \Lambda^1(\boldsymbol{C}^m) \xrightarrow{\wedge u} \Lambda^2(\boldsymbol{C}^m) \to \dots \xrightarrow{\wedge u} \Lambda^m(\boldsymbol{C}^m) \to 0$$

of vector bundle homomorphisms over $\boldsymbol{C}^m$ (exact over $u \neq 0$) into a single vector bundle isomorphism over $\boldsymbol{C}^m - \{0\}$:

$$\lambda\colon \Lambda^{\text{even}}(\boldsymbol{C}^m) \overset{\cong}{\rightarrow} \Lambda^{\text{odd}}(\boldsymbol{C}^m).$$

$$\xi \in T_0(\boldsymbol{R}^m) \qquad x \qquad \boldsymbol{R}^m$$

We already know the sequence (Λ) as the local symbol sequence of the Dolbeault complex. However, as we need to get an operator on $E = \boldsymbol{R}^m$ and not on $\boldsymbol{C}^m$ it is necessary to stress rather the real point of view. Thus we write $u \in \boldsymbol{C}^m$ as $x + i\xi \in \boldsymbol{R}^m \oplus T_0(\boldsymbol{R}^m)$, and (Λ) takes the form

$$0 \to \Lambda^0(\boldsymbol{R}^m) \otimes \boldsymbol{C} \xrightarrow{\wedge(x+i\xi)} \Lambda^1(\boldsymbol{R}^m) \otimes \boldsymbol{C} \to \dots \to \Lambda^m(\boldsymbol{R}^m) \otimes \boldsymbol{C} \to 0.$$

For $x = 0$ we get the local symbol sequence of the de Rham complex.

The main problem is now to find an elliptic operator P with $[\sigma(P)] = [\lambda]$ which is sufficiently simple for us to compute its index. Since P is supposed to be the identity at ∞, P cannot have constant coefficients, nor can it be a *differential* operator (since it is of order zero).

We first consider the case $m = 1$. Thus λ has the form

$$\boldsymbol{C} \cong \Lambda^0(\boldsymbol{R}^1) \otimes \boldsymbol{C} \xrightarrow{\cdot (x + i\xi)} \Lambda^1(\boldsymbol{R}^1) \otimes \boldsymbol{C} = \boldsymbol{C}.$$

As it stands this does not represent a pseudo-differential operator. However, we can deform λ into a map p with

$$\begin{cases} p(x, \xi) = 1 \quad \text{for } |x| \geq 1, \\ p(x, \lambda\xi) = p(x, \xi) \quad \text{for } \lambda > 0, \\ p(x, \xi) \neq 0 \quad \text{for } \xi \neq 0. \end{cases}$$

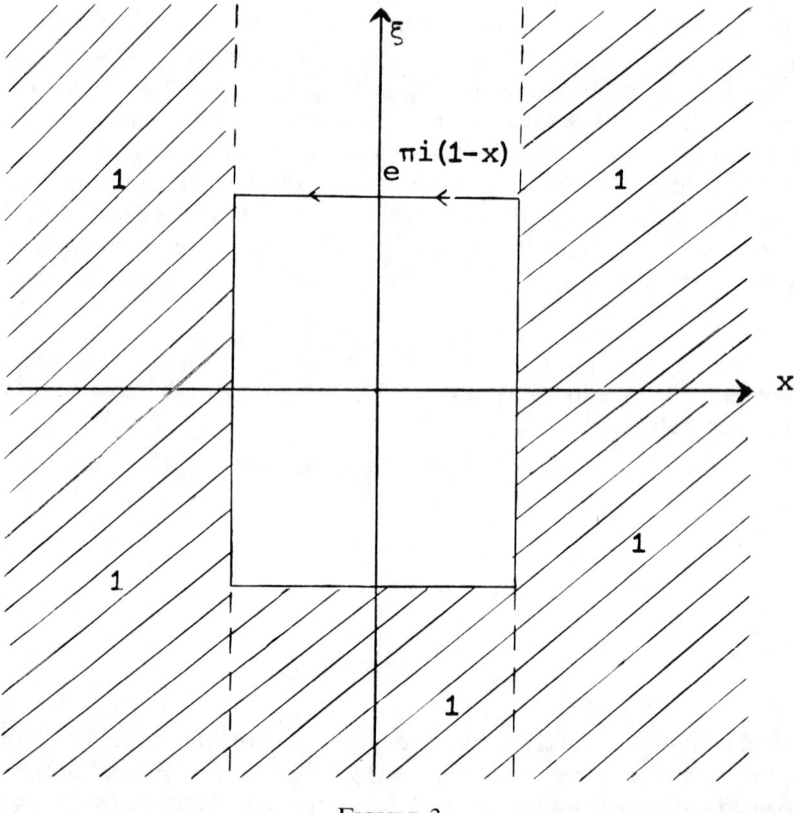

FIGURE 3

This map defines the operator we need, and it is possible to show by straightforward computation that its index is 1.

For $m > 1$ this method however is inconvenient, and we use a different technique. If m is even, we interpret $\boldsymbol{R}^m$ as the top hemisphere of S^m and extend (Λ) trivially on the bottom. Furthermore, if we deform the symbol of the de Rham complex on S^m in a way to make it constant on the equator S^{m-1}, we obtain two copies of (Λ). Thus index $(2[\lambda]) = \text{index } D^+ = \chi(S^m) = 2$, and therefore index $([\lambda]) = 1$. If m is odd, we have to combine the results for $m = 1$ and for even m in order to see also that index $([\lambda])$ is 1.

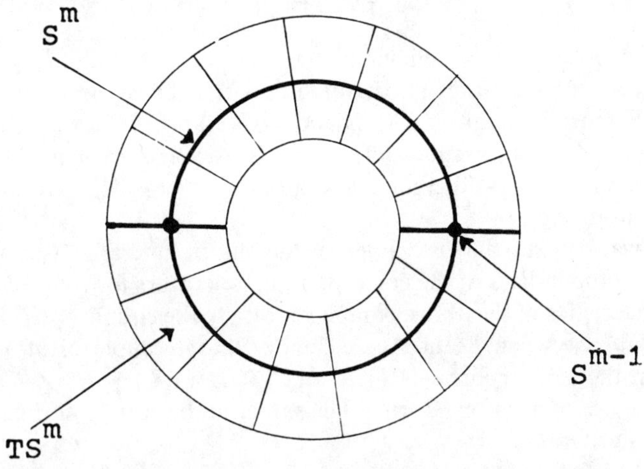

FIGURE 4

A third method of calculating index $([\lambda])$ allows us to construct an operator directly from λ without first deforming it. However, we have to introduce the weight function $e^{(|x|^2)/2}$ and the corresponding L^2-space, and we need a more general class of elliptic operators (with certain growth conditions at ∞ instead of our drastic restrictions). For example, on $\boldsymbol{R}^1$ the operator we look for is $x + d/dx$. Its symbol (in some more general sense) is $x + i\xi$, and its index in this setting turns out to be 1. While the previous methods of computing index $([\lambda])$ were rather topological, the delicacy of this last approach lies in the analysis, and we refer to [**4**] for more details.

For further applications we observe that, because of the spherical symmetry of $e^{(|x|^2)/2}$, $[\lambda]$ can be represented by an operator whose null space etc. is invariant under the action of the orthogonal group on $\boldsymbol{R}^m$. If we are careful enough with the choice of deformations, we can deduce this fact also by the more topological methods above.

(3) We now proceed to the last and most delicate step of the proof: we have to show that the homomorphism $i_3: K(TX) \to K(TN)$ preserves the analytical index. First we assume that N is trivial, i.e. $N = X \times \boldsymbol{R}^k$. Then i_3 is realized by multiplication of operators on X with the fundamental operator on $\boldsymbol{R}^k$ (or on S^k, if we prefer to work with the compact manifold $X \times S^k$). According to the

last section the index of the fundamental operator on S^k is 1. Hence the case of trivial N is sufficiently dealt with once we know that in fact quite generally the analytical index is multiplicative.

So let Y, Z be compact manifolds, and let (E) and (F) be elliptic complexes on Y and Z respectively:

$$(E): 0 \to E^0 \overset{d_E}{\to} E^1 \overset{d_E}{\to} E^2 \to \ldots$$

$$(F): 0 \to F^0 \overset{d_F}{\to} F^1 \to \ldots \quad .$$

Then their product (G) is defined by $G^i = \sum_j E^j \otimes F^{i-j}$ and $d_G = d_E \otimes 1 + (-1)^j 1 \otimes d_F$. ($d_E \otimes 1$ denotes partial differentiation by d_E on the y-variables.) If d_E, d_F are differential operators so is d_G and (G) is elliptic. If d_E, d_F are only pseudo-differential then d_G is not pseudo-differential. However, as explained by Nirenberg [2] d_G is a limit of pseudo-differential operators (if order $d_E, d_F > 0$) and everything still works out nicely.

Now we want to see that the index of the elliptic operator associated to G is the product of the indices of the corresponding operators for E and F, i.e. that the Euler characteristics of the above complexes satisfy the equality $\chi(G) = \chi(E) \cdot \chi(F)$. In the special case when E and F are the de Rham complexes of Y and Z, this follows from the isomorphism $H^*(Y \times Z; \mathbf{C}) \cong H^*(Y, \mathbf{C}) \otimes H^*(Z, \mathbf{C})$ and can be shown by the use of harmonic forms. The same can be carried through for elliptic complexes in general.

If N is not trivial we must generalize the argument above. In the general setting of fiber bundles the index of elliptic operators does not always multiply. However, in our special case we can use the strong symmetry properties of the fundamental operator on the sphere. Thus coordinate changes preserve the fact that i_3 commutes with the analytical index, and we obtain the full global commutativity of the left hand triangle in the diagram in § III(2). This completes the proof.

Let us now add some final remarks on possible extensions and further development of the index theory. First one can study the situation when a compact Lie-group acts on X and when the elliptic operator under consideration is compatible with this action. The above proof can be generalized to this situation, and one can deduce fixed point formulae connecting fixed points of the action of G with global invariants. Another generalization is to extend the index theorem to families of elliptic operators parametrised by a compact space M. In this case the index is an element of $K(M)$. Finally, for a real, skew adjoint operator one has an analog of the index with more refined properties. The dimension of the kernel equals the dimension of the cokernel and is a mod 2 homotopy invariant. An appropriate extension of the index theorem gives a topological description of this invariant.

References

1. M. F. Atiyah and I. M. Singer, *The index of elliptic operators* I, Ann. of Math. **87** (1968), 484–530.
2. L. Nirenberg, *Pseudo-differential operators*, these Proceedings, vol. 16.

3. F. Hirzebruch, *Neue topologische Methoden in der algebraischen Geometrie*, Springer-Verlag, Berlin, 1956.

4. L. Hörmander, *The index of hypoelliptic operators*, (to appear).

5. R. Palais, *Seminar on the Atiyah-Singer Index Theorem*, Ann. of Math. Studies No. 57, Princeton Univ. Press, Princeton, N.J., 1965.

Oxford University and
Institute for Advanced Study

EIGENVALUES OF THE LAPLACIAN

MARCEL BERGER

The following concerns the spectrum (i.e. the set of the eigenvalues of the Laplace operator) of two types of objects:

1. bounded domains $\mathscr{D}$ in R^n,
2. riemannian manifolds M_g, without boundary.

For the general case of a riemannian manifold with boundary, the general reference is [5]. Other references are [2] and [1].

1. **Bounded domains.** Let $\mathscr{D}$ be a bounded open subset of R^n, not necessarily simply connected, but with a very good boundary, i.e. $\forall d \in \mathscr{D}$, $\exists U$ open neighborhood of d in R^n such that, either $U \subset \mathscr{D}$, either $\exists$ a C^∞-function f on U with $U \cap \mathscr{D} = f^{-1}(]-\infty, 0[)$. The Laplace operator on $\mathscr{D}$ is $\Delta = -\Sigma(\partial^2/\partial x_i^2)$. For the boundary Dirichlet conditions $\circ|bd\mathscr{D} = 0$, being elliptic has a spectrum Spec $\mathscr{D}$ looking like Spec $\mathscr{D} = (0 < \lambda_1 < \lambda_2, \ldots, \lambda_2 < \lambda_3, \ldots \to +\infty)$, discrete and every eigenvalue being of finite multiplicity.

We are interested in the following general question: to what extent does Spec $\mathscr{D}$ characterize $\mathscr{D}$? More precisely: we write for two domains $\mathscr{D}, \mathscr{D}' \subset R^n$: $\mathscr{D} =_s \mathscr{D}'$ if Spec $\mathscr{D}$ = Spec $\mathscr{D}'$ and $\mathscr{D} = \mathscr{D}'$ if there exists an isometry r of R^n such that $r\mathscr{D} = \mathscr{D}'$.

1.1. ISOSPECTRALITY. Does $\mathscr{D} =_s \mathscr{D}'$ implies $\mathscr{D} = \mathscr{D}'$? See the musical setting in [2]. This basic question is still unsolved.

1.2. SPECTRUM VERSUS GEOMETRY. At least one can ask what information for the geometry of $\mathscr{D}$ can be gotten from Spec $\mathscr{D}$. The following results are stated for $\mathscr{D} \subset R^n$ though they were the more often historically gotten for $\mathscr{D} \subset R^2$, and the sign "$\Rightarrow$" means "determines".

1.2.1. H. WEYL(1911). Spec $\mathscr{D} \Rightarrow$ volume $\mathscr{D}$.

1.2.2. FABER-KRAHN (1923–24). Here $n = 2$: $\forall \mathscr{D} : \lambda_1 \geq \pi j^2/\text{area } \mathscr{D}$ where j is the first zero of the first Bessel function. Moreover $\lambda_1 = \pi j^2/\text{area } \mathscr{D}$ implies $\mathscr{D}$ is a disk.

1.2.3. T. CARLEMAN(1934). Spec $\mathscr{D} \Rightarrow$ area$(bd\mathscr{D})$.

COROLLARY. *The isoperimetric inequality, together with* 1.2.1 *and* 1.2.3 *implies: if $\mathscr{D}_0$ is a ball, then $\mathscr{D}_0 =_s \mathscr{D}$ implies $\mathscr{D} = \mathscr{D}_0$. Note also this can be gotten by the nth-dimensional extension of* 1.2.2.

1.2.4. KAC, MCKEAN-SINGER (1967). Spec $\mathscr{D} \Rightarrow$ number of holes of $\mathscr{D}$.

Nowadays all the above results (except 1.2.2) are gotten in one shot by the following asymptotic expansion

$$\sum_i e^{-\lambda_i t} \underset{t\to 0;\, t>0}{\sim} (4\pi t)^{-n/2}(a_0 + a_{1/2}t^{1/2} + a_1 t + a_{3/2}t^{3/2} + a_2 t^2 + \dots)$$

proved in [5]. In [5] one finds that:

$$a_0 = \operatorname{vol}\mathscr{D}, \qquad a_{1/2} = \frac{\sqrt{\pi}}{2}\operatorname{area}_{(\mathrm{bd}\mathscr{D})}, \qquad a_1 = -\tfrac{1}{6}\textstyle\int_{\mathrm{bd}\mathscr{D}} R$$

where R is the mean curvature of $\mathrm{bd}\mathscr{D}$. For $n = 2$ the integral $\int_{\mathrm{bd}\mathscr{D}} R$ of the curvature of the curve $\mathrm{bd}\mathscr{D}$ yields the number of holes of $\mathscr{D}$ by the Gauss-Bonnet theorem.

1.3. PROBLEMS. Try to help the isospectral question; find eventual counterexamples, i.e. two $\mathscr{D}, \mathscr{D}'$ with $\mathscr{D} \neq \mathscr{D}'$ and $\mathscr{D} =_s \mathscr{D}'$.

Note the spectrum of simple objects as nonrectangular parallelepipeds, simplexes, ellipsoids, are not known. For example, one would like to extend the corollary of 1.2.3 to more general classes than balls.

2. **Riemannian manifolds.** Objects now are compact riemannian manifolds M_g, without boundary, i.e. couples of a C^∞-manifold M endowed with a C^∞-riemannian structure g. Isomorphisms are *isometries* i.e. diffeomorphisms $f: M_g \to M'_{g'}$ such that $f^*g' = g$; we write $M_g = M'_{g'}$, for "M_g, $M'_{g'}$ are isometric." The Laplace-Beltrami operator on M_g is $\Delta = -\operatorname{trace}_g \operatorname{Hess}$ where the Hessian Hess is nothing but the double covariant derivative ∇^2 of a function on M_g. Being elliptic Δ has a spectrum

$$\operatorname{Spec} M_g = \{0 = \lambda_0 < \lambda_1, \dots, \lambda_1 < \lambda_2, \dots, \lambda_2 < \lambda_3, \dots \to +\infty\}$$

called the *spectrum* of our riemannian manifold M_g. We write again $M_g =_s M'_{g'}$ instead of $\operatorname{Spec} M_g = \operatorname{Spec} M'_{g'}$. Also as a convention, g_0 is going to represent the *canonical riemannian structure* of a manifold, whenever it is meaningful (and always up to a constant scalar); examples are: spheres S^n, real or complex projective spaces RP^n or CP^n, flat tori (then g_0 not unique), R^n.

2.1. ISOSPECTRALITY. For riemannian manifolds the isospectral question was answered in the negative by Milnor in 1964 [**6**]. Let $\mathscr{L}$ be a lattice in R^n; it induces a torus $R^n/\mathscr{L}$ and a flat riemannian structure $g_0/\mathscr{L}$ on it. Hence a *flat* torus $(R^n/\mathscr{L})_{g_0/\mathscr{L}}$.

2.1.1. MILNOR. $\exists$ two lattices $\mathscr{L}, \mathscr{L}'$ in R^{16} such that $(R^n/\mathscr{L})_{g_0/\mathscr{L}} =_s (R^n/\mathscr{L}')_{g_0/\mathscr{L}'}$ but $(R^n/\mathscr{L})_{g_0/\mathscr{L}} \neq (R^n/\mathscr{L}')_{g_0/\mathscr{L}'}$.

This follows directly from the following facts:

(i) Flat tori associated to lattices $\mathscr{L}, \mathscr{L}'$ are isometric if and only if $\mathscr{L}, \mathscr{L}'$ are deduced one from the other by an isometry of R^n.

(ii) For $(R^n/\mathscr{L})_{g_0/\mathscr{L}}$ the spectrum is the set of all $4\pi^2\|\xi\|^2$ where ζ runs through the dual lattice $\mathscr{L}^*_Z$ of $\mathscr{L}$ over the integers, the multiplicity of $\lambda = 4\pi^2\|\xi\|^2$ being equal to the number of $\eta \in \mathscr{L}^*_Z$ with $\|\eta\| = \|\xi\|$.

(iii) Witt found in 1939 two lattices $\mathscr{L}, \mathscr{L}'$ in R^{16} such that $\mathscr{L} \neq \mathscr{L}'$ but having always the same number of points of given norm r, this for all $r \geq 0$. By the way note:

(a) for lattices in R^2, this cannot happen. Probably neither for R^3.
(b) this phenomena happens for R^{12} (M. Kneser: [3]).
So the possibility of Milnor's type counter examples is not completely studied.

2.2. SPECTRUM VERSUS GEOMETRY. For a riemannian manifold, we denote by:

$n = \dim M_g$: the dimension of M_g
R: the scalar curvature
S: the Ricci curvature
K: the curvature tensor

and $|S|^2$, $|K|^2$ the corresponding norms for g.

2.2.1. LICHNEROWICZ [4, p. 135]. If $S \geq k \cdot g$ then $\lambda_1 \geq n/(n-1) \cdot k$.

2.2.1. OBATA [8]. If $S \geq k \cdot g$ and $\lambda_1 = (n/(n-1)) \cdot k$, then $M_g = S^n_{go}$.
Concerning λ_1 there are unpublished results by T. Aubin and J. Cheeger.

For riemannian manifolds Minakshisundaram-Pleijel found in 1949 [7] the asymptotic expansion

$$\sum_i e^{-\lambda_i t} \underset{t \to 0;\, t > 0}{\sim} (4\pi t)^{-n/2}(a_0 + a_1 t + a_2 t + \ldots)$$

with $a_i = \int_M u_i$ where $u_i : M \to R$ are functions on M which are local riemannian invariants, and defined by messy induction formulas involving normal coordinates and laplacians. Note $a_0 = 1$, hence:

2.2.2. Spec $M_g \Rightarrow \dim M$ and volume M_g,

2.2.3. (FOLK). u_1 is easy to compute: $u_1 = \frac{1}{6}R$ so that $a_1 = \frac{1}{6}\int_M R$. In particular the Gauss-Bonnet formula for $n = 2$ yields

$$\text{when } n = 2\text{: Spec } M_g \Rightarrow \chi(M)$$

where χ denotes the Euler-Poincaré characteristic of M.

2.2.4. The natural question is now if for $n = 4$ the number a_2 is or not a topological invariant. A straightforward but very unpleasant computation of u_2 can be found in [1], yielding:

$$u_2 = \frac{1}{360}(5R^2 - 2|S|^2 + 2|K|^2 + 12\Delta R)$$

hence

$$a_2 = \frac{1}{360}\int_M 5R^2 - 2|S|^2 + 2|K|^2.$$

In [5] this value of a_2 is computed in a very nice way, the basic idea being since the u_i's are riemannian invariant, one knows by general invariant stuff that u_2 has to be of the form $u_2 = \alpha \cdot R^2 + \beta|S|^2 + \gamma|K|^2 + \delta\Delta R$. Then the real coefficients α, β, γ are determined by a product argument showing they are independent of $n = \dim M$ and computed using their values for S^2_{go} and S^3_{go}.

The value gotten above for a_2 shows in particular that it is not a topological invariant: in fact when $n = 4$ the generalized Gauss-Bonnet formula reads

$$\chi(M) = \frac{1}{32\pi^2}\textstyle\int_M R^2 - 4|S|^2 + |K|^2$$

so by linear combination:

$$\textstyle\int_M R^2 + 2|S|^2 = 120a_2 - \frac{64}{3}\pi^2\chi(M).$$

Clearly $\int_M R^2 + 2|S|^2$ cannot be a topological invariant (start with a flat torus). However, the value obtained for a_2 can be used to get some uniqueness theorems of the type of the corollary in 1.2.3 (see [**1**]).

2.2.5. Suppose M_g is one of the following riemannian manifolds: S^2_{go}, RP^2_{go}, S^3_{go}, RP^3_{go}, a flat torus or a flat Klein bottle of dimension 2, then

$$M_g =_s M'_{g'} \text{ implies } M_g = M'_{g'}.$$

Also for the complex projective space and g kählerian, one has: $CP^n_{go} = {}_sCP^n_g \Rightarrow CP^n_{go} = CP^n_g$ for any n.

2.3. Problems.

2.3.1. For $n = 2$, read the topology of the compact surface M on Spec M_g. Note: $\chi(M)$ is not enough when even; and the asymptotic expansion (i.e. the a_i's) is not sufficient since a riemannian manifold M_g and one of its riemannian coverings $\widetilde{M}_{\tilde{g}}$ have the same functions u_i's (pointed out to us by M. Atiyah).

2.3.2. Read if g is kählerian on Spec M_g.

2.3.3. For $n \geq 4$ is S^n_{go} characterized by its spectrum?

2.3.4. For $n = 2$ are the M_g of constant strictly negative curvature characterized (up to isometry) by their spectrum? (See [**9**].)

2.3.5. If T_{go} is a flat torus does $T_{go} =_s M_g$ imply M_g is a flat torus for $n \geq 4$? (For $n = 3$ see [**1**].)

2.3.6. Isospectral Deformation. We end with the following question: the Milnor's type counterexamples are *discrete* in the sense that for a given dimension there is only a finite number of flat tori with the same spectrum (personal communication by M. Kneser). So we are led to look at isospectral deformations: let $g(t)$ be a one-parameter family of riemannian structures on a given manifold M:

$$\text{does "}\forall t M_{g(t)} =_s M_{g(0)}\text{" imply "}\forall t M_{g(t)} = M_{g(0)}\text{"?}$$

I don't know the answer. (See [**10**].) But let us take a look at the problem for $n = 2$ and a conformal deformation $g(t) = e^{f(t)}\cdot g(0)$. Direct computation yields $\Delta_{g(t)} = e^{-f(t)}0\cdot\Delta_{g(0)}$. So if $\phi(t)$ is a one-parameter family of functions on M such that

$$\forall t \Delta_{g(t)}\phi(t) = \lambda\cdot\phi(t) \quad \text{for a fixed } \lambda,$$

then taking the derivatives with respect to t at $t = 0$:

$$\Delta_{g(0)}\phi'(0) = \lambda \cdot \phi'(0) + \lambda f'(0)\phi(0)$$

which is possible only under the Fredholm conditions:

$$\langle f'(0), \phi(0)\psi \rangle = 0 \qquad \forall \psi \ni \Delta_{g(0)}\psi = \lambda\psi \qquad (\lambda \neq 0).$$

Applying this to the isospectral deformation and a set $\{\phi_i\}$ of orthonormal eigenfunctions of $\Delta_{g(0)}$ one gets: if $\forall t M_{g(t)} =_s M_{g(0)}$ then a necessary condition is

$$\forall \lambda_i \in \operatorname{Spec} M_{g(0)} \forall \lambda_i = \lambda_j; \ \langle f'(0), \phi_i\phi_j \rangle = 0 \qquad (\lambda_i \neq 0).$$

Denote by A_g the linear subspace of $C^\infty(M)$ generated by the products $\phi_i\phi_j$ for λ_i running through Spec M_g and $\lambda_i = \lambda_j$ and by $A_g^\perp$ its orthogonal complement. The isospectral deformation is then, for $n = 2$, related to the following questions:

Is $A_g^\perp \neq \{0\}$ for every riemannian structure g on M?
Is $A_g^\perp = \{0\}$ for every generic g?

(Note that on S^n the space $A_{g_0}^\perp$ is made up by all the spherical harmonics of odd degree.)

References

1. M. Berger, *Le spectre des variétés riemanniennes*, Revue Romaine Math. Pures et Appl. **13** (1968).

2. M. Kac, *Can one hear the shape of a drum?*, Amer. Math. Monthly **73** (1966), 1–23.

3. M. Kneser, *Lineare Relationen zwischen Darstellung-zahlen quadratischer Formen*, Math. Ann. **168** (1967), 31–39.

4. A. Lichnerowicz, *Géometrie des groupes de transformations*, Paris, 1958.

5. H. P. McKean and I. M. Singer, *Curvature and the eigenvalues of the Laplacian*, J. Differential Geometry **1** (1967), 43–69.

6. J. Milnor, *Eigenvalues of the Laplace operator on certain manifolds*, Proc. Nat. Acad. Sci., U.S.A. **51** (1964), 542.

7. S. Minakshisundaram and A. Pleijel, *Some properties of the eigenfunctions of the Laplace-operator on Riemannian manifolds*, Canad. J. Math. **1** (1949), 242–256.

8. M. Obata, *Certain conditions for a riemannian manifold to be isometric with a sphere*, J. Math. Soc. Japan, **14** (1962), 333–340.

9. H. Huber, *Zur analytischen Theorie hyperbolischer Raumformen und Bewegungsgruppen*, Math. Ann. **138** (1952).

10. S. Tanaka, *Selberg's trace formula and spectrum*, Osaka J. Math. **3** (1966).

University of Paris

ON A TOPOLOGICAL OBSTRUCTION TO INTEGRABILITY

RAOUL BOTT

1. **Introduction.** A subbundle of the tangent bundle to a C^∞-manifold is called integrable if its space of smooth sections is closed under the bracket operation. The primary purpose of this note is to show that in general subbundles of the tangent bundle need not be isomorphic to integrable ones. Thus in particular we answer a question raised by several people, see [3], for instance, as to whether every subbundle of the tangent bundle is deformable into an integrable one.

Our counterexample is based on the following

Integrability criterion. A subbundle E of a tangent bundle T is integrable only if the ring Pont T/E *generated by the real Pontrayagin classes of T/E vanishes in dimensions greater than* $2 \times \dim(T/E)$:

(1.1) $$\operatorname{Pont}^k(T/E) = 0 \quad \textit{if } k > 2\dim(T/E).$$

Note that $\operatorname{Pont}(T/E)$ may also be interpreted as $\operatorname{Pont}(T-E)$ and so depends only on the isomorphism classes of T and E. Hence if $\operatorname{Pont}(T/E)$ does not satisfy (1.1) then no subbundle of E' of T, isomorphic to E, is integrable.

The vanishing theorem (1.1) is made plausible by the following consideration: Suppose that

$$\begin{array}{c} Y \\ \downarrow \pi \\ X \end{array}$$

is a C^∞ fibering and let E be the bundle along the fibers of π. Then E is trivially integrable and fits into the exact sequence:

$$0 \to E \to T \to \pi^*TX \to 0, \qquad T = TY;$$

over Y. By naturality and the obvious vanishing of $H^*(X)$ in $\dim > \dim X$, we conclude that

(1.2) $$\operatorname{Pont}^k(T/E) = 0 \qquad \text{for } k > \dim(T/E).$$

In general an integrable E of course does not arise in this manner. However, it turns out that if one utilizes the differo-geometric construction of $\operatorname{Pont}(T/E)$ then the vanishing phenomenon (1.2) is easily seen to persist—but only in the smaller range of dimensions given by (1.1). This approach naturally yields only information about the real cohomology, and I don't know whether (1.1) is true over the integers.

The Euler class $e(T/E)$ is also not covered by this method, so that for instance I do not know whether $e(T/E)^2$ or $e(T/E)^3$ have to vanish for an integrable E. The fourth power $e(T/E)^4$ of course has to because $e(T/E)^2 \in \text{Pont}\,(T/E)$.

In the complex analytic situation one obtains perfect agreement with the case of a fibering. Indeed (1.1) is then strengthened to yield the following:

Complex integrability criterion. A holomorphic subbundle E of a holomorphic tangent bundle T is integrable only if the ring Chern (T/E), *generated by the real* Chern-*classes of* (T/E) *vanishes in dimensions greater than* $2 \times \dim_{\boldsymbol{C}}(T/E)$:

$$\text{Chern}^k(T/E) = 0, \qquad k > \dim_{\boldsymbol{R}}(T/E). \tag{1.3}$$

The proof of these propositions will be given in §2, and in §3 I will discuss the case of integrable subbundles with singularities. Here I would like to conclude with some simple corollaries to these integrability conditions which also yield the counter example alluded to earlier.

COROLLARY 1.4. *Let X be a compact analytic manifold which admits a nonvanishing holomorphic vector field. Then all the* Chern *numbers of X vanish.*

PROOF. Let E be the trivial subbundle generated by the vector-field. It is obviously integrable. Hence by (1.3) $\text{Chern}^{2n}(T/E) = 0$, $2n = \dim_{\boldsymbol{R}} X$. Because E is trivial Chern (T/E) = Chern (T) so that the corollary follows. I have noted this result earlier (see [**1**]) in a different connection.

COROLLARY 1.5. *Let T be the tangent bundle of* P_n, *the projective space of n complex dimensions, with n odd and* > 1. *Then T contains a holomorphic subbundle E of codimension* 1 *but contains no integrable holomorphic subbundle of codimension* 1.

PROOF. To construct E interpret P_n as the projective space $P(V)$ of one-dimensional subspaces of $V = \boldsymbol{C}^{n+1}$ and let

$$v \to S \to \tilde{V} \to Q \to 0$$

be the usual exact sequence of bundles over $P(V)$. (Thus $\tilde{V} = P(V) \times V$, and S is the subbundle of $\tilde{V}$ given by $S = \{(l, v) | l \in P(V), v \in V, v \in l\}$. The tangent bundle T of $P(V)$ is then well known to be given by $T = \text{Hom}\,(S, Q)$, and the dual to S in the hyperplane bundle H of $P(V)$. Now choose a basis $\{v_i\}$ of V^* and construct the *nondegenerate* 2-form;

$$\omega = v_1 \wedge v_2 + v_3 \wedge v_4 + \dots + v_n \wedge n_{n+1} \qquad (n \text{ is odd!}).$$

This form is then seen to induce a holomorphic surjection

$$T \overset{\omega_*}{\to} H^2 \to 0,$$

by the formula

$$\omega_*(t)[v] = \omega(v \wedge tv),$$

where $l \in P(V)$, $t \in T_l = \text{Hom}(l, V/l)$ and $v \in l$.

We take for E the kernel of ω_* so that

$$T/E \simeq H^2. \tag{1.6}$$

So much for the first assertion. To prove the second one assume $E \subset T$ holomorphic, integrable, and of codimension 1. Then by our criterion $c_1(T/E)^2 = 0$. Because $n > 1$, this implies $c_1(T/E) = 0$. But this contradicts the fact that $c_n(T) = c_{n-1}(E) \cdot c_1(T/E)$ is nonzero. Q.E.D.

A similar argument, using (1.1) in the form $e(T/E)^4 = 0$, yields the following:

COROLLARY 1.7. *Let T be the tangent space to P_n with n odd and > 3. Then T contains a real subbundle of codimension 2, but contains no integrable real subbundle of codimension 2.*

In the complex case one can actually strengthen (1.5) as follows:

COROLLARY 1.8. *Let X be any compact analytic manifold whose tangent bundle T is positive. Then T has no integrable analytic subbundles.*

This follows from the fact that every quotient bundle of a positive bundle is positive—see [2] for instance. Hence if E is a holomorphic subbundle of T, then T/E is positive whence $c_1^n(T/E) > 0$, $n = \dim_C X$, contradicting the integrability of E. Because complex projective space has a positive tangent bundle, this assertion strengthens 1.5.

A more direct proof of (1.8) for the complex projective spaces was pointed out to me by Joel Pasternack. He shows by a combinatorial argument that in $H^*(P_n)$, there is no nontrivial factorization of $c(T)$ into a product $c(T) = c(E) \cdot c(T/E)$ if $\text{Chern}^k(T/E) = 0$ for $k > \dim T/E$.

2. **The proof of the criteria.** Let us consider the real case alone; The complex analytic one is quite analogous. Also I will assume familiarity with the theory of real characteristic classes. Suppose then that $E \subset T$ is integrable and let $Q = T/E$. The dual Q^* is then contained in T^* and the Frobenius theorem implies that X has a coordinate covering $\{U, x_i^U\}$, $i = 1, \ldots, n$; such that on $U \in \{U\}$, Q^* is spanned by forms

$$dx_1, \ldots, dx_k, \quad k = \text{fiber} - \dim Q^*$$

where the x_i^U are the coordinates for U.

One now defines a connection D on Q^* in the following manner. Let D_U be the connection for $Q^*|U$, for which

$$D_U \cdot dx_1^U \equiv 0, \quad i = 1, \ldots, k. \tag{2.1}$$

Next choose a partition of unity λ_U for $\{U\}$ and set

$$D = \sum \lambda_U D_U.$$

It follows from (2.1) that the connection form θ_U, of D relative to the frame dx_1^U, $\ldots$, dx_k^U on U, takes the form:

$$\theta_U = \sum_V \lambda_V dg_{UV} \cdot g_{VU}, \tag{2.2}$$

where g_{UV} are the transections between the frames dx_1^U, θdx_i^V, $i = 1, \ldots k$ on $U \wedge V$: symbolizing

(2.3) $$dx^U = g_{UV} dx^V.$$

Differentiating this relation yields

(2.4) $$0 = dg_{UV} \cdot dx^V$$

from which it follows easily that the forms of θ_U are all sections of Q^*. From this it follows in turn that the *curvature forms of D all lie in the ideal, I, generated by the sections of Q^*.*

Now the ideal I clearly has the property that if $a_1, \ldots, a_{k+1}$ are any $k + 1$ elements in I then $a_1 \wedge a_2 \ldots \wedge a_{k+1} = 0$. Finally. recalling that Pont Q^* is generated by all possible symmetric polynomials applied to the curvature of D and that Pont Q^* = Pont Q, the desired criterion (1.1) follows.

Note that the Euler class of Q^* can in general not be realized in this manner. It appears as a characteristic class of the curvature only relative to a Riemannian connection.

3. **Integrable subbundles with singularities.** Suppose now that v is a holomorphic vector field on the compact analytic manifold X which vanishes only to the first order at a (necessarily) finite number of points $\{P\}$. The differential of V then induces complex linear maps $L_P \cdot T_P \to T_P$ at each of these points, and in [**1**] I gave a formula for any Chern number $c^\alpha(X)$ of X in terms of the eigenvalues of the L_P.

Precisely, if $c_i(L)$ is defined by

$$\det(1 + tL) = \sum t^i c_i(L)$$

and $c^\alpha(L) = c_1^{\alpha_1}(L) \ldots c_n^{\alpha_n}(L)$, it was shown in [**1**] that

(3.1) $$\sum_P \frac{c^\alpha(L)}{\det L} = c^\alpha(X), \quad L = L_P.$$

This formula is clearly a refinement of our Corollary 1 and leads one to conjecture that if $E \to T$ is a homomorphism of bundles whose image is an integrable subbundle of $T = TX$ outside of a singular set $\Sigma \subset X$, then the classes of $\mathrm{Pont}^k(T - E)$; $k > 2\{\dim T - \dim E\}$; should be computable in terms of local invariants of E near the singular set Σ. A first step confirming this conjecture is given by the following very recent result of P. Baum and myself which generalizes (3.1) to the case of line bundles.

Let X be a compact complex analytic manifold with tangent-bundle $T = TX$, and let

(3.2) $$E \xrightarrow{\lambda} T$$

be a holomorphic map of the line bundle E into T which vanishes only to the first order at the points $\{P\}$. The differential of λ then induces homomorphisms

$L_P : E_P \to \mathrm{End}(T_P)$ so that for every nonzero $e \in E_P$, $c^\alpha(L_P(e))$ is well defined, and if $|\alpha| = \alpha_1 + 2\alpha_i + \dots + n\alpha_n = \dim_C X$, $c^\alpha[L_P(e)]/\det L_P(e)$ is independent of $e \neq 0 \in E_P$. Denoting this ratio by $c^\alpha(L_P)/\det(L_P)$, we show

THEOREM I (BAUM-BOTT). *In the situation just described the* Chern *numbers of* $T - E$ *are related to the zeroes of* λ *by the formula*

$$\sum_{\{P\}} \frac{c^\alpha(L)}{\det L} = c^\alpha(T - E), \quad L = L_P. \tag{3.2}$$

The proof of this relation is considerably harder than the proof of (3.1) and will be reported on elsewhere. One may, of course, interpret (3.2) as generalizing (3.1) to meromorphic vector fields. Indeed the transpose of λ is a section of $T \otimes E^*$ which vanishes precisely at the zeroes of λ, so that if s is a holomorphic section of E^* which does not vanish at the zeroes of λ, then $v = \lambda^t/s$ is a meromorphic vector-field whose infinitesimal numbers $c^\alpha(L_P)/\det L_P$ are well defined and agree with the infinitesimal numbers defined for λ. Thus (3.2) expresses the relations that must exist between the local numbers at the zeroes of a meromorphic vector field of this type on X.

The complete generalization of (3.2) to higher dimensional E seems quite difficult. However, the following is not hard to prove:

THEOREM II. *Let* $\lambda : E \to T$ *be an injection outside of the submanifold* $\Sigma \subset X$, *and let its image be integrable there. Then every* $\phi \in \mathrm{Chern}^k(T - E)$, *with* $k > \dim_C(T - E)$ *is in the image of the umkehrungs homomorphism* $\iota_* : H^*(\Sigma) \to H^*(X)$.

Thus every such $\phi = \iota_* \psi$ for some $\psi \in H^{k-\sigma}(\Sigma)$, $\sigma \in \dim_R \Sigma$, and what remains to be done, is to understand the recipe for ψ in terms of ϕ and the zero behavior of λ on Σ.

REFERENCES

1. Raoul Bott, *Vector fields and characteristic numbers*, Michigan Math. J. **14** (1967), 231–244.
2. Raoul Bott and S. S. Chern, *Hermitian vector bundles and the equidistribution of the zeroes of their holomorphic sections*, Acta Math. **114** (1965), 71–112.
3. S. S. Chern, *The geometry of G-structures*, Bull. Amer. Math. Soc. **72** (1966), 167–219.

HARVARD UNIVERSITY

AN ASYMPTOTIC EXPANSION FOR THE HEAT EQUATION

PETER GREINER

Let M be a compact n-dimensional C^∞ manifold without boundary. Let E and F be two C^∞ complex vectorbundles of fiber dimension N over M. Let $P(x, D)$ be a strongly elliptic smooth linear partial differential operator of order m sending C^∞ sections of E into C^∞ sections of F. Denote by dx a density over M and by $(\cdot,\cdot)$ hermitian structures over E and F. These can be introduced locally and extended to all of M by a partition of unity. The heat operator is given by

$$L = \partial/\partial t + P(x, D) \tag{1}$$

defined on $C_0^\infty(E) \times C^\infty(R)$. Let $G(t, x, y)$ denote its Green's matrix and set

$$G(t, x) = \text{Trace}\, G(t, x, x) \tag{2}$$

$$G(t) = \textstyle\int_M G(t, x)dx. \tag{3}$$

THEOREM 1. *For each fixed* $x \in M$

$$G(t, x) \sim A_0(x)t^{-n/m} + A_1(x)t^{-n/m+1/m} + \dots \tag{4}$$

as $t \downarrow 0$. *Moreover this asymptotic expansion is uniform in* x *over* M. *As a consequence*

$$G(t) \sim t^{-n/m}\textstyle\int_M A_0(x)dx + t^{-n/m+1/m}\int_M A_1(x)dx + \dots \tag{5}$$

as $t \downarrow 0$. *The coefficients* $A_j(x)$, $j = 0, 1, 2, \dots$ *can be evaluated explicitly in terms of the coefficients of* $P(x, D)$.

The proof hinges on the construction of a very precise local parametrix for the heat operator L. Such a fundamental singularity is obtained via the methods of pseudodifferential operators. For the remainder of the fundamental solution we obtain pointwise estimates which are uniform on compact sets.

A more interesting situation occurs in the study of elliptic boundary problems. Let Ω be a precompact submanifold of M with a C^∞ boundary ω. Let $G_1, \dots, G_\mu$ be bundles on ω and let $B_1, \dots, B_\mu$ be boundary differential operators of the form

$$B_j u = \sum_{k=0}^{m-1} B_{jk}\gamma_k u \tag{6}$$

where B_{jk} is a differential operator from E to G_j in ω of order $m_j - k$ and $\gamma_k u$ is the kth normal derivative of u evaluated on ω. Our principal assumption is that B is elliptic with respect to $P(x, D) + i\tau$ for all τ with $\text{Im}\,\tau \leq 0$. For example this is satisfied by all strongly elliptic operators $P(x, D)$ with Dirichlet boundary conditions.

Again let $G(t, x, y)$ be the Green's matrix for the boundary problem (L, B) and let $G(t)$ be defined by (3). Then we have

THEOREM 2. *Let (L, B) be the above boundary problem. Then*

$$G(t) \sim C_0 t^{-n/m} + C_1 t^{-n/m+1/m} + C_2 t^{-n/m+2/m} + \dots \tag{7}$$

as $t \downarrow 0$. Furthermore the coefficients C_j, $j = 0, 1, 2, \dots$ can be evaluated explicitly in terms of the coefficients of $P(x, D)$ and B.

We have the following situation. The Green's function near ω is a sum of two terms

$$G = G' - C \tag{8}$$

where G' is the above defined local fundamental solution and C is a compensating term for the boundary. Furthermore C vanishes exponentially as $x_n \to \infty$. Hence $G(t, x)$ is only interesting for the interior of Ω. On the other hand if we integrate C over $0 \leq x_n < \infty$ we find that $G(t)$ is "essentially" obtained by integrating G' over Ω + integrating $\int_0^\infty C dx_n$ over ω. This is the decomposition we were aiming at. The construction is carried out by using Hörmander's Green's formula which reduces the problem to obtaining the inverse of an "elliptic" pseudodifferential operator on ω.

If (P, B) is a nonnegative selfadjoint boundary problem, that is

$$(P\phi, \phi) \geq c(\phi, \phi) \tag{9}$$

for some $c \geq 0$ and all $\phi \in C^\infty(E)$ for which $B\phi = 0$, then

$$G(t) = \sum_{j=1}^{\infty} \exp(-\lambda_j t) \tag{10}$$

where λ_j, $j = 1, 2, \dots$ are the eigenvalues of the self adjoint operator (P, B). A consequence of Theorem 2 is

COROLLARY 3. *Let $(-\Delta, D)$ be the negative Laplacian with Dirichlet boundary conditions in some precompact domain Ω in the plane with smooth boundary ω. Let $\mu_0, \mu_1, \mu_2, \dots$ be the eigenvalues for $(-\Delta, D)$. Then*

$$\sum_{j=0}^{\infty} e^{-\mu_j t} = \frac{|\Omega|}{4\pi t} - \frac{|\omega|}{8(\pi t)^{1/2}} + \frac{1}{6}(1 - h) + O(\sqrt{t}) \tag{11}$$

as $t \downarrow 0$, where $|\Omega|$ and $|\omega|$ denote the area and length of Ω and ω, respectively, and h is the number of holes in Ω.

This result was the motivation for this investigation. It was originally conjectured by Kac [**4**] and proved by McKean and Singer [**5**].

REFERENCES

1. R. Arima, *On general boundary value problem for parabolic equations*, J. Math. Kyoto Univ. **4** (1964), 207–243.

2. L. Hörmander, *Pseudo-differential operators and non-elliptic boundary problems,* Ann. of Math. **83** (1966), 129–209.

3. L. Hörmander, *On the Riesz means of spectral functions and eigenfunction expansions for elliptic differential operators,* Lecture at the Belfer Graduate School, Yeshiva University, November 16, 1966 (Mimeographed).

4. M. Kac, *Can one hear the shape of a drum?*, Amer. Math. Monthly **73** (1966), 1–23.

5. H. P. McKean, Jr. and I. M. Singer, *Curvature and the eigenvalues of the Laplacian,* J. Differential Geometry **1** (1967), 43–69.

6. S. Mizohata, *Sur les propriétés asymptotiques des valeurs propres pour les opérateurs elliptiques,* J. Math. Kyoto Univ. **4** (1965), 399–428.

University of Toronto

AN ANALYTIC PROOF OF THE CLASSICAL RIEMANN-ROCH THEOREM

TAKESHI KOTAKE[1]

1. In terms of vector bundles over a compact Riemann surface the Riemann-Roch theorem may be stated as follows.

THEOREM. *Let M be a compact Riemann surface of genus g_M. Let K be the canonical bundle, E a holomorphic vector bundle of fibre dimension N, and E^* the dual bundle of E, all over M. Denote by $H(M, E)$, $H(M, K \otimes E^*)$ the C-vector spaces of holomorphic sections of the bundles E and $K \otimes E^*$, the latter being the tensor product of K and E^*. Then the following formula holds:*

$$\dim H(M, E) - \dim H(M, K \otimes E^*) = c(\det E) + N(1 - g_M) \tag{1}$$

where $c(\det E)$ stands for the Chern number of the holomorphic line bundle $\det E$ *over M, namely the determinant bundle of E.*

The formula (1) takes a familiar form if we specify E to be the line bundle $\{D\}$ defined by a divisor D of M. In fact, in this case, one can identify the space $H(M, \{D\})$ with the space $L(D)$ of meromorphic functions over M, which are multiples of the inverse D^{-1} of D, and $H(M, K \otimes \{D\}^*)$ with the space $I(D)$ of meromorphic differentials over M, which are multiples of D. At the same time, the Chern number $c(\det \{D\}) = c(\{D\})$ is equal to the total degree $\deg D$ of D. Therefore, (1) gives

$$\dim L(D) - \dim I(D) = \deg D + 1 - g_M.$$

This note is devoted to a direct analytic proof of (1) in the framework of partial differential equations.

The underlying idea is very simple, and it runs as follows. Assuming a Riemannian metric in M and hermitian structures in the fibres of E, $K \otimes E^*$, one can introduce the Laplace operators Δ, $\tilde{\Delta}$, acting on the smooth sections of E, $K \otimes E^*$ respectively, so that one may think of $H(M, E)$, $H(M, K \otimes E^*)$ as eigenspaces of Δ, $\tilde{\Delta}$ with zero eigenvalue. Now if we consider the fundamental solutions $G(t, xy)$, $\tilde{G}(t, xy)$ of the heat equations $\partial/\partial t + \Delta = 0$, $\partial/\partial t + \tilde{\Delta} = 0$, then

$$\int_M \{trG(t, xx) - tr\tilde{G}(t, xx)\} dM \tag{2}$$

is easily seen to be independent of the time variable t and to be equal to $\dim H(M, E)$ $-\dim H(M, K \otimes E^*)$. Thus, the Riemann-Roch problem can be reduced to the Cauchy problem of heat equations, or specifically, to a study of fundamental

[1] This work was supported by the National Science Foundation under contract NSF–GP–5638.

solutions of these equations, which can easily be done by the use of Hilbert-Levi's parametrix method. The question here, to which we should give our attention, is to express (2) in terms of topological invariants of M and E.

2. We shall suppose the Riemann surface M to be carrying a positive definite C^∞ metric, expressed in terms of complex analytic local coordinates as $ds^2 = h(z, \bar{z})dz\, d\bar{z}$, and denote by

$$dM = \frac{i}{2}h(z, \bar{z})dz \wedge d\bar{z} \tag{3}$$

the element of surface area on M.

Let $\{U_\iota\}$ be a finite, sufficiently fine open covering of M with coordinate mapping $z_\iota : U_\iota \to C$ and let

$$g_{\iota\kappa} : U_\iota \cap U_\kappa \to \mathrm{GL}(N, C) \tag{4}$$

be the transition functions determining the holomorphic vector bundle E.

A hermitian structure in E is a C^∞ field of positive definite hermitian structure in the fibres of E. In the presentation of the bundle E by the covering and the transition functions above, it amounts to saying that attached to $\{g_{\iota\kappa}\}$ is a system $\{A_\iota\}$ of positive definite hermitian matrices A_ι, defined respectively in U_ι and varying therein smoothly, such that

$$tg_{\iota\kappa}A_\iota\bar{g}_{\iota\kappa} = A_\kappa \qquad \text{in } U_\iota \cap U_\kappa. \tag{5}$$

The construction of such matrices is locally trivial. To do so globally, we use a partition of unity subordinate to the covering $\{U_\iota\}$. No difficulty arises, since positive definite hermitian matrices of rank N constitute a convex set in $\mathrm{GL}(N, C)$.

The determinant bundle $\det E$ of E is a holomorphic line bundle over M with transition functions

$$\det g_{\iota\kappa} : U_\iota \cap U_\kappa \to C^*. \tag{6}$$

From (5) it follows that

$$\det A_\iota |\det g_{\iota\kappa}|^2 = \det A_\kappa \qquad \text{in } U_\iota \cap U_\kappa. \tag{7}$$

This clearly shows that the system $\det A = \{\det A_\iota\}$ serves to endow a hermitian structure in the bundle $\det E$. Bearing in mind that the $\det g_{\iota\kappa}$ are holomorphic, we get from (7) a closed exterior differential form of type (1, 1)

$$\partial\bar{\partial} \log(\det A) = \{\partial\bar{\partial} \log(\det A_\iota)\} \tag{8}$$

defined all over M. The form (8) is called the curvature form in $\det E$ associated with the hermitian structure just introduced, and it represents the characteristic class of the line bundle $\det E$. Integration of (8) over M then gives an analytic expression of the Chern number of $\det E$

$$c(\det F) = \frac{1}{2\pi i} \int_M \partial\bar{\partial} \log(\det A). \tag{9}$$

By a section f of E we mean a system $\{f_\iota\}$ of complex valued functions $f_\iota: U_\iota \to C^N$, subjected

$$f_\iota = g_{\iota\kappa} f_\kappa \qquad \text{in } U_\iota \cap U_\kappa \tag{10}$$

where the f_ι are considered as column vectors. The set of all such sections of class C^∞ forms a prehilbert space $C^\infty(E)$ if we define the inner product of two sections $f = \{f_\iota\}, f' = \{f'_\iota\}$ to be

$$(f, f') = \int_M {}^t(A\bar{f}) f' dM \tag{11}$$

where ${}^t(A\bar{f})f'|U_\iota = {}^t(A_\iota \bar{f}_\iota) f'_\iota$.

On $C^\infty(E) \times C^\infty(E)$ we consider a positive definite hermitian form

$$\begin{aligned} \mathfrak{D}(f, f') &= \frac{i}{2} \int_M {}^t(A\partial \bar{f}) \wedge \bar{\partial} f', \\ \mathfrak{D}(f, f') &> 0 \qquad \text{unless } f \in H(M, E) \end{aligned} \tag{12}$$

where $\partial, \bar{\partial}$ mean respectively the exterior differentiation by the complex analytic local coordinates and their complex conjugates. (12) is an analogue of the first variational form of the Dirichlet integral, familiar in potential theory. Then, an application of Stokes' theorem changes (12) into

$$(f, f') = (\Delta f, f') = (f, \Delta f'), \tag{13}$$

if we define the Laplace operator $\Delta: C^\infty(E) \to C^\infty(E)$ to be

$$\Delta = -(h_\iota^t A_\iota)^{-1} \frac{\partial}{\partial z_\iota} {}^t A_\iota \frac{\partial}{\partial \bar{z}_\iota} \qquad \text{in } U_\iota \tag{14}$$

where $h_\iota = h(z_\iota, \bar{z}_\iota)$.

The operator Δ is seen to be an elliptic partial differential operator of second degree, realized as a symmetric positive operator in the prehilbert space $C^\infty(E)$. Since M has no boundary, it therefore follows that Δ has a unique positive definite self-adjoint extension, which we denote again by Δ, in the Hilbert space $L^2(E)$, the completion of $C^\infty(E)$. Further, the operator Δ has a nonnegative discrete spectrum clustering at $+\infty$, and if we write

$$E_\lambda(\Delta) = \{f \mid f \in C^\infty(E), \Delta f = \lambda f\} \tag{15}$$

for the eigenspace of Δ with eigenvalue λ, then each of these is of finite dimension and yields a complete orthogonal decomposition of $L^2(E)$. Since in view of (13) $\Delta f = 0$ is equivalent to the holomorphy of f, we have

$$E_0(\Delta) = H(M, E). \tag{16}$$

Besides finer spectral structures, the operator Δ behaves like the usual Laplacian on compact two-dimensional Riemannian manifolds. For instance, if $(0 \leqslant)\lambda_0 \leqslant \lambda_1 \leqslant \lambda_2 \ldots$ denotes the eigenvalues and $f_0, f_1, f_2, \ldots$ the corresponding eigensections of Δ supposed to be forming a complete orthogonal basis in $L^2(F)$, then

$$\sum_{\lambda_\nu > 0} \lambda_\nu^{-2} < \infty, \tag{17}$$
$$|D^\alpha f_{\nu,\iota}(x)| < c_\alpha \lambda_\nu^{|\alpha|/2+1} \qquad (\nu = 0, 1, 2, \ldots)$$

where D^α is a differential operator of order $|\alpha|$ in z_ι, $\bar{z}_\iota$ and where the estimates are understood to hold for each component of the vector $f_{\nu,\iota} = f_\nu|U_\iota$.

The case of the bundle $\tilde{E} = K \otimes E^*$ is similar. Let K be the canonical bundle over M and let $k_{\iota\kappa}: U_\iota \cap U_\kappa \to C^*$ be its transition functions, $k_{\iota\kappa}$ being the jacobian dz_κ/dz_ι of coordinate transformation. Then the bundle $\tilde{E} = K \otimes E^*$ has the transition functions

$$k_{\iota\kappa}{}^t g_{\iota\kappa}^{-1}: U_\iota \cap U_\kappa \to \mathrm{GL}(N, C). \tag{18}$$

As a system of positive definite hermitian matrices, which gives a hermitian structure in $\tilde{E}$, one can clearly take the system

$$\tilde{A} = \{\tilde{A}_\iota\}, \quad \tilde{A}_\iota = (h_\iota{}^t A_\iota)^{-1} \qquad \text{in } U_\iota. \tag{19}$$

These being fixed, one can pass from the case of E to that of $\tilde{E}$ simply by a formal procedure; thus we define a prehilbertian structure in the space $C^\infty(\tilde{E})$ of C^∞ sections of the bundle, the Dirichlet bilinear form in $C^\infty(\tilde{E})$ and the Laplace operator $\tilde{\Delta}: C^\infty(\tilde{E}) \to C^\infty(\tilde{E})$. $\tilde{\Delta}$ is an elliptic operator

$$\tilde{\Delta} = -A_\iota \frac{\partial}{\partial z_\iota}(h_\iota A_\iota)^{-1} \frac{\partial}{\partial \bar{z}_\iota} \qquad \text{in } U_\iota \tag{20}$$

realized as a positive definite symmetric operator in the prehilbertian space $C^\infty(\tilde{E})$; its eigenspace $E_0(\tilde{\Delta})$ with zero eigenvalue consists of holomorphic sections of the bundle $\tilde{E} = K \otimes E^*$, that is,

$$E_0(\tilde{\Delta}) = H(M, K \otimes E^*). \tag{21}$$

We shall write $(0 \leqslant)\, \tilde{\lambda}_0 \leqslant \tilde{\lambda}_1 \leqslant \tilde{\lambda}_2 \ldots$ for the eigenvalues and $\tilde{f}_0, \tilde{f}_1, \tilde{f}_2, \ldots$ for the eigensections of $\tilde{\Delta}$.

After these preparations we can now state the formula (1) equivalently as

THEOREM′.

$$\dim E_0(\Delta) - \dim E_0(\tilde{\Delta}) = \frac{1}{2\pi i} \int_M \partial\bar{\partial} \log(\det A) + N(1 - g_M). \tag{1′}$$

3. The problem is now to compute the integer $\dim E_0(\Delta) - \dim E_0(\tilde{\Delta})$. For this purpose we shall make use of heat equations associated with the Laplace operators $\Delta, \tilde{\Delta}$. The integer in question will then have an integral expression by means of the fundamental solutions of these heat equations. The following lemma, which reflects in a sense the formal similarity of Δ and $\tilde{\Delta}$, is crucial in this respect.

LEMMA 1. *The operators $\Delta, \tilde{\Delta}$ have the same positive spectra, along with the multiplicity of each eigenvalue. In other words, if $\lambda > 0$ is an eigenvalue and $E_\lambda(\Delta)$ the* *associated eigenspace of* Δ, *then this* λ *is also an eigenvalue of* $\tilde{\Delta}$ *and the dimension of the corresponding eigenspace* $E_\lambda(\tilde{\Delta})$ *is the same as that of* $E_\lambda(\Delta)$.

PROOF. Let $E_\lambda(\Delta) \ni f|U_\iota = f_\iota$, so that $\Delta f = \lambda f$, or

$$-(h_\iota {}^t A_\iota)^{-1} \frac{\partial}{\partial z_\iota} {}^t A_\iota \frac{\partial f_\iota}{\partial \bar{z}_\iota} = \lambda f_\iota \qquad \text{in each } U_\iota.$$

From this,

$$-A_\iota \frac{\partial}{\partial z_\iota}(h_\iota A_\iota)^{-1} \frac{\partial \tilde{f}}{\partial \bar{z}_\iota} = \lambda \tilde{f}_\iota \qquad \text{in each } U_\iota$$

where $C^\infty(\tilde{E}) = \tilde{f}|U_\iota = \tilde{f}_\iota = A_\iota \partial \bar{f}_\iota / \partial z_\iota$, that is, $\tilde{\Delta}\tilde{f} = \lambda \tilde{f}$, or $\tilde{f} \in E_\lambda(\tilde{\Delta})$. But here $\tilde{f} = 0$ if and only if $f = 0$ since $\lambda \neq 0$ by hypothesis. Thus the mapping $f \to \tilde{f}$ is an isomorphism of $E_\lambda(\Delta)$ into $E_\lambda(\tilde{\Delta})$, which implies that $\dim E_\lambda(\Delta) \leqslant \dim E_\lambda(\tilde{\Delta})$. Meanwhile, by a similar consideration, we can obtain $\dim E_\lambda(\tilde{\Delta}) \leqslant \dim E_\lambda(\Delta)$. Hence the lemma.

Let E_x be the fibre of E over $x \in M$ and E_y that of E^* over $y \in M$. Then the collection of $E_x \otimes E_y$, $(x, y) \in M \times M$ may be naturally endowed with a bundle structure to yield a holomorphic vector bundle $E \hat{\otimes} E^*$ over $M \times M$ of fibre dimension N^2, namely the exterior tensor product of the bundles E and E^* over M.

Consider now the heat equation

$$Q = \partial/\partial t + \Delta = 0 \qquad (t > 0). \tag{22}$$

By its fundamental solution we then mean the function $R_+ \ni t \to G(t, xy) \in C^\infty(E \hat{\otimes} E^*)$, which generates the solutions of the Cauchy problem of (22) in the sense that for any $f \in C^\infty(E)$

$$(Gf)(t, x) = \textstyle\int_M G(t, xy) f(y) dM(y) \tag{23}$$

is regular for $t \geqslant 0$, has initial value $(Gf)(0, x) = f(x)$ and satisfies (22). In our case where the Laplace operator Δ is realized as a positive self-adjoint operator in $L^2(E)$, it can be thought of as the kernels of the operators $e^{-t\Delta}$, $t > 0$, these forming a semigroup of bounded selfadjoint operators in $L^2(E)$ generated by Δ, and it can be expressed, in terms of the eigenfunction expansion of Δ, by

$$G(t, xy) = \textstyle\sum_\nu e^{-t\lambda_\nu} f_\nu(x) \otimes {}^t(A\bar{f}_\nu)(y) \tag{24}$$

where $f_\nu(x) \otimes {}^t(A\bar{f}_\nu)(y)$ is to be regarded as the $N \times N$ matrix $f_{\nu,\iota}(x){}^t(A_\kappa \bar{f}_{\nu,\kappa})(y)$ in each $U_\iota \times U_\kappa \ni (x, y)$. Notice here that in view of (17) $G(t, xy)$ is of class C^∞ for $t > 0$.

Let us now consider (24) on the diagonal of $M \times M$ and take its trace. We then obtain a smooth function on M

$$\operatorname{tr} G(t, xx) = \textstyle\sum_\nu e^{-t\lambda_\nu} {}^t(A\bar{f}_\nu)(x) f_\nu(x) \tag{25}$$

depending on the parameter $t \in R_+$, and its integration over M gives

$$\int_M \operatorname{tr} G(t, xx)\, dM(x) = \sum_\nu e^{-t\lambda_\nu} \qquad (t > 0) \tag{26}$$

since the f_ν are normalized.

Likewise, letting $\tilde{E} \hat{\otimes} \tilde{E}^*$ be the exterior product bundle over $M \times M$ of $\tilde{E} = K \otimes E^*$ and its dual $\tilde{E}^* = K^* \otimes E$, we can express, in terms of the eigenfunction expansion of $\tilde{\Delta}$, the fundamental solution $R_+ \ni t \to \tilde{G}(t, xy) \in C^\infty(\tilde{E} \hat{\otimes} \tilde{E}^*)$ of the heat equation

$$\tilde{Q} = \partial/\partial t + \tilde{\Delta} = 0 \qquad (t > 0) \tag{27}$$

as

$$\tilde{G}(t, xy) = \sum_\nu e^{-t\lambda_\nu} \tilde{f}_\nu(x) \otimes {}^t(\tilde{A}\bar{\tilde{f}}_\nu)(y), \tag{28}$$

and from this

$$\begin{aligned} \operatorname{tr} \tilde{G}(t, xx) &= \sum_\nu e^{-t\tilde{\lambda}_\nu} {}^t(\tilde{A}\bar{\tilde{f}}_\nu)(x) \tilde{f}_\nu(x), \\ \int_M \operatorname{tr} \tilde{G}(t, xx)\, dM(x) &= \sum_\nu e^{-t\tilde{\lambda}_\nu}. \end{aligned} \tag{29}$$

Now consider the difference between (26) and (29). Then, since $\dim E_0(\Delta) = \sum_{\lambda_\nu = 0} 1$, $\dim E_0(\tilde{\Delta}) = \sum_{\tilde{\lambda}_\nu = 0} 1$, and since the terms corresponding to the positive eigenvalues in those expressions mutually cancel out by virtue of Lemma 1, there follows

LEMMA 2.

$$\dim E_0(\Delta) - \dim E_0(\tilde{\Delta}) = \int_M (\operatorname{tr} G - \operatorname{tr} \tilde{G})(t, xx) dM(x)$$

independent of $t > 0$.

4. We now proceed to compute the limit value of $\operatorname{tr} G(t, xx) - tr\ \tilde{G}(t, xx)$ when $t \in R_+$ tends to zero. The consideration is local.

We may suppose the covering $\{U_\iota\}$ to be chosen fine enough so that relative to the metric ds^2 the geodesic distance $r = r(xy)$ between two points $x, y \in M$ is well defined inside each of the U_ι. Now, letting V be an open subset of U_ι with $V \subset \bar{V} \subset U_\iota$ and letting $p \in C_0^\infty(U_\iota)$ with $p|V \equiv 1$, we define

$$\begin{aligned} \gamma &= (\pi t)^{-1} p(x) p(y) \exp(-r^2(xy)/t) \qquad (t > O), \\ \Gamma &= \gamma \cdot I \qquad (I = \text{unit} \in \mathrm{GL}(N, C)) \\ Q\Gamma &= (\partial/\partial t + \Delta_x)\Gamma(t, xy), \\ J &= \Gamma + \sum_{\mu \geq 1} (-1)^\mu \Gamma \bigstar (Q\Gamma)^\mu \end{aligned} \tag{30}$$

where

$$\begin{aligned} \Gamma \bigstar (Q\Gamma)^\mu &= \int_0^t ds \int_M \Gamma(s, x\cdot)(Q\Gamma)^\mu(t - s, \cdot y) dM, \\ (Q\Gamma)^{\mu+1} &= Q\Gamma \bigstar (Q\Gamma)^\mu. \end{aligned}$$

It is then easy to observe that

(a) the series J converges and represents a smooth function $R_+ \ni t \to J(t, xy) \in C_0^\infty(U_\iota \times U_\iota)$;

(b) if $j^{(m)}$, $m = 0, 1, 2, \ldots$ denotes the first m-terms of J and $D^{\nu,\alpha}$ a differential operator of order ν in t and of order $|\alpha|$ in a local coordinate in $U_\iota \times U_\iota$, then there exists a constant $c = c_{\nu,\alpha}$ such that for $0 < t \leqslant 1$,

$$|D^{\nu,\alpha}(J - J^{(m)})| \leqslant ct^{m-2\nu-|\alpha|-2/2}\exp(-r^2/2t)$$

where in the left-hand side the absolute value of the matrix is meant to be the maximum of the absolute values of all of its components. This convention will be agreed on hereafter too when the estimate of matrix-valued functions is in concern;

(c) the function $Jf = \int_M J(t, xy)f(y)dM(y), f \in C^\infty(U_\iota)$ satisfies

$$(31) \qquad \begin{aligned} Q(Jf)(t, x) &= (\Theta f)(t, x) \quad \text{for } t > 0, \\ (Jf)(0, x) &= (p^2 f)(x) \end{aligned}$$

where $(\Theta f)(t, x) = \int_M \Theta(t, xy)f(y)dM$ with

$$(32) \qquad \Theta(t, xy) = \textstyle\sum_{\mu \geq 1}(-1)^{\mu-1}(1 - p^2(x))(Q\Gamma)^\mu(t, xy).$$

Here, in view of $p|V \equiv 1$ and (b), Θ vanishes identically unless $x \in U_\iota - V$ and further, if x remains in a compact subset W of V, we have

$$|D^{\nu,\alpha}\Theta| \leqslant c\exp(-c'/t) \qquad (0 < t \leqslant 1)$$

with suitable constants $c = c_{\nu,\alpha}$ and c', the latter depending on dist $(W, \partial V)$.

From this there follows

LEMMA 3. *Let G_ι be the restriction of the fundamental solution G (24) over $U_\iota \times U_\iota$ and let W be any compact subset in $V \subset U_\iota$. Then there exist constants c, c' such that*

$$|(G_\iota - J)(t, xy)| \leqslant c\exp(-c'/t)$$

provided $x, y \in W$, $0 < t \leqslant 1$, where c' depends on dist $(W, \partial V)$.

PROOF. If we consider Jf in (b) as the solution of the inhomogeneous Cauchy problem (31) for Q and write it, according to Duhamel, in an integrated form by use of the fundamental solution G, we obtain

$$(G_i - J)(t, xy) = -\int_0^t ds \int_{U_\iota - V} G_i(s, x\cdot)(\Theta)(t - s, \cdot y)dM$$

for $x, y \in W$. The lemma is then a trivial consequence of the estimates (17) and (c).

We call J the parametrix in U_ι of Q. By (14), (20), one sees that the differential operators Q, $\tilde{Q}$ have the same principal part. Therefore, to construct the parametrix $\tilde{J}$ in U_ι for $\tilde{Q}$, one can take the same Γ as its first approximation and proceed just as above, simply replacing Q with $\tilde{Q}$. The above results then hold as such in this case too.

The Lemma 3 being thought of as a quantitative statement of the pseudo-local properties of the solutions of heat equations, it holds in a much more general situation, and plays an important role in the spectral analysis of elliptic operators. The generalization is thanks to Bergandal-Gårding.

We now have enough information to prove

LEMMA 4. *Let the notation be as above. Let $x \in U_\iota$ and W be a compact neighborhood of x in U_ι. Then we have*

$$(\operatorname{tr} G - \operatorname{tr} \tilde{G})(t, xx) = -(\pi h_\iota)^{-1}(\partial^2(h_\iota^{N/2} \det A_\iota)/\partial z_\iota \partial \bar{z}_\iota)(x) + O(t^{1/2})$$

uniformly in $x \in W$ and for $0 < t \leqslant 1$.

PROOF. Choose an open set V in U_ι so that $W \subset V \subset \bar{V} \subset U_\iota$, and construct as before the parametrices $J, \tilde{J}$ of $Q, \tilde{Q}$ in U_ι. Then, (b) and Lemma 3 entail

$$\operatorname{tr} G - \operatorname{tr} \tilde{G} = -\operatorname{tr} \Gamma \bigstar (Q - \tilde{Q})\Gamma + \{\operatorname{tr} \Gamma \bigstar (Q\Gamma)^2 - \operatorname{tr} \Gamma \bigstar (\tilde{Q}\Gamma)^2\} + O(t^{1/2})$$

where the estimate is uniform in $x \in W$ and for $0 < t \leqslant 1$.

But here on the right-hand side, if one makes explicit the second member by use of (14), (20), and cancels out the naughty terms therein, it can easily be shown that

$$\operatorname{tr} \Gamma \bigstar (Q\Gamma)^2 - \operatorname{tr} \Gamma \bigstar (\tilde{Q}\Gamma)^2 = O(t^{1/2})$$

uniformly in $x \in W$ and for $0 < t \leqslant 1$. Moreover, if we write in U_ι

$$\begin{aligned} Q - \tilde{Q} &= -h_\iota^{-1} l_\iota \partial/\partial \bar{z}_\iota \\ l_\iota &= {}^t A_\iota^{-1} \partial^t A_\iota/\partial z_\iota - h_\iota A_\iota \partial(h_\iota A_\iota)^{-1}/\partial z_\iota \\ \operatorname{tr} l_\iota &= 2\partial \log (h_\iota^{N/2} \det A_\iota)/\partial z_\iota \end{aligned} \tag{33}$$

and note the symmetry $\Gamma(t, xy) = \Gamma(t, yx) \in C_0^\infty(U_\iota \times U_\iota)$ for fixed $t > 0$, it immediately follows by partial integration that

$$\operatorname{tr} \Gamma \bigstar (Q - \tilde{Q})\Gamma = \tfrac{1}{2} \operatorname{tr} \Gamma \bigstar h_\iota^{-1} (\partial l_\iota/\partial \bar{z}_\iota)\Gamma.$$

Thus, in view of (30) and (33),

$$\begin{aligned} \operatorname{tr} G - \operatorname{tr} \tilde{G} &= -\tfrac{1}{2} \operatorname{tr} \Gamma \bigstar h_\iota^{-1}(\partial l_\iota/\partial \bar{z}_\iota)\Gamma + O(t^{1/2}) \\ &= -\tfrac{1}{2}\gamma \bigstar h_\iota^{-1} \operatorname{tr} (\partial l_\iota/\partial \bar{z}_\iota)\gamma + O(t^{1/2}) \\ &= -(2\pi h_\iota)^{-1} \operatorname{tr} (\partial l_\iota/\partial \bar{z}_\iota)(x) + O(t^{1/2}) \\ &= -(\pi h_\iota)^{-1} \partial^2 \log (h_\iota^{N/2} \det A_\iota)/\partial z_\iota \partial \bar{z}_\iota + O(t^{1/2}) \end{aligned}$$

uniformly in $x \in W$ and for $0 < t \leqslant 1$. This proves the lemma.

The Theorem is now an immediate consequence of the Lemma 4. In fact, let $\sum_\iota \zeta_\iota = 1$ be a C^∞ partition of unity subordinate to the covering $\{U_\iota\}$. Then, from the results obtained above, there follows, letting $t \to 0$,

$$\begin{aligned}\dim H(M, E) - \dim H(M, K \otimes E^*) &= \dim E_0(\Delta) - \dim E_0(\tilde{\Delta}) \\ &= \textstyle\sum_\iota \int_M \zeta_\iota (\operatorname{tr} G - \operatorname{tr} \tilde{G})(t, xx) dM(x) \\ &= \frac{1}{2\pi i} \sum_\iota \int_M \zeta_\iota \frac{\partial^2 \log (h_\iota^{N/2} \det A_\iota)}{\partial z_\iota \partial \bar{z}_\iota} dz_\iota \wedge d\bar{z}_\iota \qquad (34) \\ &= \frac{1}{2\pi i} \int_M \partial\bar{\partial} \log (\det A) + \frac{N}{4\pi i} \int_M \partial\bar{\partial} \log h.\end{aligned}$$

The formula (34) holds for any hermitian, holomorphic vector bundle E over M. Hence, in (34), specifying in particular the bundle E to be a trivial line bundle I over M, for which we may obviously suppose $A_\iota = I \in C$, we obtain incidentally the Gauss-Bonnet formula

$$\chi(M) = 2(1 - g_M) = \frac{1}{2\pi i} \int_M \partial\bar{\partial} \log h \qquad (35)$$

for the Euler characteristic $\chi(M)$ of M, since $\dim H(M,I) = 1$ and $\dim H(M, K) = g_M =$ genus of M.

If we combine (34) and (35), we shall thus have completed the proof of the theorem.

5. We should like to mention in passing that the Atiyah-Singer theorem, concerning the index of elliptic operators acting on the sections of vector bundles over compact manifolds, can also be conceived of in the same context as that considered above. In fact, if $L: C^\infty(E) \to C^\infty(\tilde{E})$ is an elliptic operator, E, $\tilde{E}$ being hermitian vector bundles of the same fiber dimension over a compact Riemannian manifold M, and if L^* denotes the metric adjoint of L, then it can easily be shown that we have in this case an analogue of the Lemma 1, and thus that the ind (L), which is by definition $\dim \ker (L) - \dim \ker (L^*)$, can be expressed, in terms of the fundamental solutions $G(t, xy)$, $\tilde{G}(t, xy)$ of the heat equations $\partial/\partial t + L^*L$, $\partial/\partial t + LL^*$, as $\operatorname{ind}(L) = \lim_{t\to 0} \int_M (\operatorname{tr} G - \operatorname{tr} \tilde{G})(t, xx) dM(x)$. Furthermore, it can be shown (by Hörmander and by the author, both unpublished) that if $\dim M = m$, $\operatorname{ord}(L) = n$, then for small t there holds the following asymptotic expansion: $\operatorname{tr} G(t, xx) \sim t^{-m/n} \sum_{\mu \geq 0} c_\mu(x) t^{\mu/n}$, with coefficients $c_\mu \in C^\infty(M)$, and correspondingly for $\operatorname{tr} \tilde{G}(t, xx)$ with $\tilde{c}_\mu \in C^\infty(M)$, computable from the expression of the differential operators L, L^*. With these facts, we can obtain for the index a computational formula $\operatorname{ind}(L) = \int_M (c_n - \tilde{c}_n)(x) dM(x)$, which however is too hopelessly complicated as such to have any geometric meaning except when the dimension of the basic manifold is small. But if there were to be any hope of giving an analytic proof of the Atiyah-Singer theorem in this way, it seems as if it would lie in finding a way of somehow involving invariant theoretical considerations in the computation of the coefficients of the above asymptotic expansions, as was done by McKean-Singer in their resolution of the Kac conjecture, yet keeping in mind that what interests us here is not the individual coefficients c_n, $\tilde{c}_n$, but rather their difference $c_n - \tilde{c}_n$.

As for the elliptic approach to this subject, we would refer the reader to **[8]** and **[11]** below.

REFERENCES

1. S. Agmon, *Lectures on elliptic boundary value problems,* Van Nostrand, Princeton, N.J., 1966.

2. M. F. Atiyah and I. M. Singer, *The index of elliptic operators on compact manifolds,* Bull. Amer. Math. Soc. **69** (1963), 422–433.

3. G. Bergandal, *Convergence and summability of eigenfunction expansions connected with elliptic differential operators,* Medd. fr. Lunds Univ. Mat. Sem. **15** (1959).

4. S. S. Chern, *Complex Manifolds,* Lecture notes, Chicago University (1956).

5. S. D. Eidelman, *On fundamental solutions of parabolic systems,* Mat. Sb. N. S. **38** (**80**) (1956), 51–92. (Russian)

6. L. Gårding, *On the asymptotic properties of the spectral function belonging to a self-adjoint semi-bounded extension of an elliptic differential operator,* Kungl. Fysiogr. Sallsk. i Lund Förh. **24** (1954), 1–18.

7. F. Hirzebruch, *Neue topologische Methoden in des algebraische Geometrie,* Springer, Berlin, 1956.

8. T. Kotake and M. S. Narasimhan, *Regularity theorems for fractional powers of a linear elliptic operator,* Bull. Soc. Math. France **90** (1962), 449–471.

9. H. P. McKean and I. M. Singer, *Curvature and the eigenvalues of the Laplacian,* J. Differential Geometry **1** (1967), 43–69.

10. S. Minakshisundaram and A. Pleijel, *Some properties of the eigenfunctions of the Laplace-operator on Riemannian manifolds,* Canad. J. Math. **1** (1949), 242–256.

11. R. T. Seeley, *Complex powers of an elliptic operator*, Proc. Sympos. Pure Math., vol. 10, Amer. Math. Soc., Providence, R.I., 1967, pp. 288–307.

12. H. Weyl, *Die Idee der Riemannschen Fläche,* Teubner, Leipzig, 1955.

MASSACHUSETTS INSTITUTE OF TECHNOLOGY

THE EIGENVALUES OF THE LAPLACE OPERATOR FOR THE EXTERIOR PROBLEM

PETER D. LAX[1] AND RALPH S. PHILLIPS[2]

Let D be a compact domain in $\boldsymbol{R}^n$, with a smooth boundary; denote by $\lambda_1, \lambda_2, \ldots$ the eigenvalues of the Laplace operator on D, subject to Dirichlet boundary condition. Much is known about the connection between geometric properties of D and the set of eigenvalues λ. In this note we describe briefly some connections of D with the set of eigenvalues of the Laplacian over the exterior of D, for n odd.

At first this sounds nonsensical since the spectrum of the Laplace operators over the exterior is continuous; there are no proper eigenfunctions and eigenvalues. Indeed the eigenfunctions we have in mind are highly improper: they grow like $\exp\{\text{const}|x|\}$ as $|x| \to \infty$. However they satisfy an appropriate version of the radiation condition at ∞.

What is the significance of these offbeat eigenfunctions? It turns out that they enter the asymptotic description of solutions of the wave equation in the exterior domain. More precisely, subject to a geometric condition which will be described below, the following is true:

Let $u(x, t)$ be a solution of the wave equation in the exterior, vanishing on the boundary of D, whose initial data are zero outside a bounded set. Then u has the asymptotic expansion

$$u(x, t) = \sum a_j \exp(\mu_j t) v_j(x)$$

valid as $t \to \infty$, uniformly for bounded set of x, where μ_j and v_j depend only on D, satisfy

$$(\mu_j^2 - \Delta)v_j = 0, \qquad v_j = 0 \quad \text{on } \partial D,$$

and v_j satisfies the radiation condition. The numbers μ_j are complex, arranged so that

$$0 > \operatorname{Re} \mu_1 \geq \operatorname{Re} \mu_2 \geq \ldots \to -\infty.$$

This has been established rigorously for D convex by Ludwig and Morawetz (see [2] and [3]). In [1] we have conjectured that it is true if and only if D is noncapturing, in the following sense:

Consider all reflected rays in the exterior of D, i.e. straight line segments which change their direction at the boundary of D according to the law of reflection. D is called noncapturing if the length of all reflected rays inside a sphere containing D is less than some constant.

[1] Sponsored in part by AT(30-D-1480.

[2] Sponsored in part by NSF-GP-8857 and AF-F44620-68-C-0054.

In [**4**] we have studied the purely decaying modes, i.e. those with real and negative μ_j. Quite surprisingly these exhibit the same kind of monotonicity as the classical interior eigenvalues, i.e. each μ_j decreases if D is increased. This leads, via comparison with spheres, to estimates of the distribution of the μ_j.

A similar result holds for the Neumann boundary condition.

For background material, see Chapter V of our book [**1**].

Bibliography

1. P. D. Lax and R. S. Phillips, *Scattering theory*, Academic Press, New York, 1967.

2. C. S. Morawetz and D. Ludwig, *The generalized Hughens' Principle for reflecting bodies*, Comm. Pure Appl. Math. (to appear).

3. R. S. Phillips, *A remark on the previous paper of D. Ludwig and C. S. Morawetz*, Comm. Pure Appl. Math. (to appear).

4. P. D. Lax and R. S. Phillips, *Decaying modes for the wave equation in the exterior of an obstacle*, Comm. Pure Appl. Math. (to appear).

Courant Institute of Mathematical Sciences
Stanford University

PSEUDO-DIFFERENTIAL OPERATORS

L. NIRENBERG[1]

1. The notion of pseudodifferential operator has grown in recent years out of attempts to describe an algebra of linear operators large enough to contain all differential operators and, whenever possible, their inverses. Of course, it is important here that the new operators be close enough to differential operators so that it is convenient to compute with them. The general functional calculus enables one to define rather general functions of operators, but it is often difficult to investigate their properties.

Our aim in these lectures is to present a class of pseudodifferential operators—one of several occurring in the current literature—and to give fairly simple and informal derivations of their basic properties, including their invariance on manifolds, as preparation for the lectures of M. Atiyah.

Let M be a smooth differential manifold, and let $L: C_0^\infty(M) \to C^\infty(M)$ be a linear operator (or L may be a linear operator acting on C^∞ sections of one vector bundle over M to another). We say that L is a local operator if for each $x \in M$ the value $Lu(x)$ depends only on the values of u near x or, equivalently, if supp $Lu \subset$ supp u for every $u \in C_0^\infty(M)$. Thus every differential operator is local, and according to a theorem of J. Peetre these are the only local operators. In the sequel we shall consider operators $L: C_0^\infty(M) \to C^\infty(M)$, which extend to map the space $\mathscr{E}'(M)$ of distributions with compact support in M into the space $\mathscr{D}'(M)$ of all distributions on M. Such an operator will be called *pseudolocal* if sing supp $Lu \subset$ sing supp u for each $u \in \mathscr{E}'(M)$, where sing supp v is the complement of the open set on which v is C^∞. Differential operators are clearly pseudolocal; and if $K \in C(\boldsymbol{R}^2)$ is of class C^∞ off the diagonal, then the operator

$$Lu(x) = \int K(x, y)u(y)dy, \quad u \in C_0^\infty$$

is pseudolocal but not local.

We shall write $x = (x_1, \ldots, x_n)$ for the coordinate in $\boldsymbol{R}^n$ and $\xi = (\xi_1, \ldots, \xi_n)$ for the dual coordinate. For an n-tuple $\alpha = (\alpha_1, \ldots, \alpha_n)$ of nonnegative integers we write

$$|\alpha| = \alpha_1 + \ldots + \alpha_n, \qquad \alpha! = \alpha_1! \cdot \ldots \cdot \alpha_n!;$$
$$x^\alpha = x_1^{\alpha_1} \cdot \ldots \cdot x_n^{\alpha_n}, \qquad \xi^\alpha = \xi_1^{\alpha_1} \cdot \ldots \cdot \xi_n^{\alpha_n};$$
$$\partial_x^\alpha = (\partial/\partial x_1)^{\alpha_1} \cdot \ldots \cdot (\partial/\partial x_n)^{\alpha_n},$$
$$\partial_\xi^\alpha = (\partial/\partial \xi_1)^{\alpha_1} \cdot \ldots \cdot (\partial/\partial \xi_n)^{\alpha_n};$$
$$D_x^\alpha = (-i)^{|\alpha|} \partial_x^\alpha.$$

[1] Notes by W. Sweeney. The author was supported under grant AF–49(638)–1719 with the Air Force Office of Scientific Research. Reproduction in whole or in part is permitted for any purpose of the United States Government.

Our operators will be defined with the aid of Fourier transform; we recall the definition, relative to a local coordinate x on a patch containing the support of u:

$$\tilde{u}(\xi) = (2\pi)^{-n}\int e^{-i\xi\cdot x}u(x)dx, \quad u\in C_0^\infty(\boldsymbol{R}^n)$$

$$(1) \qquad \begin{aligned} u(x) &= \int e^{i\xi\cdot x}\tilde{u}(\xi)d\xi,\\ \|\tilde{u}\|_{L_2} &= \|u\|_{L_2},\\ (D^\alpha u)^\sim(\xi) &= \xi^\alpha\tilde{u}(\xi). \end{aligned}$$

For any real number s we define

$$\|u\|_s = \{\int|\tilde{u}(\xi)|^2(1+|\xi|^2)^s d\xi\}^{1/2}$$

for $u\in C_0^\infty(\boldsymbol{R}^n)$, and we denote by $H_s(\boldsymbol{R}^n)$ the completion of $C_0^\infty(\boldsymbol{R}^n)$ in this norm. Then $H_s(\boldsymbol{R}^n)$ is a Hilbert space; and when s is a nonnegative integer, $H_s(\boldsymbol{R}^n)$ consists of all $u\in L_2(\boldsymbol{R}^n)$ with $\partial_x^\alpha u\in L_2(\boldsymbol{R}^n)$ for $0\le|\alpha|\le s$. It is clear that $H_s(\boldsymbol{R}^n)$ $\subset H_r(\boldsymbol{R}^n)$ if $s\ge r$, and $\bigcap_s H_s(\boldsymbol{R}^n)\subset C^\infty(\boldsymbol{R}^n)$. For a compact manifold M (without boundary) we may use a finite covering $\{U_j\}$ of M by coordinate patches and a subordinate partition of unity $\{\rho_j\}$ to define

$$H_s(M) = \{u\in\mathscr{D}'(M)|\rho_j u\in H_s(\boldsymbol{R}^n) \text{ for each } j\}, \quad \|u\|_s = \{\textstyle\sum_j\|\rho_j u\|_s^2\}^{1/2}.$$

We shall see later on that different choices of $\{U_j\}$ and $\{\rho_j\}$ define the same space $H_s(M)$ and norms $\|\cdot\|_s$ which are equivalent. If M is compact and if $s>r$, then, as is well known, bounded sets in $H_s(M)$ are precompact in $H_r(M)$.

A linear operator $L: C_0^\infty(M)\to C^\infty(M)$ is said to be of order m if it extends to a bounded operator $L: H_s(M)\to H_{s-m}(M)$ for each real s. If L is of order $<m$ and if M is compact, then by the previous property the extension $L: H_s(M)\to H_{s-m}(M)$ is a compact operator. In particular, if L has order $-\infty$, then $L: H_s(M)\to H_r(M)$ is always compact, and by Sobolev's lemma $L(H_s(M))\subset C^\infty(M)$. We shall write $\mathscr{L}_{-\infty}$ for the class of operators of order $-\infty$.

To represent our operators in terms of the Fourier transform consider first a differential operator $p(x,D) = \sum_{|\alpha|\le m}a_\alpha(x)D^\alpha$ on $\boldsymbol{R}^n$. The formulas in (1) yield

$$\begin{aligned} p(x,D)u(x) &= \textstyle\sum_\alpha a_\alpha(x)\int e^{i\xi\cdot x}\xi^\alpha\tilde{u}(\xi)d\xi\\ &= \int e^{i\xi\cdot x}p(x,\xi)\tilde{u}(\xi)d\xi, \end{aligned}$$

where the polynomial $p(x,\xi) = \sum_{|\alpha|\le m}a_\alpha(x)\xi^\alpha$ is called the symbol of $p(x,D)$. To define pseudodifferential operators we shall use the same formula

$$(2) \qquad Pu(x) = p(x,D)u(x) = \int e^{ix\cdot\xi}p(x,\xi)\tilde{u}(\xi)d\xi$$

with a larger class of functions $p(x,\xi)$. Not every function $p(x,\xi)$ leads to a pseudolocal operator; for instance if $p(x,\xi) = e^{i\xi\cdot x_0}$, x_0 fixed, then P is a "shift" operator and hence not pseudolocal. The pseudolocal character of P will be reflected in the behavior for large $|\xi|$ of p and its derivatives; we will require rapid decay at infinity for high ξ derivatives. We call $p\in C^\infty(\boldsymbol{R}^n\times\boldsymbol{R}^n)$ a symbol of order m and write $p \in S^m$ if for every compact $K \subset \boldsymbol{R}^n$ and for every α, β there exists a constant $c = c_{\alpha\beta K}$ such that

(3) $$|\partial_\xi^\alpha D_x^\beta p(x, \xi)| \leq c(1 + |\xi|)^{m-|\alpha|}$$

holds for all $(x, \xi) \in K \times \boldsymbol{R}^n$. For simplicity we shall always assume that symbols have compact support in the x variable.

THEOREM 1. *If $p \in S^m$, then* (2) *defines a linear operator $P: C_0^\infty(\boldsymbol{R}^n) \to C^\infty(\boldsymbol{R}^n)$.*

PROOF. If $u \in C_0^\infty(\boldsymbol{R}^n)$, then $|\tilde{u}|$ decreases faster than any power of $|\xi|$ as $|\xi|$ becomes large; thus for each integer N and for each u and β the estimate

$$|D_x^\beta p(x, \xi)\tilde{u}(\xi)| \leq c(1 + |\xi|)^m(1 + |\xi|)^{-N}$$

holds for all x, ξ. These estimates justify differentiation under the integral sign in (2); the resulting integrals are absolutely convergent, and the desired result follows immediately.

It will be convenient later on to have alternate representation formulas for the pseudodifferential operator P. If we substitute the definition of Fourier transform into (2), then we obtain

(4) $$Pu(x) = \int\int e^{i\xi\cdot(x-z)}p(x, \xi)u(z)dz d\xi$$

and

(5) $$Pu(x) = \int K(x, x - y)u(y)dy,$$

where

(5′) $$K(x, z) = (2\pi)^{-n}\int e^{i\xi\cdot z}p(x, \xi)d\xi.$$

Also

(6) $$(Pu)\tilde{}(\eta) = \int \tilde{p}(\eta - \xi, \xi)\tilde{u}(\xi)d\xi,$$

where $\tilde{p}$ is the Fourier transform of p with respect to the first argument.

The kernel K given by (5′) will in general be singular at $z = 0$, and in the earlier work it was the representation (5) which was most used, giving rise to the expression "singular integral operators".

THEOREM 2. *If $p \in S^m$, then P has order m.*

PROOF. For the sake of simplicity we prove the inequality $\|Pu\|_s \leq c\|u\|_{s+m}$ only in the case $s = m = 0$. Since p has compact support in x and since (3) holds with $|\alpha| = m = 0$ and large $|\beta|$, an estimate

$$|\tilde{p}(\eta - \xi, \xi)| \leq c_N(1 + |\xi - \eta|)^{-N}$$

holds for each integer N and each $(\xi, \eta) \in \boldsymbol{R}^n \times \boldsymbol{R}^n$.

Hence by (6)

$$|(Pu)\tilde{}(\eta)| < c\int(1 + |\xi - \eta|)^{-n-3}|\tilde{u}(\xi)|d\xi$$

and

$$\|Pu\|_{L_2} = \|(Pu)^{\sim}\|_{L_2} \leq c\|(1 + |\xi|)^{-n-3}\|_{L_1}\|\tilde{u}\|_{L_2} \leq c'\|u\|_{L_2},$$

as required.

With our operator P extended as a map $H_s \to H_{s-m}$ we shall next establish the pseudolocal property, using the representation (5).

Assume, for the moment, that $m < -n$. Then the function $K(x, z)$ in (5′) is continuous; in fact, if m is large negative, then $K(x, z)$ is easily seen to be rather smooth. For $z \neq 0$, $K(x, z)$ belongs, in fact, to C^∞. To see this suppose that $z_1 \neq 0$ and note that repeated integration by parts yields

$$K(x, z) = (i)^N \int \frac{e^{i\xi \cdot z}}{z_1^N}\Big(\frac{\partial}{\partial \xi_1}\Big)^N p(x, \xi)d\xi \tag{7}$$

for any positive integer N. Hence

$$|\partial_x^\beta \partial_z^\gamma K(x, z)| \leq \text{constant} \int (| + |\xi|^{|\gamma|})(|z_1|^{-N} + |z_1|^{-N-|\gamma|})|\partial_x^\beta(\frac{\partial}{\partial \xi_1})^N p(x, \xi)|d\xi$$

and the integral converges for N sufficiently large. Thus $K(x, z)$ is of class C^∞ for $z \neq 0$.

For arbitrary m, let λ be an integer with $2\lambda > n + m$, and write

$$\begin{aligned} Pu(x) &= \int e^{i\xi \cdot x} p(x, \xi)\tilde{u}(\xi)d\xi \\ &= \int e^{i\xi \cdot x} p(x, \xi)(1 + |\xi|^2)^{-\lambda}(1 + |\xi|^2)^\lambda \tilde{u}(\xi)d\xi \\ &= \int e^{i\xi \cdot x} p(x, \xi)(1 + |\xi|^2)^{-\lambda}((1 - \Delta)^\lambda u)^{\sim}(\xi)d\xi. \end{aligned}$$

Since $p(x, \xi)(1 + |\xi|^2)^{-\lambda}$ is of order $m - 2\lambda < -n$, the argument just given yields

$$Pu(x) = \int K_\lambda(x, x - y)(1 - \Delta)^\lambda u(y)dy \tag{8}$$

with a continuous kernel $K_\lambda(x, z)$ which is C^∞ for $z \neq 0$.

THEOREM 3. *A pseudodifferential operator P is pseudolocal.*

PROOF. We have to prove sing supp $Pu \subset$ sing supp u for any $u \in \mathscr{E}'(M)$. Suppose that u is C^∞ in a neighborhood of x_0. Let $\zeta \in C_0^\infty$ be such that u is C^∞ on supp ζ and $\zeta \equiv 1$ on a neighborhood of x_0, and set

$$u = \zeta u + (1 - \zeta)u = u_1 + u_2.$$

Since $u_1 \in C_0^\infty$ we know that $Pu_1 \in C^\infty$. Next,

$$Pu_2 = \int K_\lambda(x, x - y)(1 - \Delta)^\lambda u_2(y)dy;$$

if x is near x_0, then x is away from the support of u_2. We may then multiply K_λ by a C_0^∞ function ψ of y which is one on the support of u_2 and zero on a neighborhood of x_0. The function u_2 may be expressed in the form $u_2 = (1 - \Delta)^N v$, where v is continuous; and we may now integrate by parts, throwing all the derivatives on ψK_λ. The resulting expression may be differentiated under the integral sign, for

x close to x_0, and we conclude that Pu is C^∞ in a neighborhood of x_0, completing the proof.

We shall consider pseudodifferential operators P modulo the class $\mathscr{L}_{-\infty}$ of infinitely smoothing operators. In many instances more precise knowledge of P is unnecessary. For example, let L be a linear operator which has a left inverse Q mod $\mathscr{L}_{-\infty}$. Then $QL = I + T$, where $T \in \mathscr{L}_{-\infty}$, and

$$\text{sing supp } QLu = \text{sing supp } (I + T)u = \text{sing supp } u$$

for $u \in \mathscr{E}'(\boldsymbol{R}^n)$. If Q is pseudolocal, then

$$\text{sing supp } u \subset \text{sing supp } Lu$$

for $u \in \mathscr{E}'(\boldsymbol{R}^n)$. Operators having this property are called hypoelliptic. We have proved that L is hypoelliptic if it has a pseudolocal left inverse mod $\mathscr{L}_{-\infty}$. The problem of characterizing all hypoelliptic differential operators is still not solved.

Next, we consider sums of operators of decreasing orders (analogous to terms of different orders for differential operators). If $\{p_j\}$ is a sequence of symbols, p_j in S^{m_j}, and if $m_j \searrow -\infty$, then there exists $p \in S^{m_0}$ such that

$$p - \sum_{j=0}^{N-1} p_j \in S^{m_N} \tag{9}$$

for each $N \geq 1$. In fact, such a p may be given explicitly by

$$p(x, \xi) = \sum_{j=0}^{\infty} \phi(\xi/t_j) p_j(x, \xi),$$

where $\phi \in C^\infty(\boldsymbol{R}^n)$ is identically 0 for $|\xi| < 1$ and identically 1 for $|\xi| > 2$. If $\{t_j\}$ is a rapidly increasing sequence $\to \infty$ (note for any fixed (x, ξ) the sum is finite) then it is easy to prove that $p \in S^{m_0}$. By Theorem 2 the pseudodifferential operator $P - \sum_0^N P_j$ has order m_{N+1}. We shall write $p \sim \sum p_j$ to signify that (9) holds for each N.

2. We now turn to more interesting properties of the operators: that they are closed under composition and may be invariantly defined on a manifold. We recall that the functions u may be vector valued and the symbols may be matrices.

THEOREM 4. *Let P and Q be pseudodifferential operators with symbols $p \in S^{m_1}$ and $q \in S^{m_2}$, respectively. Then $R = QP$ is a pseudodifferential operator with symbol $r(x, \xi) \in S^{m_1+m_2}$, and one has the analogue of Leibniz's formula*:

$$r(x, \xi) \sim \sum_\alpha \frac{1}{\alpha!} \partial_\xi^\alpha q(x, \xi) \cdot D_x^\alpha p(x, \xi). \tag{10}$$

Observe that the various terms on the right have decreasing orders of growth in ξ.

PROOF. Using (4) for Q and (2) for P, we obtain

$$\begin{aligned} Ru(x) &= QPu(x) \\ &= (2\pi)^{-n} \int\int e^{i\xi\cdot(x-y)} q(x,\xi) \int e^{iy\cdot\eta} p(y,\eta)\tilde{u}(\eta)d\eta dy d\xi \\ &= \int r(x,\eta) e^{i\eta\cdot x}\tilde{u}(\eta)d\eta, \end{aligned}$$

where

$$r(x,\eta) = (2\pi)^{-n} \int\int q(x,\xi)p(y,\eta)e^{i(\xi-\eta)\cdot(x-y)}dy d\xi.$$

It is easy to verify directly that $r \in S^{m_1+m_2}$, and by Taylor's formula

$$r(x,\eta) = (2\pi)^{-n} \int\int \Big(\sum_{|\alpha|\le N} \frac{1}{\alpha!}\partial_\eta^\alpha q(x,\eta)(\xi-\eta)^\alpha\Big)\cdot p(y,\eta)e^{i(\xi-\eta)\cdot(x-y)}dy d\xi + r_N.$$

Integrating first with respect to y and then ξ, one verifies directly that

$$(2\pi)^{-n} \int\int (\xi-\eta)^\alpha p(y,\eta)e^{i(\xi-\eta)\cdot(x-y)}dy d\xi = D_x^\alpha p(x,\eta).$$

Hence

$$r(x,\eta) = \sum_{|\alpha|\le N} \frac{1}{\alpha!}\partial_\eta^\alpha q(x,\eta)D_x^\alpha p(x,\eta) + r_N.$$

To complete the proof one checks by direct estimates that $r_N \in S^{m_1+m_2-N}$.

We remark that if the symbols are matrices, then the order of composition in (10) is important. In the scalar case, however, it does not matter, and we see from (10) that $PQ - QP$ will then have order $m_1 + m_2 - 1$.

THEOREM 5. *If P is a pseudodifferential operator, there exists a pseudodifferential operator P^* such that*

$$\langle Pu, v\rangle = \langle u, P^* v\rangle$$

for $u, v \in C_0^\infty(\mathbf{R})^n$. Moreover, if p^* denotes the symbol of P^*, then

$$p^*(x,\xi) \sim \sum_\alpha \frac{1}{\alpha!} D_x^\alpha \partial_\xi^\alpha \overline{p(x,\xi)^t}. \tag{11}$$

The proof of Theorem 5 uses Taylor's formula as above, and is contained in the remark below.

Theorems 4 and 5 show that the pseudodifferential operators form an algebra with conjugation modulo $\mathscr{L}_{-\infty}$. By (10) and (11) the same is true for the class of pseudodifferential operators whose symbols are sums of terms which are positively homogeneous in ξ for $|\xi| >$ constant.

In our definition of pseudodifferential operator $p(x, D)$ we have, loosely speaking, allowed the D operators to act first—the operations of multiplying by functions of x then following; this is clear from the form (2) for differential operators. The adjoint of the operator reverses in some sense the order (as well as taking complex conjugation). On taking the product of two operators one again mixes these operators, and formulas (11) and (10) show how to undo this mixing. In (12) below we consider more general forms of our operators—analogous to first multiplying by functions of x, then by functions of D, then by functions of x again. It is convenient to consider still more general "mixings" of this kind, and this more general formulation is carried out in Friedrichs [7].

Turning now to the consideration of our operators on a manifold, we have to show that under a C^∞ change of variables the class of operators is invariant. The proof we give is due to M. Kuranishi (to appear) and will make use of the representation (4). It is convenient first to consider more general representations.

REMARK. Consider an operator

$$Qu(x) = \int e^{i\xi\cdot(x-z)}q(x,\xi,z)u(z)dzd\xi \tag{12}$$

where $q \in C^\infty(\boldsymbol{R}^n \times \boldsymbol{R}^n \times \boldsymbol{R}^n)$ has compact support in x and z and satisfies conditions analogous to (3): for all α, β, γ

$$\left|\partial_\xi^\alpha D_x^\beta D_z^\gamma q(x,\xi,z)\right| \le c_{\alpha,\beta,\gamma}(1+|\xi|)^{m-|\alpha|}. \tag{12'}$$

Then Q is in fact a pseudodifferential operator with symbol $p(x,\xi) \in S^m$ which satisfies

$$p(x,\xi) \sim \sum_\alpha \frac{1}{\alpha!}\partial_\xi^\alpha D_z^\alpha q(x,\xi,z)\big|_{z=x}. \tag{13}$$

PROOF. Carrying out the dz integration in (12), we obtain

$$Qu(x) = \iint e^{i\xi\cdot x}\tilde{q}(x,\xi,\xi-\eta)\tilde{u}(\eta)d\eta d\xi,$$

where $\tilde{q}(x,\xi,\zeta)$ is the Fourier transform with respect to z and thus

$$p(x,\eta) = \int e^{i(\xi-\eta)\cdot x}\tilde{q}(x,\xi,\xi-\eta)d\xi$$

$$= \int e^{i\zeta\cdot x}\tilde{q}(x,\zeta+\eta,\zeta)d\zeta. \tag{13'}$$

It is easy to check that $p(x,\eta)$ is in S^m. Furthermore (13) follows easily from (13′) (or also directly from (12)) by expanding

$$\tilde{q}(x,\zeta+\eta,\zeta) \sim \sum_\alpha \frac{1}{\alpha!}\partial_\eta^\alpha \tilde{q}(x,\eta,\zeta)\zeta^\alpha$$

and integrating.

In order to prove invariance it suffices, with the aid of a partition of unity, to consider functions with small support in a coordinate patch.

Let $x = \chi(y)$ give a change of coordinate on $U \subset \boldsymbol{R}^n$ and let P be a pseudodifferential operator whose symbol satisfies $p(x,\xi) = 0$ when $x \notin U$. Let Q be the operator defined by

$$Qv(y) = P(v\cdot\chi^{-1})(\chi(y)).$$

THEOREM 6. *Q is a pseudodifferential operator with symbol*

$$q(y,\eta) \sim p(\chi(y), [(\partial\chi/\partial y)^t]^{-1}\eta) + \textit{lower order terms}. \tag{14}$$

PROOF. According to (4) we have

$$Pu(x) = (2\pi)^{-n} \iint e^{i\xi\cdot(x-z)} p(x, \xi) u(z) dz d\xi$$

for $u \in C_0^\infty(U)$. Writing $z = \chi(w)$ and $v(w) = u(\chi(w))$, we obtain

$$Pu(\chi(y)) = (2\pi)^{-n} \iint e^{i\xi\cdot(\chi(y)-\chi(w))} p(\chi(y), \xi)\cdot|\partial\chi/\partial w| v(w) dw d\xi,$$

where $|\partial\chi/\partial w|$ denotes the determinant of the Jacobian matrix $\partial\chi/\partial w$. Now $\chi(y) - \chi(w) = H(y, w)\cdot(y - w)$ where $H(y, w)$ is a nonsingular matrix for w close to y and $H(w, w) = \partial\chi/\partial w$. Thus

$$Pu(\chi(y)) = (2\pi)^{-n} \iint e^{i\xi\cdot H(y,w)(y-w)} p(\chi(y), \xi)\cdot|\partial\chi/\partial w| v(w) dw d\xi,$$

and, setting $\zeta = H(y, w)^t\xi$,

$$\begin{aligned} Pu(\chi(y)) &= (2\pi)^{-n} \iint e^{i\zeta\cdot(y-w)} p(\chi(y), [H(y, w)^t]^{-1}\zeta)\cdot \\ &\qquad |\partial\chi/\partial w| \psi(w)\cdot \frac{1}{|H(y, w)|} v(w) dw d\zeta \\ &= (2\pi)^{-n} \iint e^{i\zeta(y-w)} q(y, \zeta, w) v(w) dw\, d\zeta, \end{aligned}$$

where

$$q(y, \zeta, w) = p(\chi(y), [H(y, w)^t]^{-1}\zeta)|\partial\chi/\partial w|\cdot\frac{1}{|H(y, w)|}\psi(w)$$

and $\psi \in C_0^\infty(\boldsymbol{R}^n)$ is identically 1 on a neighborhood of supp v. It is easy to verify that q satisfies (12′) and, applying the Remark, we obtain p of the form (13) of which the leading part is shown in Theorem 6, and the proof is complete.

Theorem 6 shows that we may use local coordinates to define the notion of pseudodifferential operator on a manifold. This may be done with the aid of suitable partitions of unity ϕ_i, ψ_j: we first multiply the function by ϕ_i, apply the operator in a local coordinate patch containing its support, then multiply by ψ_j and sum over i and j. The formula in Theorem 6 shows how the symbol transforms under coordinate change. (For operators P mapping sections of one vector bundle to another, one must also take into account the effect on p of a change of basis in the fibre.) The same proof shows that the subalgebra of operators whose symbols are sums of homogeneous terms is also well defined on a manifold; in **[8]** L. Hörmander has given a purely intrinsic characterization of this subalgebra.

If P is a pseudodifferential operator, then its symbol $p(x, \xi)$ may be recaptured from the action of the operator in the following way: Let $f \in C_0^\infty$ be identically one in a neighborhood of the point x. Then by (4)

$$\begin{aligned} &e^{-i\xi\cdot x} P(f e^{i\xi\cdot x})(x) \\ &= \iint e^{i(\eta-\xi)\cdot(x-z)} p(x, \eta) f(z) dz d\eta \\ &= \iint e^{i\zeta\cdot(x-z)} p(x, \xi + \zeta) f(z) dz d\zeta \\ &= \iint e^{i\zeta\cdot(x-z)} \sum_{|\alpha|\le N} \frac{1}{\alpha!} \partial_\xi^\alpha p(x, \xi) \zeta^\alpha f(z) dz d\zeta + r_N \\ &= \sum_{|\alpha|\le N} \frac{1}{\alpha!} \partial_\xi^\alpha p(x, \xi) D_x^\alpha f(x) + r_N. \end{aligned}$$

Since $f \equiv 1$ near x, we see that this expression

$$= p(x, \xi) + r_N,$$

and for large N one sees easily that r_N decays in ξ faster than any given power at infinity.

3. We shall now study elliptic operators on a compact manifold (without boundary). Let P be a pseudodifferential operator whose symbol is a square matrix $p \in S^m$. We say that $p(x, \xi)$ is elliptic if there exists a constant c such that $p(x, \xi)$ is invertible for $|\xi|$ large and if its norm satisfies

$$|p(x, \xi)^{-1}| \leq c(1 + |\xi|)^{-m}$$

for $x \in K$ and $|\xi| \geq c$. The notion of ellipticity does not depend on the choice of coordinates, and hence it makes sense to speak of an elliptic pseudodifferential operator on a manifold.

We claim that the elliptic operators are invertible (modulo $\mathscr{L}_{-\infty}$) in the algebra of pseudodifferential operators. In fact, to find a left inverse for $P = p(x, D)$ we must find q such that

$$\sum_{\alpha} \frac{1}{\alpha!} \partial_\xi^\alpha q(x, \xi) D_x^\alpha p(x, \xi) \sim 1.$$

To do this we set $q \sim \sum q_j$, where $q_j \in S^{-m-j}$, and solve

$$\sum_{\alpha, j} \frac{1}{\alpha!} \partial_\xi^\alpha q_j(x, \xi) D_x^\alpha p(x, \xi) \sim 1$$

by solving the equations

$$q_0(x, \xi) p(x, \xi) = 1,$$

$$q_1(x, \xi) p(x, \xi) + \sum_{|\alpha|=1} \partial_\xi^\alpha q_0(x, \xi) D_x^\alpha p(x, \xi) = 0,$$

$$q_2(x, \xi) p(x, \xi) + \sum_{|\alpha|=1} \partial_\xi^\alpha q_1(x, \xi) D_x^\alpha p(x, \xi) + \sum_{|\alpha|=2} \frac{1}{\alpha!} \partial_\xi^\alpha q_0(x, \xi) D_x^\alpha p(x, \xi) = 0,$$

$$\vdots \qquad\qquad \vdots \qquad\qquad \vdots$$

successively for $q_0, q_1, q_2, \ldots$. A similar construction yields a right inverse Q' for P. Since $Q \equiv QPQ' \equiv Q'$ modulo $\mathscr{L}_{-\infty}$, we see that Q is also a right inverse for P.

If P is any elliptic pseudodifferential operator of order $m \geq 0$ on a compact manifold M, then one verifies with the aid of a partition of unity that

$$\|Pu\|_{L_2} + \|u\|_{L_2} \geq c\|u\|_m$$

for $u \in C^\infty(M)$, where $\|\cdot\|_m$ s the norm previously defined in terms of local coordinates. In fact the norms on the left and right are equivalent. To see this let

Q be a pseudodifferential operator of order $-m$ such that $QP = I + T$, $T \in \mathscr{L}_{-\infty}$. Then

$$\begin{aligned}\|u\|_m &\leq \|(I + T)u\|_m + \|Tu\|_m \\ &\leq \|QPu\|_m + \|Tu\|_m \\ &\leq c(\|Pu\|_{L_2} + \|u\|_{L_2}) \quad \text{for } y \in C^\infty(M).\end{aligned}$$

Thus the norm $\|\cdot\|_m$ is equivalent to $\|P\cdot\|_{L_2} + \|\cdot\|_{L_2}$, and it follows that the space $H_m(M)$ is well defined, independent of any particular covering—as claimed earlier. The same must be true for $H_{-m}(M)$ which is dual to $H_m(M)$.

Since an elliptic operator has an inverse in our algebra, it is necessarily hypoelliptic—as shown in § 1. The same is true for an overdetermined elliptic system, *i.e.*, one in which the symbol has more rows than columns and has maximal rank for large $|\xi|$—thus the square matrix

$$\overline{p(x, \xi)^t}\, p(x, \xi)$$

is elliptic in our previous sense. To see that this overdetermined system is hypoelliptic, let u be a solution of $Pu = f$, and let Q be the operator corresponding to the symbol $\overline{p(x, \xi)^t}$. Then $QPu = Qf$, and since Q is pseudolocal, Qf is C^∞ wherever f is C^∞, and u is then also C^∞ there since QP is elliptic in our previous sense and hence hypoelliptic.

Returning to the case of a square matrix p and P an operator on a compact manifold M, we claim that as a map of H_s to H_{s-m} it is a Fredholm operator. Recall that a Fredholm operator between two Banach spaces is an operator T which has closed range $R(T)$ of finite codimension, and a finite dimensional null space $N(T)$. The index of such an operator is defined by dim $N(T)$ $-$ codim $R(T)$, or

$$\text{index } T = \dim N(T) - \dim N(T^*).$$

We shall use the following facts about index:

(1) if T is Fredholm and C is compact, then index $(T + C) =$ index T;

(2) if T_1 and T_2 are Fredholm, then index $T_1T_2 =$ index T_1 + index T_2;

(3) if T_t, $0 \leq t \leq 1$, is a family of Fredholm operators, continuous in the operator norm topology, then index $T_0 =$ index T_1.

It follows from classical theory that T is a Fredholm operator if and only if it has a right inverse and a left inverse modulo compact operators. Hence, if P is an elliptic pseudodifferential operator with symbol in S^m on a compact manifold M, then for each s the following map is a Fredholm operator:

$$P: H_s(M) \to H_{s-m}(M). \tag{15}$$

As an illustration of the use of elliptic pseudodifferential operators we prove the following result, in which we assume that for $|\xi| = 1$, $\lim_{\lambda\to\infty} \lambda^{-m} p(x, \lambda\xi)$ exists uniformly.

THEOREM 7. *Let P be an elliptic pseudodifferential operator acting on complex valued functions on a compact manifold M of dimension > 2. Then for each s the operator* (15) *has index* 0.

PROOF (R. SEELEY [**15**]). Our aim is to deform the given operator P into one whose index is known. Assume at first that P has order 0, and let $p(x, \xi)$ be the symbol of P in some coordinate patch. Then index P depends only on the behavior of $p(x, \xi)$ for ξ in a neighborhood of ∞, because when $p(x, \xi)$ is changed for ξ in a bounded set, then $p(x, D): H_s(M) \to H_{s-m}(M)$ changes only by a compact operator. Now for large R we can deform $p(x, \xi)$ to $p(x, R\xi/|\xi|)$ for $|\xi| > R$ through elliptic symbols. Since the index does not change under this deformation, we may assume that $p(x, \xi)$ is homogeneous in ξ for large ξ. In fact, in virtue of (14), we may consider $p(x, \xi)$ to be an invariantly defined function on the cotangent bundle $T^*(M)$; this is because the part of $p(x, \xi)$ which does not transform this way is of lower order and hence defines a compact operator: $H_s(M) \to H_{s-m}(M)$. After choosing a Riemannian metric on M, $p(x, \xi)$ becomes a well defined map on the sphere bundle $ST^*(M)$. The rest of the proof is essentially contained in the following lemma.

LEMMA. *Assume that for each fixed* x, $\arg p(x, \xi)$ *can be considered as a continuous function on the unit* ξ *sphere in* T^*_x. *Then* $p(x, \xi)$ *can be deformed through elliptic symbols to a symbol which is independent of* ξ.

PROOF. We first perform the deformation for fixed x. We write Σ^{n-1} for the unit sphere in T^*_x and choose a family K_t of functions on $\Sigma^{n-1} \times \Sigma^{n-1}$ such that

(1) $K_t(\xi, \eta)$ depends only on the Riemannian distance $d(\xi, \eta)$;

(2) $\int K_t(\xi, \eta) dv(\eta) = 1$, where dv = volume element on Σ^{n-1};

(3) $K_1(\xi, \eta) = 1/\mathrm{vol}\,\Sigma^{n-1}$

(4) $K_t(\xi, \eta)$ is continuous in (t, ξ, η) for $t > 0$;

(5) as $t \to 0$, $K_t(\xi, \eta)$, as a function of η, tends to the δ function at ξ.

The required deformation is given by:

$$|p_t(x, \xi)| = \int K_t(\xi, \eta)|p_0(x, \eta)| dv(\eta);$$
$$\arg p_t(x, \xi) = \int K_t(\xi, \eta) \arg p_0(x, \eta) dv(\eta);$$
$$p_t = |p_t| e^{i \arg p_t}$$

This deformation is well defined for every x. Although $\arg p(x, \xi)$ is continuous, it is determined only up to an additive multiple of 2π and hence so is $\arg p_t(x, \xi)$. However, it is clear that p_t is well defined and hence the lemma is proved.

Since $\dim M > 2$, the unit ξ sphere for every fixed x is simply connected and so the lemma may be applied. We then conclude that index P is the same as the index of the operator which consists of multiplication by $p_1(x) \neq 0$. Since the index of this operator is 0, the proof is complete in case $m = 0$. To treat the general case one chooses an elliptic pseudodifferential operator Q of order $-m/2$ in order to apply the previous argument to Q^*QP, which has order 0. Since Q^*Q is selfadjoint, index $Q^*Q = 0$ and $0 =$ index $Q^*QP =$ index $Q^*Q +$ index $P =$ index P, as required.

The index of elliptic operators in general has been determined by Atiyah and Singer (see [1] for a recent presentation).

Before proceeding we point out a deficiency of the class of pseudodifferential operators given above. If M and N are manifolds and if P is a pseudodifferential operator on M, then we can define

$$P' : C_0^\infty(M \times N) \to C^\infty(M \times N) \tag{16}$$

by merely letting P act on the first variable. This map need not be a pseudodifferential operator as defined here because

$$|\partial_\xi^\alpha D_x^\beta p(x, \xi)| \leq c(1 + |\xi| + |\eta|)^{m - |\alpha|}$$

cannot hold for large $|\alpha|$ unless $\partial_\xi^\alpha D_x^\beta p(x, \xi)$ vanishes identically.

One way to widen the class of pseudodifferential operators is to consider the class P of linear operators $P' : C_0^\infty(M) \to C^\infty(M)$ such that for some m and each $s \in \boldsymbol{R}$

$$P' : H_s(M) \to H_{s-m}(M)$$

is the limit in the norm topology of operators $q(x, D)$ with $q(x, \xi) \in S^m$. For $m > 0$ the map P' in (16) has this property as is not difficult to verify (see [**1**]).

This suggests the question: how is the norm of P, as a map of H_s to H_{s-m}, related to the symbol? This is apparently not a very suitable question; a more appropriate result is expressed in the following (which is formulated merely for $m = 0$):

THEOREM 8. *If P is a pseudodifferential operator on a compact manifold with symbol in S^0, then as a map of L_2 into L_2 it satisfies*

$$\|\hat{P}\| - \inf_C \|P + C\| = \overline{\lim_{\xi \to \infty}} \sup_x |p(x, \xi)| = \lambda, \tag{17}$$

where $\|\cdot\|$ denotes the norm of the operator as a map of L_2 into L_2, and the infimum is taken over all compact operators C on L_2; $|p(x, \xi)|$ represents the norm of the matrix $p(x, \xi)$ relative to orthonormal frames if P acts on sections of a Hermitian vector bundle.

From (14) we see that the right-hand side is well defined, independent of the particular coordinates.

If we restrict ourselves to symbols which are sums of homogeneous functions of ξ for $|\xi|$ large and let P be the class of operators, as above, which are limits in the norm topology of such special pseudodifferential operators as maps of H_s into H_{s-m} for all s, then it follows from Theorem 8 that to each such operator we may associate in a one to one way a *continuous* symbol $p(x, \xi)$ which is homogeneous in ξ for large $|\xi|$.

Before proving Theorem 8 we shall first prove (for the case $m = 0$):

Gårding's inequality: Consider a pseudodifferential operator Q on a compact manifold M whose symbol $q(x, \xi) \in S^0$ is a Hermitian positive semidefinite matrix. Then for every real s and $\varepsilon > 0$ there is a constant $C = C(\varepsilon, s)$ such that for every $u \in L_2(M)$

(18) $$\operatorname{Re}\langle Qu, u\rangle_s + \varepsilon\langle u, u\rangle_s \geq -C\|u\|^2_{s-1/2},$$

where $\langle\cdot,\cdot\rangle_s$ represents H_s scalar product for u (possibly in a Hermitian vector bundle).

PROOF. We shall consider only the case $s = 0$. For $\varepsilon > 0$ we may write $q(x, \xi) + \varepsilon I = r^2(x, \xi)$; here I is the unit matrix—with $r(x, \xi)$ a Hermitian matrix symbol in S^0. If R is the operator corresponding to r, then its adjoint R^* differs from R by an operator of order -1 (see Theorem 5). Furthermore, $Q + \varepsilon I - R^*R$ is also of order -1. Therefore

$$\operatorname{Re}(Qu, u) + \varepsilon(u, u) - (R^*Ru, u) \geq -\text{constant } \|u\|^2_{-1/2}$$

and since $(R^*Ru, u) = (Ru, Ru) \geq 0$, the result follows.

COROLLARY. *With λ as defined in* (17), *for every $\varepsilon > 0$ there is a constant $C(\varepsilon)$ such that*

(17′) $$\|Pu\|^2_0 \leq (\lambda + \varepsilon)^2\|u\|^2_0 + C(\varepsilon)\|u\|^2_{-1/2}.$$

PROOF. We may modify p on a set of compact support in ξ so as to make it satisfy

$$|p(x, \xi)| \leq \lambda + \varepsilon/2$$

for all x, ξ. The new operator differs from the preceding one by an operator of order -1, and thus (17′) will be unaffected. Then the symbol of the operator $(\lambda + \varepsilon)^2 - P^*P$ will (aside from lower order terms) be a positive definite Hermitian matrix, and (17′) then follows from (18).

We mention that a sharp form of Gårding's inequality (18) (and hence also of (17′)), namely with $\varepsilon = 0$ and C some constant, also holds; see [**13**] and, for a simple proof, [**7**] and Calderón's lectures in [**18**].

PROOF OF THEOREM 8. The inequality

$$\|\hat{P}\| \geq \varlimsup_{\xi\to\infty} \sup_x |p(x, \xi)| = \lambda$$

is easily established by considering the action of $P + C$ on functions of the form

$$|\xi^k|^{n/4}u((x - x_0)|\xi^k|^{1/2})e^{i\xi^k x}$$

for a suitable sequence $\xi^k \to \infty$ and with $u \in C_0^\infty$, $u \not\equiv 0$; and we shall prove only the other inequality

(17″) $$\|\hat{P}\| \leq \lambda.$$

To this end let $\phi(\xi)$ be a C_0^∞ symbol, $0 \leq \phi \leq 1$, which is equal to one on a large ball $|\xi| \leq R$ (in each of a finite number of coordinate systems covering M). Then the operator $\phi(D)$ on $L_2(M)$ is compact and hence so is $P\phi(D)$. Then according to (17′) we have

$$\begin{aligned}\|P(I - \phi(D))u\|^2_0 &\leq (\lambda + \varepsilon)^2\|(I - \phi(D))u\|^2_0 \\ &\quad + C(\varepsilon)\|(I - \phi(D))u\|^2_{-1/2} \\ &\leq (\lambda + \varepsilon)^2\|u\|^2_0 + C(\varepsilon)\|(I - \phi(D))u\|^2_{-1/2}.\end{aligned}$$

If we choose R very large then it is clear from the definition of $\|\cdot\|_{-1/2}$ that

$$\|(I - \phi(D))u\|^2_{-1/2} \leq (\text{arbitrarily small constant})\|u\|_0^2$$

so that

$$\|Pu - P\phi(D)u\|_0^2 \leq (\lambda + 2\varepsilon)^2\|u\|_0^2.$$

Since $P\phi(D)$ is compact, the inequality (17″) follows.

4. The class of pseudodifferential operators that we have presented is unfortunately too restricted for many purposes. One of the aims in considering such general operators is to include inverses of a wide class of differential operators. However, the only invertible operators in our algebra are the elliptic ones. The heat operator in, say, one space variable

$$-\partial^2/\partial x_1^2 + \partial/\partial x_2$$

has symbol $\xi_1^2 + i\xi_2$ and its inverse is a well behaved operator, however its symbol $q = (\xi_1^2 + i\xi_2)^{-1}$ does not belong to any of our classes S^m. One merely has

$$\partial_\xi^\alpha q \leq c(1 + |\xi|)^{-1-|\alpha|/2}.$$

The classes of symbols S^m that we have considered are special cases of more general classes $S^m_{\rho,\delta}$ considered in [10] by Hörmander. These are symbols p satisfying

$$|\partial_\xi^\alpha D_x^\beta p(x, \xi)| \leq C_{\alpha,\beta}(1 + |\xi|)^{m-\rho|\alpha|+\delta|\beta|}, \quad 0 \leq \delta < \rho \leq 1.$$

If we also have $1 - \rho \leq \delta < \rho \leq 1$ then the corresponding operators may be invariantly defined on a manifold, and all the results that we have proved above hold for these operators. Indeed all the proofs that we have given extend to this wider class with one exception. Our proof of Theorem 2, that these operators are of order m, will no longer work; it will merely show that the operators are of finite order. The proof in [**10**(3.3)] that the order is m is rather delicate; it makes use of a special partition of unity in the ξ space. We remark that the inverse of the heat operator is still not contained within this wider class. For further work on parabolic equations we refer to [**17**]. H. Kumano-go (to appear) has considered a certain class of operators in which the norms $\|\ \ \|_s$ are modified by replacing $1 + |\xi|^2$ in their definition by another function $\lambda(\xi)$ having certain similar properties but having different order of growth in different ξ-directions.

We mention, next, that many operators arising in the solution of differential equations are not pseudolocal. For instance, if L is a hyperbolic operator, say the wave operator

$$\partial^2/\partial t^2 - \sum \partial^2/\partial x^{i2},$$

the operator P mapping the Cauchy data, u and $\partial u/\partial t$, at time $t = 0$ to their values at time T is not pseudolocal. Very recently, in a very interesting way, Hörmander [**11**] has introduced a wider class of operators, no longer pseudolocal, in order to study hyperbolic equations. These have the form

$$Pu(x) = (2\pi)^{-n} \int e^{i\phi(x,\xi,z)} p(x, \xi, z)u(z)\,dz\,d\xi$$

where p is a symbol satisfying (12′) but where ϕ is a *nonlinear* function satisfying certain conditions—as a special case ϕ may be homogeneous in ξ of degree one for $|\xi| \geq 1$ and, when restricted to the sphere bundle $|\xi| = 1$, it has no zero which is a critical point in (y, ξ) if x is fixed or in (x, ξ) if y is fixed. Although the operator P is no longer pseudolocal it is possible to study the "spread" of the singular support. With the aid of such operators one obtains close approximations to solutions of the initial value problem for hyperbolic equations—employing a method which is a familiar one in geometric optics.

It is clear that still more general operators will play an important role in the future developments.

In this report we have had, of necessity, to confine ourselves only to the simplest and most basic properties of pseudodifferential operators. The treatment has been based, to a great extent, on a part of the paper [**10**] by Hörmander. We have omitted any discussion of boundary value problems, as developed by Vishik and Eskin and others. References to them may be found in the other papers in the bibliography which we have kept rather small. In [**9**] there is, in particular, a reduction of elliptic boundary value problems to elliptic pseudodifferential problems on the boundary. Many interesting applications have been given by Calderón, see [**3–5**]. In [**16**] Seeley has derived an analytic formula for the index of elliptic operators with the aid of pseudodifferential operators; analogous formulas were obtained independently by Hörmander and Kotake. Calderón [**6**] has also described an analytic expression for the index. Many of the recent results can be found summarized in [**5**], [**17**] and [**18**].

Appendix

As an illustration of the use of pseudodifferential operators we shall present a simple treatment of the initial value problem for hyperbolic differential equations (see Calderón [**3**]). For simplicity we shall confine ourselves to a first order system of N equations for N unknowns $u = (u^1, \ldots, u^N)$ on a compact n-dimensional manifold M, with u depending also on a time variable, t:

$$Lu = D_t + Pu = f \tag{19}$$

with initial data

$$u(x, 0) = 0. \tag{20}$$

Here $D_t = (1/i)\,\partial/\partial t$ and, for each t, P is assumed to be a first order differential operator on M whose symbol is an $N \times N$ matrix depending also on t as a parameter, $0 \leq t \leq T$, and with C^∞ coefficient matrices:

$$p(t, x, \xi) = \sum_1^n a_j(t, x)\xi_j + a_0(t, x).$$

(The argument given here extends also to an operator P which for each t is a pseudodifferential operator acting on N-vectors on M, with symbol $p(t, x, \xi) \in S^1$.)

It is a known fact that in order for the initial value problem to be well posed it is necessary that all eigenvalues of the leading symbol matrix

$$a(t, x, \xi) = \sum_1^n a_j(t, x)\xi_j$$

be *real*, for every (x, ξ). This is, however, not sufficient; a sufficient condition is that for every $\xi \neq 0$ the eigenvalues be *real* and *distinct*. The result that we present here is a more general one. We shall assume that for every (x, ξ) with $|\xi| = 1$, i.e. on the sphere bundle, the matrix $a = a(t, x, \xi)$ is similar to a hermitian matrix, i.e. there exists a nonsingular $N \times N$ matrix $r = r(t, x, \xi)$ such that

$$rar^{-1} \text{ is hermitian} \tag{21}$$

—this clearly implies that the eigenvalues of a are real. But we shall require more: that the matrix $r(t, x, \xi)$ can be chosen so as to vary *smoothly* (C^∞) with respect to t, x, ξ, and so that $r(t, x, \xi)$, when extended as homogeneous in ξ of degree zero for $|\xi| > 1$, may be regarded as a symbol in S^0 for each t. If the eigenvalues of a are real and *distinct* for each $\xi \neq 0$ then it is not difficult to see that such a matrix r may be found. If, however, the eigenvalues are not distinct a *smooth* symmetrizer r may not exist.

Assuming the existence of such a smooth symmetrizer $r(t, x, \xi)$ we shall prove

THEOREM 9. *For every $f(t, x) \in C^\infty(M \times [0, T])$ there exists a unique C^∞ solution of the initial value problem* (19), (20).

The proof of Theorem 9 is based on the following estimate in terms of our norms $\| \quad \|_s$ on M. If we merely wanted a solution in L_2 it would suffice to take $s = 0$; to obtain a C^∞ solution it is convenient to keep s arbitrary.

BASIC INEQUALITY. For every real s there is a constant $C(s)$ (depending also on T) such that if $u(x, 0) = 0$ then

$$\|u(\cdot, t)\|_s \leq C(s) \int_0^t \|Lu(\cdot, \tau)\|_s d\tau, \quad 0 \leq t \leq T. \tag{22}$$

Inequality (22) yields the uniqueness of the solution.

PROOF OF (22). Let $r^*(t, x, \xi) = \overline{r(t, x, \xi)^t}$ be the adjoint matrix to r. From (21) we see that $r^*rar^{-1} = r^*ra$ is hermitian. The matrix r^*r is positive definite hermitian. For each t let $Q = Q(t)$ be the corresponding pseudodifferential operator acting on N-vectors on M. By Theorem 5, Q is an operator of order zero, and differs from its adjoint in H_s by an operator of order -1. By adding to Q an operator of order -1 we may suppose, in virtue of Gårding's inequality (18), that for each t,

$$\operatorname{Re} \langle Q(t)u, u \rangle_s \geq c \langle u, u \rangle_s \tag{23}$$

for some positive constant c.

With A denoting the differential operator $A = \sum_1^n a_j D_j$, we obtain by direct calculation

(24) $$D_t\langle Qu, u\rangle_s = \langle QLu, u\rangle_s + \langle Qu, Lu\rangle_s - \langle QAu, u\rangle_s - \langle Qu, Au\rangle_s - i\langle Q_t u, u\rangle_s - \langle Qa_0 u, u\rangle_s - \langle Qu, a_0 u\rangle_s$$

where $Q_t = \partial/\partial t(Q)$ is, for each t, a pseudodifferential operator of order zero. Since the leading part r^*ra of the symbol of QA is hermitian it follows from Theorem 5 that for each t, the operators $QA - A^*Q^*$ and $Q^*A - A^*Q$ are of order zero (here * refers to the adjoint as pseudodifferential operators) so that, as one readily verifies

$$|\mathrm{Im}\langle QAu, u\rangle_s| + |\mathrm{Im}\langle Qu, Au\rangle_s| \leq C\|u\|_s^2$$

for some constant C. Here, and in the following, C is used to denote various constants which do not depend on u. Consequently if we now integrate (24) with respect to t, and take imaginary parts, we find, since Q is of order zero

$$\tfrac{1}{2}\mathrm{Re}\langle Qu, u\rangle_s(t) \leq -\mathrm{Im}\int_0^t (\langle QLu, u\rangle_s + \langle Qu, Lu\rangle_s)(\tau)d\tau + C\int_0^t \|u(\cdot, \tau)\|_s^2 d\tau$$

$$\leq C\int_0^t (\|u(\cdot, \tau)\|_s\|Lu(\cdot, \tau)\|_s + \|u(\cdot, \tau)\|_s^2)d\tau.$$

From (23) we obtain

(25) $$\|u(\cdot, t)\|_s^2 \leq C\int_0^t (\|u(\cdot, \tau)\|_s\|Lu(\cdot, \tau)\|_s + \|u(\cdot, \tau)\|_s^2)d\tau;$$

it is now a simple exercise to derive (22) from (25).

PROOF OF THEOREM 9. Taking

$$\{u, v\} = \int_0^T \langle u(\cdot, t), v(\cdot, t)\rangle dt,$$

as the L_2 scalar product on $M \times [0, T]$, where $\langle , \rangle$ denotes L_2 scalar product on M, the formal adjoint L^* of L is defined by the relation

$$\{Lu, v\} = \{u, L^*v\} \quad \text{if } u(x, 0) = v(x, T) = 0$$

and is again an operator of the same type as L; in fact $L - L^*$ is of order zero. The adjoint boundary condition is $u = 0$ on $t = T$, and we may therefore suppose that the analogue of (22) also holds for the adjoint problem, i.e., if $v(x, T) = 0$,

(22′) $$\|v(\cdot, t)\|_s \leq C(s)\int_t^T \|L^*v(\cdot, \tau)\|_s d\tau.$$

As a consequence of (22′) we have

(26) $$\int_0^T \|v(\cdot, t)\|_s^2 dt \leq C^2T^2\int_0^T \|L^*v(\cdot, \tau)\|_s^2 d\tau.$$

Now the space $L_2(H_s)$ of square integrable functions in t with values in H_s may be regarded as the dual (with respect to the bilinear relation $\{\ \ \}$) of the space of square integrable functions in t with values in H_{-s}. Hence, if f is given, then, for any C^∞ function $v(x, t)$ with $v(x, T) = 0$, we have

$$\begin{aligned} |\{v,f\}| &\leq \int_0^T \| f(\cdot,\tau)\|_{-s} \|v(\cdot,\tau)\|_s d\tau \\ &\leq [\int \| f(\cdot,\tau)\|_{-s}^2 d\tau]^{1/2} [\int \|v(\cdot,\tau)\|_s^2 d\tau] \\ &\leq C^2 T [\int \| f\|_{-s}^2 d\tau]^{1/2} [\int \|L^* v\|_s^2 d\tau]^{1/2} \end{aligned}$$

in virtue of (26). Thus the linear functional $\{v,f\}$ of v is continuous in the norm $[\int \|L^*v\|_s^2 d\tau]^{1/2}$ and hence, by the usual representation theorem, admits the representation

$$\{v,f\} = \int_0^T \langle L^*v, u\rangle dt \tag{27}$$

where $u(x, t)$ is a square integrable (vector) function of t with values in H_{-s}. The function u is therefore a distribution solution of

$$Lu = f \tag{28}$$

belonging to $L_2(H_{-s})$. Solving for $D_t u$ in the differential equation (28) we see that $D_t u \in L_2(H_{-s-1})$ and, differentiating successively, we find that $D_t^k u \in L_2(H_{-s-k})$. If we now choose s to be a large negative integer we see that the resulting function u has square integrable derivatives of large order in $M \times (0, T)$. Hence u is a smooth function in $M \times [0, T]$. From the equation (27) it follows finally, by Green's theorem, that on $t = 0$, $\langle v, u\rangle = 0$ for all C^∞ functions v vanishing at $t = T$. Hence we conclude that $u(x, 0) = 0$. Thus u is a solution of the initial value problem. Since, by (22), the solution is unique, and since we can take $-s$ arbitrarily large, it follows that the solution belongs to C^∞ in $M \times [0, T]$.

Friedrichs and Lax have treated the initial value problem for a more general class of equations and initial and end surfaces; their work is contained in Friedrichs [7, Part 2, §22].

BIBLIOGRAPHY

1. M. F. Atiyah and I. M. Singer, *The index of elliptic operators*, I. Ann. of Math., (1968) 484–530.

2. A. P. Calderón and A. Zygmund, *Singular integral operators and differential equations*, Amer. J. Math. **79** (1957), 289–309.

3. A. P. Calderón, *Integrales singulares y sus applicaciones a ecuaciones diferenciales hyperbolicas*, Cursos y seminorias de matematica, Fasc. 3, Univ. of Buenos Aires (1960).

4. ———, *Existence and uniqueness theorems for systems of partial differential equations*, Sympos. Fluid Dynamics, Univ. of Maryland, Inst. for Fluid Dynamics, 1961.

5. ———, *Singular integrals*, Bull. Amer. Math. Soc. **3** (1966), 427–465.

6. ———, *The analytic calculation of the index of elliptic equations*, Proc. Nat. Acad. Sci. (1967), 1193–1194.

7. K. O. Friedrichs, *Lecture notes on pseudodifferential operators*, New York University, 1968.

8. L. Hörmander, *Pseudodifferential operators*, Comm. Pure Appl. Math. **18** (1965), 501–517.

9. ———, *Pseudodifferential operators and nonelliptic boundary problems,* Ann. of Math. **83** (1966), 129–209.

10. ———, "Pseudo-differential operators and hypoelliptic equations" in *Singular integrals*, Proc. Sympos. Pure Math., Vol. 10, Amer. Math. Soc., Providence, R.I., 1968.

11. ———, *The spectral function of an elliptic operator*, Acta Math. **121** (1968), 193–218.

12. J. J. Kohn and L. Nirenberg, *An algebra of pseudodifferential operators,* Comm. Pure Appl. Math. **18** (1965), 269–305.

13. P. D. Lax and L. Nirenberg, *On stability for difference schemes; a sharp form of Gårding's inequality,* Comm. Pure Appl. Math. **19** (1966), 473–492.

14. R. Palais, *Seminar on the Atiyah-Singer index theorem,* Ann. of Math. Studies No. 47, Princeton Univ. Press, Princeton, N.J., 1965.

15. R. T. Seeley, *Regularization of singular integral operators on compact manifolds,* Amer. J. Math. **83** (1961), 265–275.

16. ———, "Complex powers of an elliptic operator" in *Singular integrals,* Proc. Sympos. Pure Math., Vol. 10, Amer. Math. Soc., Providence, R.I., 1968, 288–307.

17. *Singular integrals,* Proc. Sympos. Pure Math., Vol. 10, Amer. Math. Soc., Providence, R.I., 1968.

18. C.I.M.E. course on pseudodifferential operators, Stresa, Italy, August 1968, Proceedings to appear.

COURANT INSTITUTE OF MATHEMATICAL SCIENCES

ON THE CALCULUS OF SYMBOLS FOR PSEUDO-DIFFERENTIAL OPERATORS

BENT E. PETERSEN[1]

Let M be a smooth paracompact manifold and let E, F and G be smooth complex vector bundles on M. If P is a linear differential operator from E to F, if s is a smooth section of E and if g is a smooth real-valued function on M then $e^{-i\lambda g}P(e^{i\lambda g}s)$ is a polynomial in λ. In [2] L. Hörmander studies a class of continuous linear operators from compactly supported smooth sections of E to smooth sections of F for which the expression $e^{-i\lambda g}P(e^{i\lambda g}s)$ may be developed in an asymptotic series in descending powers of λ. This class in particular contains the differential operators, the parametrices of elliptic differential operators, and the operators which admit smooth kernels. We will call these operators pseudo-differential operators.

In [2] Hörmander obtains the calculus relative to coordinates for the coefficients of the asymptotic series under natural operations on the pseudo-differential operators (e.g. composition, inverse, transpose). In the case of differential operators the coefficient of the highest order term in the polynomial $e^{-i\lambda g}P(e^{i\lambda g}s)$ is just the symbol $\sigma(P)$ of P evaluated at dg. Therefore we also call the coefficients of the asymptotic series of a pseudo-differential operator P the symbols of P. Our purpose here is to describe a coordinate free calculus for these symbols as obtained in [**4**]. Other treatments of similar results may be found in W. Shih [**5**] and A. P. Calderón [**1**].

I wish to thank I. M. Singer for suggesting this problem and for his help.

1. **Pseudo-differential operators.** A sequence (z_n), $n = 0, 1, 2, \ldots$, of *distinct* complex numbers will be called a sequence of exponents if

$$\operatorname{Re} z_0 \geq \operatorname{Re} z_1 \geq \ldots \qquad \text{and} \qquad \limsup \operatorname{Re} z_n = -\infty$$

where Re denotes the real part. Then a pseudo-differential operator is a continuous linear map

$$P:\Gamma_c^\infty(E) \to \Gamma^\infty(F)$$

such that there is a sequence of exponents (z_n) such that the following conditions are satisfied:

If $s \in \Gamma_c^\infty(E)$ and if K is a compact subset of $\Gamma^\infty(1)$ such that whenever g is in K then $dg \neq 0$ on supp s, then there is an asymptotic expansion

$$e^{-i\lambda g}P(e^{i\lambda g}s) \sim \sum_{k=0}^{\infty} P_k(s, g)\lambda^{z_k}$$

[1] Research supported in part by the National Research Council of Canada.

as $\lambda \to +\infty$, uniformly for $g \in K$. That is, for each integer $N > 0$ the set

$$\{\lambda^{-z_N}[e^{-i\lambda g}P(e^{i\lambda g}s) - \sum_{k<N} P_k(s,g)\lambda^{z_k}] : \lambda \geq 1, g \in K\}$$

is a bounded subset of $\Gamma^\infty(F)$. Here $\Gamma^\infty(G)$ denotes the set of smooth sections of G and $\Gamma_c^\infty(G)$ consists of those sections in $\Gamma^\infty(G)$ which are compactly supported. 1 is the trivial *real* line bundle on M, dg is the exterior derivative of g and supp s is the support of s. The sets $\Gamma^\infty(G)$ and $\Gamma_c^\infty(G)$ are given the usual Schwartz topologies.

The set $\mathscr{P}(E, F)$ of pseudo-differential operators is a left module over the ring of smooth functions on M. The asymptotic expansion associated to a pseudo-differential operator is unique, and if the expansion is identically zero we say that P has order $-\infty$. The operators of order $-\infty$ form a submodule $\mathscr{P}_{-\infty}(E, F)$ of $\mathscr{P}(E, F)$ and consist precisely of the continuous linear operators from $\Gamma_c^\infty(E)$ to $\Gamma^\infty(F)$ which have smooth kernels.

2. **The symbols.** Let $J^n(G)$ denote the nth jet bundle of G and j_n the nth jet extension map [3]. Let T^* be the cotangent bundle of M. Let

$$\psi_n : J^{n+1}(1) \to J^n(T^*)$$

be the unique morphism associated to the differential operator

$$j_n d : \Gamma^\infty(1) \to \Gamma^\infty(J^n(T^*)),$$

i.e. if g is a smooth function then $\psi_n j_{n+1}(g) = j_n(dg)$. Then the sequence

$$0 \to 1 \to J^{n+1}(1) \xrightarrow{\psi_n} J^n(T^*)$$

is exact where the first morphism maps a constant to the $n + 1$ jet of the corresponding constant section. It follows that the image Ψ^n of ψ_n is a smooth vector subbundle of $J^n(T^*)$ and we obtain a short exact sequence

$$0 \to 1 \to J^{n+1}(1) \to \Psi^n \to 0$$

which is split by the natural projection of $J^{n+1}(1)$ onto 1.

We denote by Ψ_0^n the open submanifold of Ψ^n complementary to the kernel of the natural projection of Ψ^n onto T^*. in particular $\Psi_0^0 = T^* - (0)$. Let

$$p_n : \Psi_0^n \to M$$

be the projection of the bundle Ψ_0^n. Then if $n \geq 0$ is an integer and z is a complex number we define

$$\mathrm{Smbl}_z^n(E, F)$$

to be the set of morphisms $\sigma : p_n^*(J^n(E)) \to p_n^* F$ which are positively homogeneous of degree z over the fibres of Ψ_0^n. Functions on M lift to functions on Ψ_0^n which are positively homogeneous of degree zero. Hence $\mathrm{Smbl}_z^n(E, F)$ has a natural structure as a module over the ring of smooth functions on M. We define

$\mathrm{Smbl}_z(E, F) = \mathrm{Smbl}_z^0(E, F)$. Then the definition of $\mathrm{Smbl}_z(E, F)$ is the same as the one given in [3].

THEOREM. [4], [5]. *If $P \in \mathscr{P}(E, F)$ with exponents (z_k) and if n_k is defined to be the largest integer such that $0 \leq n_k \leq \mathrm{Re}(z_0 - z_k)$ then there exists a unique*

$$\sigma_k(P) \in \mathrm{Smbl}_{z_k}^{n_k}(E, F)$$

such that if $s \in \Gamma_c^\infty(E)$, $g \in \Gamma^\infty(1)$ and $dg \neq 0$ on supp s *then*

$$P_k(s, g) = \sigma_k(P) \cdot j_{n_k}(dg) \cdot j_{n_k}(s).$$

Now let D be a covariant derivative on E and let ∇ be a covariant derivative on T^*. Then the pair (D, ∇) induces kth order differential operators

$$D^{(k)} : \Gamma^\infty(E) \to \Gamma^\infty(\mathscr{S}^k T^* \otimes E)$$

which are kth total differentials in the sense that the associated morphisms

$$J^k(E) \to \mathscr{S}^k T^* \otimes E$$

are splittings of the exact jet bundle sequences

$$0 \to \mathscr{S}^k T^* \otimes E \to J^k(E) \to J^{k-1}(E) \to 0$$

(see [3]). The exterior derivative d in particular is a covariant derivative on 1 and hence the pair (d, ∇) induces kth total differentials

$$\partial^{(k)} : \Gamma^\infty(1) \to \Gamma^\infty(\mathscr{S}^k T^*).$$

These total differentials are related by a Leibnitz rule [4]

$$D^{(n)}(fs) = \sum_{k=0}^{n} \binom{n}{k} \partial^{(k)}(f) \cdot D^{(n-k)}(s)$$

where $f \in \Gamma^\infty(1)$, $s \in \Gamma^\infty(E)$, and where the indicated product is the symmetric product.

It follows then that relative to a pair of covariant derivatives (D, ∇) we have an induced map

$$\mathrm{Smbl}_z^n(E, F) \to \mathrm{Smbl}_z(E, F)$$

which maps $\sigma \in \mathrm{Smbl}_z^n(E, F)$ to the composition

$$T^* - (0) \to \Psi_0^n \xrightarrow{\sigma} \mathrm{Hom}(J^n(E), F) \to \mathrm{Hom}(E, F)$$

where the first map comes from the splittings induced by $\partial^{(k)}$, $k = 2, 3, \ldots, n + 1$, and the last map comes from the splittings induced by $D^{(k)}$, $k = 1, 2, \ldots, n$. By applying this map to the symbols of a pseudo-differential operator we obtain a linear map

$$\tau : \mathscr{P}(E, F) \to \mathscr{F}(E, F)$$

where $\mathscr{F}(E, F)$ is the module of all formal sums

$$\sum_{k=0}^{\infty} \tau_k$$

where $\tau_k \in \mathrm{Smbl}_{z_k}(E, F)$, (z_k) being any sequence of exponents.

THEOREM. [4]. *The sequence*

$$0 \to \mathscr{P}_{-\infty}(E, F) \to \mathscr{P}(E, F) \xrightarrow{\tau} \mathscr{F}(E, F) \to 0$$

is exact for any choice of (D, ∇).

3. **The calculus.** Let ∇ be a covariant derivative on T^*, D_E a covariant derivative on E, and D_F a covariant derivative on F. Let $D_F^{(n)}$ be the total differentials induced by the pair (D_F, ∇). Let $P \in \mathscr{P}(E, F)$ with exponents (z_k), $Q \in \mathscr{P}(F, G)$ with exponents (w_k), $f \in \Gamma_c^\infty(1)$ and let $R = Q f P \in \mathscr{P}(E, G)$. Let

$$\tau(P) = \sum_{k=0}^{\infty} \tau_k(P) \qquad \text{relative to } (D_E, \nabla),$$

$$\tau(Q) = \sum_{k=0}^{\infty} \tau_k(Q) \qquad \text{relative to } (D_F, \nabla),$$

$$\tau(R) = \sum_{k=0}^{\infty} \tau_k(R) \qquad \text{relative to } (D_E, \nabla).$$

THEOREM. [4]. *We have*

$$\sum_{k=0}^{\infty} \tau_k(R) = \sum_{n,j,k=0}^{\infty} \frac{i^{-n}}{n!} \tau_k(Q)^{(n)} \circ [D_F^{(n)}{}_* f \tau_j(P)]$$

where equality of these formal sums means equality of the parts homogeneous of the same degree and where

$$D_F^{(n)}{}_* f \tau_j(P) \in \mathrm{Smbl}_{z_j}(E, \mathscr{S}^n T^* \otimes F)$$

is defined to be the image of the composition $D_F^{(n)} \circ f \tau_j(P)$ *under the map*

$$\mathrm{Smbl}_{z_j}^n(E, \mathscr{S}^n T^* \otimes F) \to \mathrm{Smbl}_{z_j}(E, \mathscr{S}^n T^* \otimes F)$$

induced by the pair (D_E, ∇), *and where*

$$\tau_k(Q)^{(n)} \in \mathrm{Smbl}_{w_k - n}(\mathscr{S}^n T^* \otimes F, G)$$

is the nth Fréchet derivative of $\tau_k(Q)$ *along the fibers of* T^*.

Let P be as above. Let Ω be the volume bundle of M, i.e. the trivial line bundle whose smooth sections are the smooth densities on M, and let $E' = \mathrm{Hom}(E, \Omega)$. Then there is a unique $P' \in \mathscr{S}(F', E')$ such that

$$\textstyle\int_M \langle u, Ps \rangle = \int_M \langle P'u, s \rangle$$

if $u \in \Gamma_c^\infty(F')$ and $s \in \Gamma_c^\infty(E)$, [2]. Moreover P' has the same exponents as P. In addition to the covariant derivatives above, choose a covariant derivative $D_{F'}$, on F', and let

$$\tau(P') = \sum_{k=0}^{\infty} \tau_k(P') \qquad \text{relative to } (D_{F'}, \nabla).$$

THEOREM. [4]. *We have*

$$\sum_{k=0}^{\infty} \tau_k(P') = \sum_{j,n=0}^{\infty} \frac{i^{-n}}{n!} (D_E^{(n)})' * (\tau_j(P)^{(n)})'$$

where $*$ is defined as before but relative to $(D_{F'}, \nabla)$, where $(D_E^{(n)})'$ is the transpose of the differential operator $D_E^{(n)}$, and where $(\tau_j(P)^{(n)})'$ is defined as the composition

$$T^* - (0) \xrightarrow{-1} T^* - (0) \xrightarrow{\tau} \operatorname{Hom}(\mathscr{S}^n T^* \otimes E, F) \xrightarrow{A} \operatorname{Hom}(F', \mathscr{S}^n T \otimes E')$$

where $\tau = \tau_j(P)^{(n)}$ and A is the map $\operatorname{Hom}(\cdot, \Omega)$.

If $E = F = G = 1$, $D_E = D_F = D_{F'} = d$, and if M is an open submanifold of Euclidean space and ∇ is the covariant derivative on T^* which kills the exterior derivatives of the coordinate functions, then the formulae given in this section reduce to those given by Hörmander [2].

REFERENCES

1. A. P. Calderón, *Singular integrals,* Bull. Amer. Math. Soc. **72** (1966), 427–465.

2. L. Hörmander, *Pseudo-differential operators,* Comm. Pure Appl. Math. **18** (1965), 501–517.

3. R. S. Palais, et al., *Seminar on the Atiyah-Singer index theorem,* Princeton Univ. Press, Princeton, N.J., 1965.

4. B. E. Petersen, Ph.D. Thesis, Massachusetts Institute of Technology, Cambridge, Mass., 1968.

5. W. Shih, *On the symbol of a pseudo-differential operator,* Bull. Amer. Math. Soc. **74** (1968), 657–659.

OREGON STATE UNIVERSITY

UNE REMARQUE SUR LES OPERATEURS INTEGRAUX

WEISHU SHIH

Le but de cette remarque est d'associer à une certaine classe d'équations intégrales dont le noyau $k_t(x, y)$ dépend d'un parametre $t \in B$, un invariant topologique $e(k)$ qui se trouve dans le groupe $K^1(B)$. Cet invariant peut être considéré comme la contre partie de l'indice d'une famille d'opérateurs différentiels elliptiques qui se trouve dans $K^0(B)$. Nous étudierons plus tard le calcul de cet invarient.

Soit X un fibré de fibre X_0 une variété différentiable orientée compacte de base un espace topologique B et de groupe structure un sous groupe des difféomorphismes respectant l'orientation:

$$f: X \to B.$$

On notera $X_t = f^{-1}(t)$, $t \in B$ les fibres de X. Une métrique riemannienne est donnée sur l'espace tangent le long des fibres $T_B(X)$ de X. Soit

$$p: E \to X$$

un fibré vectoriel complexe sur X munie d'une structure hermitienne. Désignons par p_i^*E $i = 1, 2$ le fibré induit sur le produit fibré $X\ x_B\ X$ par les projections canoniques $p_i: X\ x_B X \to X$. Alors une équation intégrale est donnée par son noyau

$$k \in \operatorname{Hom}(p_1^*E, p_2^*E),$$

qui est un morphisme de fibrés vectoriel bas Xx_BX. Notons

$$C_B^\infty(E) \to B$$

le fibré dont la fibre au dessus d'un point $t \in B$ s'identifie à l'espace des sections: $C^\infty(E_t)$ du fibré vectoriel E_t (restriction de E à X_t). Alors k définit un morphisme de fibré

$$\tilde{k}: C_B^\infty(E) \to C_B^\infty(E)$$

par

$$\tilde{k}(g)(y) = \textstyle\int_{X_t} k_t(x, y)(g(x))d\omega_t$$

où $g \in C^\infty(E_t)$ et $d\omega_t$ est la forme d'orientation de X_t.

La théorie classique du déterminant de Fredholm peut se généraliser comme suit. Désignons par

$$k \circ \Delta \in \operatorname{End}(E) = \Delta^* \operatorname{Hom}(p_1^*E, p_2^*E)$$

la section du fibré d'endomorphismes de E obtenue par la composition de k avec l'application diagonale $\Delta: X \to Xx_BX$, et soit $(k \circ \Delta)^n$ sa n-ième puissance dans End(E). Rappelons que la fonction trace "tr" est bien définie sur End(E), on peut ainsi considérer les fonctions continues sur B définies par

$$\sigma_n(t): t \to \int_{X_t} \mathrm{tr}(k \circ \Delta)^n d\omega_t \qquad n = 2, 3 \ldots.$$

Alors le déterminant de Fredholm de k est défini par la fonction

$$\det_F(k)(t) = \sum_{n=0} \frac{(-1)^n}{n!} A_n(t)$$

où $A_0(t) = 1$, et pour $n \geq 1$

$$A_n(t) = \det \begin{pmatrix} 0, n-1, & 0 \ldots & 0 & 0 \\ \sigma_2(t), 0 & n-2, \ldots & 0 & 0 \\ & & 0 & 1 \\ \sigma_n(t), \sigma_{n-1}(t), & & \sigma_2(t), & 0 \end{pmatrix}$$

On a alors le

THEOREME DE FREDHOLM. *La condition nécessaire et suffisante pour que l'opérateur integral*

$$1 + \tilde{k}: C_B^\infty(E) \to C_B^\infty(E)$$

soit un isomorphisme est que: $\det_F(k)(t) \neq 0$ *pour tout* $t \in B$.

EXEMPLE. Soit H l'opérateur de la projection sur les formes harmoniques le long des fibres de X. Alors de déterminant de Fredholm de son noyau, qui est une forme double, n'est pas nul. On peut en particulier, prendre l'espace de base B comme l'espace des structures riemanniennes sur une variété compacte donnée.

Donnons nous un noyau k dont le déterminant de Fredholm ne soit pas nul, alors la restriction de l'isomorphisme $1 + \tilde{k}$ à chaque fibre est un opérateur du type: identité plus un opérateur compact, donc définit un élément de $K^1(B)$ de l'espace de base B

$$e(k) \in K^1(B)$$

que l'on peut interpréter comme un analogue de l'indice pour un opérateur intégral de noyau k. Les propriétés élémentaires e.g. somme de Whitney, se vérifient aussitôt et nous étudierons plus tard les relations avec l'autre invariant.

REFERENCES

1. M. F. Atiyah and I. M. Singer, *The index of elliptic operators*: I, III, Ann. of Math. (2) **87** (1968), 484–530, 546–604.

2. M. F. Atiyah and F. Hirzebruch, *Vector bundles and homogeneous spaces,* Proc. Sympos. Pure Math., Vol. 3, American Mathematical Society, Providence, R.I., 1961, pp. 7–38.

3. R. S. Palais, *On the homotopy type of certain groups of operators,* Topology **3** (1965), 1–9.

4. G. De Rham, *Variétés différentiables. Formes, courants, formes harmoniques,* Actualités Sci. Ind., No. 1222 = Publ. Inst. Math. Univ. Nancago III, Hermann et Cie, Paris, 1955.

5. F. Smithies, *The Fredholm theory of integral equations,* Duke Math. J. **8** (1941), 107–130.

6. Weishu Shih, *Fiber cobordism and the index of a family of elliptic differential operators,* Bull. Amer. Math. Soc. **72** (1966), 984–991.

INSTITUTE FOR ADVANCED STUDY

A DIFFERENTIAL GEOMETRIC APPROACH TO CHARACTERISTICS

ROBERT B. GARDNER[1]

In this paper I will consider the C^∞-theory of nonlinear systems of partial differential equations in the spirit of the geometric approach of the French school of the late nineteenth century [2] [3]. As such we view a system of rth order p.d.e. for maps

$$f: M \to N$$

between manifolds M and N, as a differential manifold Σ together with a one-to-one differentiable map

$$i: \Sigma \to J^r(M, N),$$

where $J^r(M, N)$ is the manifold of all r-jets of maps from M to N.

As an example let me consider the parabolic Monge-Ampere equation for maps

$$f: R^2 \to R.$$

We will introduce coordinates x, y in the source, z in the target, $p = \partial z/\partial x$, $q = \partial z/\partial y$ in the fibers of the 1-jets, and $r = \partial^2 z/\partial x^2$, $s = \partial^2 z/\partial x \partial y$, $t = \partial^2 z/\partial y^2$ in the fibers of the 2-jets. As such the equation is given by the abstract manifold R^7, which we coordinatize by (x, y, z, p, q, u, v) and the one-to-one differentiable map

$$i: R^7 \to J^2(R^2, R)$$

given by

$$i(x, y, z, p, q, u, v) = (x, y, z, p, q, f_1 + vu^2, f_2 + vu, f_3 + v)$$

where $f_i = f_i(x, y, z, p, q)$ for $1 \leq i \leq 3$. Classically this is just the equation

$$(r - f_1)(t - f_3) - (s - f_2)^2 = 0.$$

This example shows that we may not assume that the mapping i has maximal rank and still include all the classical equations of p.d.e., and is the reason why we have not assumed i to be an imbedding.

A *solution* to a system

$$i: \Sigma \to J^r(M, N)$$

is a map

$$f: M \to N$$

[1] National Science Foundation Grants GP-8735 and GP-5455.

such that the r-graph of f lies on the image of Σ. The r-graph is the mapping

$$j^r(f): M \to J^r(M, N)$$

defined by

$$j^r(f)(p) = j^r_p(f)$$

where $j^r_p(f)$ is the r-jet of f at p.

The basic idea of the geometric method is to replace the problem of finding maps f, by the problem of finding submanifolds of Σ whose images under the mapping i are r-graphs of solutions. That is to find a manifold M' and a mapping σ such that

$$\sigma: M' \to \Sigma$$

and

$$i \circ \sigma(M') = j^r(f)(M)$$

for some mapping

$$f: M \to N.$$

If $\sigma: M' \to \Sigma$ is such a submanifold we will say that it gives rise to a solution. A foundational result in the theory is the following:

THEOREM 1. *There is a pfaffian system $\Omega^r(M, N)$ defined on $J^r(M, N)$ such that a submanifold of* $\dim M$

$$\sigma: M' \to \Sigma$$

gives rise to a solution if and only if

(1) $$(i \circ \sigma)^* \Omega^r(M, N) = 0$$

and

(2) $$\sigma_* T(M') \cap V_F(\Sigma) = 0$$

where

$$V_F(\Sigma) = \text{kernel}\,(\pi^{r,r-1}{}_* \circ i_*)$$

and $\pi^{r,r-1}$ is the natural projection from the r-jets to the $(r-1)$ *jets.*

Specifically

$$\Omega^r(M, N)|_{j^r_p(f)} = \pi^{r,r-1*}[j^{r-1}(f)_* T_p(M)]^\perp.$$

In our example the 1-graph of a function has equations

$$z = f, \qquad p = \partial f/\partial x, \qquad q = \partial f/\partial y$$

and hence the lift of the annihilator of its tangent space gives

$$(3)\quad \Omega^2(R^2, R) = \pi^{2,1*}\begin{cases} dz - \partial f/\partial x\, dx - \partial f/\partial y\, dy \\ dp - \partial^2 f/\partial x^2 dx - \partial^2 f/\partial x\,\partial y\, dy \\ dq - \partial^2 f/\partial x\,\partial y\, dx - \partial^2 f/\partial y^2 dy \end{cases} = \begin{cases} dz - pdx - qdy \\ dp - rdx - sdy. \\ dq - sdx - tdy \end{cases}$$

The vector field system for this example is simply

$$V_F(R^7) = \{\partial/\partial u, \partial/\partial v\}.$$

As a result of Theorem 1, the study of solutions of systems of p.d.e. is reduced to the study of integral manifolds of the pfaffian system

$$\Omega(\Sigma) = i^*\Omega^r(M, N)$$

which is generally not completely integrable.

In various papers culminating in his book *Les systems differentiels exterieurs et leurs applications geometrique*, E. Cartan developed a method for extending integral manifolds of pfaffian systems to larger integral manifolds. This is based on the notion of the Cauchy characteristic system of Σ which is defined as follows:

$$\text{Char}\,\Omega(\Sigma) = \{X \in \Gamma(\Sigma) | X \in \Omega(\Sigma)^\perp, X \lrcorner\, d\Omega(\Sigma) \subset \Omega(\Sigma)\}$$

where $\Gamma(\Sigma)$ is the module of vector fields on Σ, $\lrcorner$ is the adjoint of left exterior multiplication, and $d\Omega(\Sigma)$ is the set of all exterior derivatives of elements of $\Omega(\Sigma)$. This vector field system can be characterized as the annihilator of the minimal enveloping subspace of the closed ideal generated by $\Omega(\Sigma)$, it is completely integrable, and has the property that $\Omega(\Sigma)$ is invariant along its integral submanifolds [**4**]. As such if we consider initial data

$$\begin{array}{ccc} \Sigma & \xrightarrow{i} & J^r(M, N) \\ \Delta \uparrow & & \downarrow \alpha \\ N & & M \end{array}$$

which thus satisfies

$$(4)\qquad \alpha \circ i \circ \Delta \text{ is one-to-one and differentiable}$$

and

$$(5)\qquad \Delta^*\Omega(\Sigma) = 0,$$

and if $\{X^1, \ldots, X^k\}$ defines a completely integrable subsystem of $\text{Char}\,\Omega(\Sigma)$ transversal to $V_F(\Sigma) + \text{Im}\,\Delta_*$, then the mapping

$$\sigma \cdot N\ X\ R^k \to \Sigma$$

defined by

$$\sigma(u, t^1, \ldots, t^k) = \exp_{\Delta(u)} t^1 X^1 + \ldots + t^k X^k,$$

where the right-hand side is the flow of the vector field $t^1 X^1 + \ldots + t^k X^k$ at time 1 from the point $\Delta(u)$, is an integral manifold of dimension equal to the dimension of N plus k.

Unfortunately except in special cases this vector field system is always zero. For example in the case of the parabolic Monge-Ampere equation

$$\operatorname{Char} \Omega(R^7) = 0.$$

The theory of nonintegrable pfaffian systems is in its infancy, but examples indicate that the structure of integral manifolds depends heavily on the structure of various subsystems [**1**]. As such we introduce a generalization of Cauchy characteristic which depends on a subsystem.

DEFINITION. Let $J \subset \Omega(\Sigma)$ be a subsystem, then the J-characteristic system of Σ is the vector field system

$$\operatorname{Char}(\Omega(\Sigma), dJ) = \{X \in \Gamma(\Sigma) | x \in \Omega(\Sigma)^{\perp}, X \lrcorner dJ \subset \Omega(\Sigma)\}.$$

This vector field system may be characterized as the annhilator of the minimal enveloping subspace of the ideal which is 1-generated by $\Omega(\Sigma)$, and 2-generated by the exterior derivatives of elements of J.

The study of integral curves of such a system leads to the introduction of the J-focal system.

DEFINITION. Let $J \subset \Omega(\Sigma)$ be a subsystem, then the J-focal system of Σ is the pfaffian system

$$F(J) = \{Z \lrcorner dJ | \, Z \in V_F(\Sigma)\}.$$

It follows from the definitions that

$$F(J) \subset [\operatorname{Char}(\Omega(\Sigma), dJ)]^{\perp}.$$

The significance of this subsystem $F(J)$ is given by the following theorem.

THEOREM 2. *Let $i:\Sigma \to J^r(M, N)$ be a system of rth order p.d.e. where i_* has maximal rank, then for any submanifold*

$$\sigma : M' \to \Sigma$$

giving rise to a solution, we have

$$\sigma^* F(J) = \sigma^*[\operatorname{Char}(\Omega(\Sigma), dJ)]^{\perp}.$$

As a result an integral curve of $F(J)$ which is not an integral curve of $\operatorname{Char}(\Omega(\Sigma), dJ)$ can never be tangent to a submanifold giving rise to a solution.

Now I would like to consider the application of the theory of J-characteristics to the Cauchy problem. Since the results are local we will restrict to the case of nonlinear systems of p.d.e. for maps between euclidean spaces

$$i : \Sigma \to J^r(R^m, R^n)$$

with initial data

$$\Delta : R^{m-1} \to \Sigma$$

satisfying

(4) $\alpha \circ i \circ \Delta$ is one-to-one and differentiable

and

(5) $$\Delta^*\Omega(\Sigma) = 0.$$

If we introduce the mapping

$$j_0 : R^{m-1} \to R^{m-1}XR$$

by $j_0(u) = (u, 0)$, then the Cauchy problem is to find a submanifold

$$\sigma : R^{m-1}XR \to \Sigma$$

which gives rise to a solution and makes the diagram

$$\begin{array}{ccc} R^{m-1}XR & \xrightarrow{\sigma} & \Sigma \\ & {\scriptstyle j_0} \nwarrow \quad \nearrow {\scriptstyle \Delta} & \\ & R^{m-1} & \end{array}$$

commutative.

The Cauchy method for solving this problem is to find a vector field X on Σ such that

$$\sigma(u, t) = \exp_{\Delta(u)} tX$$

is a solution.

As such it is natural to ask for necessary and sufficient conditions on vector fields X, depending on the initial data, such that

$$\sigma(u, t) = \exp_{\Delta(u)} tX$$

gives rise to a solution.

A naive answer to this question is that

(6) $X \in \Omega(\Sigma)^{\perp}$ with $X \neq 0$

(7) for fixed t, $(\exp tX)^*\Omega(\Sigma) \subset \ker \Delta^*$

(8) X is transversal to $V_F(\Sigma) + \operatorname{im} \Delta_*$

are such conditions. The problem is that condition (7) is uncomputable. This may be overcome by the following notion.

DEFINITION A vector field on Σ is k-stable if

(9) $X \in \Omega(\Sigma)^{\perp}$ with $X \neq 0$

and

(10) the sequence of Lie derivatives

$$\Omega(\Sigma) \subset D_X^1\Omega(\Sigma) = \{\mathscr{L}_X\Omega(\Sigma)\} \subset \ldots \subset D_X^k\Omega(\Sigma) = \{\mathscr{L}_X D_X^{k-1}\Omega(\Sigma)\}$$

is locally finitely generated and noetherian.

If the sequence (10) is noetherian and hence stabilizes at some finite stage, we will write

$$D_X\Omega(\Sigma) = \bigcup_k D_X^k\Omega(\Sigma).$$

Let us note that any well behaved vector field in $\Omega(\Sigma)^\perp$ will be k-stable for some k.

With this preparation we may state the main theorem which solves the existence of solutions of the Cauchy problem.

THEOREM 3. *A k-stable vector field X defines a solution of the Cauchy problem if and only if*

(11) $$X \text{ is transversal to } V_F(\Sigma) + \operatorname{Im}\Delta_*$$

and

(12) $$\Delta^* D_X\Omega(\Sigma) = 0.$$

Let us remark that condition (10) is only a general position condition. The real effect of the initial data is completely separated out in condition (11) which is a condition completely determined by the initial data. Finally this condition is actually computable, and hence the solvability of the Cauchy problem is reduced to the study of k-stable vector fields.

If we make an additional assumption we may reduce the amount of computations needed in order to check condition (12). This assumption is the following:

(A) Every solution of Σ arises from the orbit of a J-characteristic vector field. That is, there is a vector field Y on every submanifold

$$\sigma: M \to \Sigma$$

which gives rise to a solution, such that $\sigma_* Y \in \operatorname{Char}(\Omega(\Sigma), dJ)$.

Under this assumption it suffices to study the k-stable vector fields in $\operatorname{Char}(\Omega(\Sigma), dJ)$ which are generally fewer than the set of all k-stable vector fields.

Finally I want to mention a way to intrinsically find subsystems J of $\Omega(\Sigma)$. The idea is to take advantage of the nonintegrability of $\Omega(\Sigma)$ in order to build tensors on $\Omega(\Sigma)$ and then let J be an invariant subspace associated to that tensor.

As an example let us consider a general second order p.d.e. for one unknown function in two independent variables,

$$F(x, y, z, p, q, r, s, t) = 0$$

given by

$$i: \Sigma \to J^2(R^2, R).$$

Then we may define a symmetric bilinear form

$$\langle\, , \rangle : \Omega(\Sigma) \otimes \Omega(\Sigma) \to C^\infty(\Sigma)$$

by

$$d\phi \wedge d\psi \wedge w^1 \wedge w^2 \wedge w^3 = \langle \phi, \psi \rangle v_7$$

where $\phi, \psi \in \Omega(\Sigma)$, v_7 is a volume element on Σ, and w^1, w^2, w^3 is any basis of $\Omega(\Sigma)$. This definition is well defined up to multiplication of the form by a nonzero scalar factor. If we choose the basis of $\Omega(\Sigma)$ given in equation (3), then the matrix of this form is

$$\begin{pmatrix} 0 & 0 & 0 \\ 0 & F_t & -\frac{1}{2}F_s \\ 0 & -\frac{1}{2}F_s & F_r \end{pmatrix}.$$

Thus if we let

$$\Delta = F_r F_t - \tfrac{1}{4}F_s^2$$

we recover the classical partitioning of these equations into types as invariants of the bilinear form $\langle\,,\rangle$.

If $\Delta = 0$, then the equation is parabolic and there is a unique two-dimensional maximal totally isotropic subsystem which we can let serve as J.

If $\Delta < 0$, then the equation is hyperbolic, and there are two two-dimensional maximal totally isotropic subsystems which we can let serve for J.

In any of these cases if

$$\Delta : R \to \Sigma$$

is a J-characteristic curve such that

$$\alpha \circ i \circ \Delta : R \to R^2$$

is one-to-one and differentiable, and has image defined as the locus of

$$\gamma(x, y) = 0$$

then

$$0 = F_r(\partial\gamma/\partial x)^2 + F_s(\partial\gamma/\partial x)(\partial\gamma/\partial y) + F_t(\partial\gamma/\partial y)^2$$

and hence this construction produces the Monge characteristic curves as a special case.

In the original example of the parabolic Monge-Ampere equations

$$J = \begin{cases} w^1 = dz - pdx - qdy \\ \pi^2 = dp - udq - (f_1 - f_2u)dx - (f_2 - f_3u)dy \end{cases}$$

and these equations are geometrically divided into two further types [**1**].

Type I characterized by $d\pi^2 \wedge d\pi^2 \wedge w^1 \wedge \pi^2 = 0$,

and

Type II characterized by $d\pi^2 \wedge d\pi^2 \wedge w^1 \wedge \pi^2 \neq 0$.

In the case of the Type I equations the J-characteristic system is easy to calculate and gives

$$\begin{aligned}\operatorname{Char}(\Omega(\Sigma), dJ) = \{\partial/\partial v, -\partial/\partial x + u\partial/\partial y + (uq - p)\partial/\partial z \\ - (f_1 - f_2 u)\partial/\partial p - (f_2 - f_3 u)\partial/\partial q \\ - (u^3 \partial f_3/\partial p + \partial f_1/\partial q - u(2\partial f_2/\partial q - \partial f_1/\partial p) \\ - u^2(2\partial f_2/\partial p - \partial f_3/\partial q))\partial/\partial u\}.\end{aligned}$$

The detailed knowledge of the two generators can be used to make estimates on how far a solution can be extended. As an example, if the coefficient of the term in $\partial/\partial u$ vanishes, which happens if

$$f_i = f_i(x, y, z), \qquad 1 \le i \le 3,$$

and we consider initial data along $x = 0$ given by

$$\Delta(y) = (0, y, z(y), p(y), q(y), u(y), v(y))$$

then it is easy to prove that a solution can only exist for

$$0 \le x < |1/u'(y)|.$$

As a concrete example we take

$$(r - x)t - s^2 = 0$$

with data

$$\Delta(y) = (0, y, y^2/2, y^2/2, y, y, 1).$$

This has the solution

$$z = 1/2(y^2/(1 - x)) + x^3/6$$

which cannot be extended beyond $x = |1/u'(y)| = 1/1 = 1$.

References

1. E. Cartan, *Les systems de Pfaff a cinq variables*, Ann. Ec. Normale **27** (1910), 109–192.

2. ———, *Les systems differentiels exterieurs et leurs applications geometrique*. Hermann, Paris, 1945.

3. E. Goursat, *Leçons sur l'integration des equations aux derivees partielles du second ordre a deux variables independentes*, 2 vols., Hermann, Paris, 1898.

4. S. Sternberg, *Lectures on differential geometry*, Prentice-Hall, Englewood Cliffs, N.J., 1964.

Columbia University and
Courant Institute of Mathematical Sciences

FORMAL THEORY OF OVERDETERMINED LINEAR PARTIAL DIFFERENTIAL EQUATIONS

HUBERT GOLDSCHMIDT [1]

We are interested in studying the inhomogeneous linear partial differential equation

$$(1) \qquad De = f$$

on a manifold X. In general, this equation cannot be solved if f is given arbitrarily. For example, the inhomogeneous equation $de = f$, where e is a function on $\boldsymbol{R}^n$ and f is a 1-form, implies $df = 0$. As in this case, if the equation (1) is regular, there is a condition which must be imposed on f, a formal integrability condition, $D'f = 0$, which may always be expressed in terms of a linear differential operator D' of finite order. If f_x is the formal power series expansion of f at a point x, there are conditions on f_x necessary for the existence of a formal power series e_x such that $De_x = f_x$. These conditions are obtained by repeatedly differentiating the equation. The purpose of the formal theory of differential equations is to construct a systematic theory of these formal obstructions and of the existence of formal solutions.

We give an account of some of the main results of [**1**], [**2**], [**3**] and [**5**]. §2 contains theorems on the existence of formal solutions and compatibility conditions. §3 is concerned with elliptic equations and the Spencer sequence.

The author wishes to thank C. J. Henrich for writing notes on which this paper is based.

1. **Differential operators.** Let us define a linear differential operator as follows. Let E, F denote vector bundles over a connected n-dimensional manifold X. A linear map $D: C^\infty(E) \to C^\infty(F)$ from the C^∞-sections of E to the C^∞-sections of F is a *differential operator of order* k from E to F if, for any point $x_0 \in X$, the action of D may be expressed as

$$(Ds)(x) = \sum_{|\alpha| \le k} a_\alpha(x)(\partial^\alpha s)(x)$$

in terms of a local coordinate on X and local frames for E and F on a neighborhood of x_0 in X. Here ∂^α is the usual multi-index notation and $a_\alpha(x)$ is a matrix depending differentiably on x. We define the bundle of k-jets of sections of E as

$$J_k(E) = \bigcup_{x \in X} J_k(E)_x$$

[1] This work was supported in part by the National Science Foundation grant GP 7461.

where $J_k(E)_x$ is the quotient of the space of sections of E by the subspace of sections which vanish to order $k + 1$ at x. The k-jet $j_k(s)(x)$ at $x \in X$ corresponding to a section s is determined by the values at x of s and all its derivatives of order $\leq k$ with respect to any local coordinate system. The vector bundle structure of $J_k(E)$ is determined by the requirement that the map $j_k(s): x \mapsto j_k(s)(x)$ be a differentiable section of $J_k(E)$, whenever $s \in C^\infty(E)$. There is a natural projection π_{k-1} of $J_k(E)$ onto $J_{k-1}(E)$ sending the k-jet $j_k(s)(x)$ into the $(k - 1)$-jet $j_{k-1}(s)(x)$ which it determines. The kernel of π_{k-1} can be naturally identified with the space of kth order homogeneous E-valued polynomials on the tangent bundle T of X, or with $S^kT^* \otimes E$. Here T^* is the cotangent bundle and S^kT^* denotes the kth symmetric product of T^*. The map $j_k: C^\infty(E) \to C^\infty(J_k(E))$ is a differential operator of order k, which is universal in the sense that, for any differential operator D of order k from E to F, there is a unique vector bundle morphism $\phi: J_k(E) \to F$ such that $D = \phi \circ j_k$. We shall denote the sheaves of germs of sections of E, F by $\mathscr{E}$, $\mathscr{F}$ respectively and also denote by D the map from $\mathscr{E}$ to $\mathscr{F}$ induced by the differential operator D.

We shall study the formal solutions of equation (1), where $e \in C^\infty(E)$, $f \in C^\infty(F)$, by examining ϕ and maps which are deduced from ϕ.

A solution $u \in J_k(E)_x$ of the equation $\phi u = f(x)$, where $f \in C^\infty(F)$, $x \in X$, is a truncated Taylor series of order k satisfying the differential equation (1) at x. We are interested in the series which satisfy (1) to higher order at x, that is, in the truncated Taylor series of order $k + l$ satisfying (1) to order l at x, which can be defined by the process of prolongation. The operator $j_l \circ D$ from E to $J_l(F)$ is a differential operator of order $k + l$. It therefore can be expressed in the form $j_l \circ D = p_l(\phi) \circ j_{k+l}$ where $p_l(\phi): J_{k+l}(E) \to J_l(F)$ is a vector bundle morphism, called the lth *prolongation* of ϕ. The equation

$$p_l(\phi)(u) = j_l(f)(x), \qquad u \in J_{k+l}(E)_x, \quad x \in X,$$

is obtained by differentiating (1) up to l times in all possible ways. A solution u of this equation is a solution of (1) to order l at x.

If $D = j_k: C^\infty(E) \to C^\infty(J_k(E))$ and $\phi = \mathrm{id}_k$ is the identity map of $J_k(E)$, the prolongation

$$p_l(\mathrm{id}_k): J_{k+l}(E) \to J_l(J_k(E))$$

satisfies $p_l(\mathrm{id}_k) j_{k+l}(s) = j_l(j_k(s))$ and is a monomorphism of vector bundles.

Let us define $R_{k+l} = \ker p_l(\phi)$; it is the space of $(k + l)$th order solutions of the homogeneous equation. In general, R_{k+l} is a family of vector spaces over X; in case their dimensions are constant, it is a vector bundle. If R_k is a vector bundle, then it determines R_{k+l} completely and we call R_{k+l} the lth prolongation of R_k. One may simply define a homogeneous differential equation as a subbundle R_k of $J_k(E)$.

The restriction of $p_l(\phi)$ to the subbundle $S^{k+l}T^* \otimes E$ of $J_{k+l}(E)$ gives rise to a map $\sigma_l(\phi)$, the symbol of $p_l(\phi)$. One can see that $\sigma_l(\phi)$ is determined by $\sigma(\phi) = \sigma_0(\phi)$, the symbol of ϕ. If we let $g_{k+l} = \ker \sigma_l(\phi)$, we have the commutative diagram of exact sequences:

$$\begin{array}{ccccc}
 & & 0 & & 0 \\
 & & \downarrow & & \downarrow \\
0 \to g_{k+l+1} & \longrightarrow & S^{k+l+1}T^*\otimes E & \xrightarrow{\sigma_{l+1}(\phi)} & S^{l+1}T^*\otimes F \\
\downarrow & & \downarrow & & \downarrow \\
0 \to R_{k+l+1} & \longrightarrow & J_{k+l+1}(E) & \xrightarrow{p_{l+1}(\phi)} & J_{l+1}(F) \\
\downarrow \pi_{k+l} & & \downarrow \pi_{k+l} & & \downarrow \pi_l \\
0 \to R_{k+l} & \longrightarrow & J_{k+l}(E) & \xrightarrow{p_l(\phi)} & J_l(F) \\
 & & \downarrow & & \downarrow \\
 & & 0 & & 0
\end{array}$$

We shall also use the notations $p_l(D)$, $\sigma_l(D)$, $\sigma(D)$ for $p_l(\phi)$, $\sigma_l(\phi)$, $\sigma(\phi)$ respectively.

The projective limit of the bundles $J_k(E)$ is denoted by $J_\infty(E)$; one may describe its members as formal power series with "coefficients" in E. The subset $R_\infty = \text{pr lim } R_{k+l}$ of $J_\infty(E)$ is the set of formal solutions of the homogeneous equation and the kernel of $p_\infty(\phi) = \text{pr lim } p_l(\phi)$, the formal operator corresponding to D. One does not know a priori that $R_\infty \neq 0$. We shall say that $\phi: J_k(E) \to F$ is formally integrable if, for all $l \geq 0$, R_{k+l} is a vector bundle and

$$\pi_{k+l}: R_{k+l+1} \to R_{k+l}$$

is surjective. Similarly, we say that a subbundle R_k of $J_k(E)$ is formally integrable if there exists a formally integrable vector bundle morphism ϕ from $J_k(E)$ to some other vector bundle over X such that $R_k = \ker \phi$. A formally integrable operator has many formal solutions, if $R_k \neq 0$; therefore we shall want to find sufficient conditions for ϕ to be formally integrable. It is for such operators that one can prove estimates leading to analytic existence theorems.

We shall conclude this section by constructing the symbol cohomology, first introduced by Spencer (see [**1**], [**2**] and [**6**]), For $P \in S^mT^*$, define

$$\delta P \in T^* \otimes S^{m-1}T^*$$

to be the exterior derivative of P considered as a homogeneous polynomial on T. Then δ can be extended to a map

$$\delta: \Lambda^jT^* \otimes S^mT^* \to \Lambda^{j+1}T^* \otimes S^{m-1}T^*$$

by setting

$$\delta(\omega \otimes u) = (-1)^j \omega \Lambda \delta u, \qquad \omega \in \Lambda^jT^*, \quad u \in S^mT^*.$$

We obtain a complex

$$0 \to S^mT^* \otimes E \xrightarrow{\delta} T^* \otimes S^{m-1}T^* \otimes E \xrightarrow{\delta} \cdots \to \Lambda^nT^* \otimes S^{m-n}T^* \otimes E \to 0$$

which is exact for $m \geq 1$. One can verify that

$$\delta(\Lambda^jT^* \otimes g_{k+l+1}) \subset \Lambda^{j+1}T^* \otimes g_{k+l},$$

so we have a complex

$$(2)\qquad \begin{aligned} 0 \to g_{k+l} &\xrightarrow{\delta} T^* \otimes g_{k+l-1} \xrightarrow{\delta} \Lambda^2 T^* \otimes g_{k+l-2} \to \cdots \\ &\to \Lambda^l T^* \otimes g_k \xrightarrow{\delta} \Lambda^{l+1} T^* \otimes S^{k-1} T^* \otimes E \end{aligned}$$

(we set $g_{k-1} = S^{k-1}T^* \otimes E$). We denote the cohomology of this complex at $\Lambda^j T^* \otimes g_m$ by $H^{m,j}$. These are the *symbol cohomology groups* of ϕ. Sequence (2) is automatically exact in the first two places, i.e. $H^{m,0} = H^{m,1} = 0$ for $m \geq k$. We shall say that g_k is *2-acyclic* if $H^{k+l,2} = 0$ for $l \geq 0$, and *involutive* if $H^{k+l,j} = 0$ for $l \geq 0$, $j \geq 0$. These definitions are meaningful since g_{k+l} depends only on the family of subspaces g_k of $S^k T^* \otimes E$.

2. **Formal integrability and compatibility conditions.**

THEOREM 1. *Let $\phi: J_k(E) \to F$ be a vector bundle morphism. Then sufficient conditions for ϕ to be formally integrable are*
(i) *R_{k+1} is a vector bundle;*
(ii) *$\pi_k: R_{k+1} \to R_k$ is surjective;*
(iii) *g_k is 2-acyclic.*
Moreover, under these conditions, there exist a vector bundle G and a vector bundle morphism $\psi: J_1(F) \to G$ such that the sequence

$$\mathscr{E} \xrightarrow{\phi \circ j_k} \mathscr{F} \xrightarrow{\psi \circ j_1} \mathscr{G}$$

is formally exact in that the sequence

$$J_\infty(E) \xrightarrow{p_\infty(\phi)} J_\infty(F) \xrightarrow{p_\infty(\psi)} J_\infty(G)$$

is exact; in fact the sequences

$$J_{k+l+1}(E) \xrightarrow{p_{l+1}(\phi)} J_{l+1}(F) \xrightarrow{p_l(\psi)} J_l(G)$$

are exact.

Let $G = \operatorname{coker} p_1(\phi)$ and let ψ be the natural map. By hypothesis (i), G is a vector bundle. Now hypothesis (ii) can be shown to imply the exactness of the sequence

$$S^{k+1}T^* \otimes E \xrightarrow{\sigma_1(\phi)} T^* \otimes F \xrightarrow{\sigma(\psi)} G.$$

It then follows that the cohomology of the sequence

$$S^{k+2}T^* \otimes E \xrightarrow{\sigma_2(\phi)} S^2 T^* \otimes F \xrightarrow{\sigma_1(\psi)} T^* \otimes G$$

is isomorphic to $H^{k,2}$. Consequently one can construct a curvature map $\kappa: R_{k+1} \to H^{k,2}$ such that the sequence

$$R_{k+2} \xrightarrow{\pi_{k+1}} R_{k+1} \xrightarrow{\kappa} H^{k,2}$$

is exact and which therefore expresses the obstruction to extending a solution of order $k + 1$ to a solution of order $k + 2$. Now hypothesis (iii) applies. That R_{k+2} is a vector bundle is established by a continuity argument; and induction completes the first part of the theorem. The second part is an easy consequence of the above steps. For details of the proof, see [**1**].

The conclusions of Theorem 1 are satisfactory, but the hypotheses are not: on the face of it, verifying (iii) takes an infinite number of steps. However, Elie Cartan gave a "finite" definition of involutiveness which is equivalent to ours, as was shown by Serre (see [**4**, Appendix]). Also, we have

THEOREM 2. *There is an integer k_0, determined by n, k and the rank of E, such that if $k_1 \geq k_0$, the following two conditions are sufficient for R_{k_1} to be formally integrable*:

(i) *R_{k_1+1} is a vector bundle;*

(ii) *$\pi_{k_1} : R_{k_1+1} \to R_{k_1}$ is surjective.*

The proof of Theorem 2 consists in a lemma which states that there exists an integer k_0 for which $H^{k_0+l,j} = 0$ for all $j, l \geq 0$; then Theorem 1 applies.

We shall say that a vector bundle morphism $\phi : J_k(E) \to F$ is *regular* if for all $l \geq 0$, the map $p_l(\phi) : J_{k+l}(E) \to J_l(F)$ is of constant rank. Similarly, we shall call $R_k = \ker \phi$ a regular equation if R_{k+l} is a vector bundle for $l \geq 0$. For regular operators, the compatibility conditions can be found in a satisfactory way, as given by

THEOREM 3. *Let $D = \phi \circ j_k$, where ϕ is a regular morphism. Then there are a vector bundle G and an epimorphism of vector bundles $\phi' : J_l(F) \to G$ determining a differential operator $D' = \phi' \circ j_l$ such that*

(i) *for each $r \geq 0$, the sequence*

$$J_{k+l+r}(E) \xrightarrow{p_{l+r}(\phi)} J_{l+r}(F) \xrightarrow{p_r(\phi')} J_r(G)$$

is exact; the sequence $\mathscr{E} \xrightarrow{D} \mathscr{F} \xrightarrow{D'} \mathscr{G}$ is formally exact in that the sequence

$$J_\infty(E) \xrightarrow{p_\infty(\phi)} J_\infty(F) \xrightarrow{p_\infty(\phi')} J_\infty(G)$$

is exact;

(ii) *the operator D' is minimal in the sense that for any morphism $\phi'' : J_m(F) \to H$ such that the sequences*

$$J_{k+m+r}(E) \xrightarrow{p_{m+r}(\phi)} J_{m+r}(F) \xrightarrow{p_r(\phi'')} J_r(H)$$

are exact, we have $m \geq l$ and there is a differential operator $\bar{D}$ from G to H of order $m - l$ satisfying $\phi'' \circ j_m = \bar{D} \circ D'$;

(iii) *if D'_1 is any differential operator from F to G_1 satisfying conditions* (i) *and* (ii) *like D', then there is an isomorphism $\psi : G \to G_1$ of vector bundles such that $D'_1 = \psi \circ D'$.*

Let $G = \operatorname{coker} p_l(\phi)$, for some l and ϕ' be the natural map. For verification of the statements see [2] and [3]. In fact, the regularity of ϕ is almost a necessary condition for the existence of compatibility conditions of the above type (see [2]).

We shall say that a section f of F satisfies the *compatibility conditions* for D if $D'f = 0$, where D' is given in Theorem 3. This condition clearly does not depend on the choice of D'.

Some equations are not formally integrable, nor are any of their prolongations. Thus $\pi_{k+l}: R_{k+l+m} \to R_{k+l}$ is not surjective, for some l, $m \geq 0$. Otherwise expressed, the equations of order $k + l + m$ obtained by prolongation imply further relations among the derivatives of order $k + l$. The next theorem states that for some equations one can assemble these new relations together with the equations obtained by prolongation into a new formally integrable differential equation with the same solutions as the old one. Let us say that $\phi: J_k(E) \to F$ or that $D = \phi \circ j_k$ is *completely regular* if ϕ is regular and the maps

$$\pi_{k+l}: R_{k+l+m} \to R_{k+l}$$

are of constant rank for all l, $m \geq 0$.

THEOREM 4 (SEE [2]). *Let D be a completely regular differential operator from E to F. Then there are integers l_0, $m_0 \geq 0$ such that the equation $R^{(l_0)}_{k+m_0} = \pi_{k+m_0}(R_{k+m_0+l_0})$ is a formally integrable equation with the same solutions and formal solutions as R_k; its rth prolongation is $R^{(l_0)}_{k+m_0+r} = \pi_{k+m_0+r}(R_{k+m_0+l_0+r})$.*

Moreover, we have the following

COROLLARY (SEE [3]). *If the hypotheses of Theorem 4 are satisfied, then there are a vector bundle F_1 and a differential operator P from F to F_1 of order $m_0 + l_0$ such that*

(i) *$D_1 = P \circ D$ is an operator of order $k + m_0$; its associated equation is the formally integrable equation $R^{(l_0)}_{k+m_0} \subset J_{k+m_0}(E)$;*

(ii) *if $f \in C^\infty(F)$ satisfies the compatibility conditions for D, then $Pf \in C^\infty(F_1)$ satisfies the compatibility conditions for D_1 and the solutions and formal solutions of the inhomogeneous equations $De = f$ and $D_1 e = Pf$ are the same.*

We have the more precise relation of P and the compatibility conditions.

THEOREM 5 (SEE [3]). *Let D be a completely regular differential operator of order k from E to F. Let P be as in the above corollary. Let D' be a differential operator from F to G satisfying condition* (i) *of Theorem 3, and let D'_1 be any differential operator from F_1 to G having the same relation to $D_1 = P \circ D$. Then there is an integer m and a differential operator Q from G to $J_m(G_1)$ such that*

(i) *the diagram*

$$\begin{array}{ccccc} \mathscr{E} & \xrightarrow{D} & \mathscr{F} & \xrightarrow{D'} & \mathscr{G} \\ \downarrow & & \downarrow P & & \downarrow Q \\ \mathscr{E} & \xrightarrow{D_1} & \mathscr{F}_1 & \xrightarrow{j_m \circ D'_1} & \mathscr{J}_m(G_1) \end{array}$$

commutes; hence if $f \in C^\infty(F)$ satisfies $D'f = 0$, then $D'_1(Pf) = 0$;

(ii) *the operator P induces an isomorphism from the cohomology of the complex* $\mathscr{E} \xrightarrow{D} \mathscr{F} \xrightarrow{D'} \mathscr{G}$ *to that of the complex* $\mathscr{E} \xrightarrow{D_1} \mathscr{F}_1 \xrightarrow{D'_1} \mathscr{G}_1$.

3. **The Spencer sequence.** We have seen that we may reduce the study of completely regular differential operators to the formally integrable case. In this section, we construct the sophisticated Spencer sequence of a formally integrable equation according to [1].

THEOREM 6. *Let $R_k \subset J_k(E)$ be a formally integrable equation; assume that g_{k+1} is involutive. Let $\mathscr{S}$ be the sheaf of germs of solutions of R_k.*

(i) *There is a complex*

$$0 \to \mathscr{S} \xrightarrow{j_k} \mathscr{C}^0 \xrightarrow{D^0} \mathscr{C}^1 \xrightarrow{D^1} \mathscr{C}^2 \xrightarrow{D^2} \cdots \tag{3}$$

which is uniquely determined (up to isomorphism) by the properties

(a) $C^0 = R_k$;

(b) *D^i is a first-order differential operator from C^i to C^{i+1} whose symbol $\sigma(D^i): T^* \otimes C^i \to C^{i+1}$ is an epimorphism;*

(c) *the sequence (3) is formally exact and the sequences*

$$0 \to R_{k+m} \to J_m(C^0) \xrightarrow{p_{m-1}(D^0)} J_{m-1}(C^1) \xrightarrow{p_{m-2}(D^1)} J_{m-2}(C^2) \to \cdots \to J_{m-l}(C^l)$$

are exact at R_{k+m} for $m \geq 1$ and at $J_{m-i}(C^i)$ for $m \geq i + 1$.

(ii) *This sequence has the following properties:*

(a) *we have $C^l \cong \Lambda^l T^* \otimes R_k/\delta(\Lambda^{l-1}T^* \otimes g_{k+1})$ and $\sigma(D^l)$ is induced by the multiplication map $T^* \otimes \Lambda^l T^* \to \Lambda^{l+1}T^*$; in particular $C^l = 0$ for $l > n$;*

(b) *the sequence of sheaves*

$$0 \to \mathscr{S} \xrightarrow{j_k} \mathscr{C}^0 \xrightarrow{D^0} \mathscr{C}^1$$

is exact.

(iii) (*due to Quillen* [5], *see also* [2]). *Let D be any differential operator of order k from E to F whose associated equation is R_k. Let $D' = \phi' \circ j_l$ be any operator of order l from F to G such that the sequences*

$$J_{k+l+r}(E) \xrightarrow{p_{l+r}(\phi)} J_{l+r}(F) \xrightarrow{p_r(\phi')} J_r(G)$$

are exact, for $r \geq 0$. Then the cohomology of the sequence $\mathscr{C}^0 \xrightarrow{D^0} \mathscr{C}^1 \xrightarrow{D^1} \mathscr{C}^2$ is isomorphic to the cohomology of the sequence $\mathscr{E} \xrightarrow{D} \mathscr{F} \xrightarrow{D'} \mathscr{G}$.

OUTLINE OF PROOF OF (i) (SEE [1]). Define C^1 to be the cokernel of the natural monomorphism $R_{k+1} \to J_1(C^0)$ and let the natural projection be $\rho_0: J_1(C^0) \to C^1$. Then show that the sequences

$$0 \to R_{k+m} \to J_m(C^0) \xrightarrow{p_{m-1}(\rho_0)} J_{m-1}(C^1)$$

are exact for $m \geq 1$. For $l > 1$, recursively define C^l and ρ_{l-1} as the cokernel of $p_1(\rho_{l-2}): J_2(C^{l-2}) \to J_1(C^{l-1})$ and the natural projection. The hypothesis that

g_{k+1} is involutive implies that the sequence

$$J_{m+2}(C^{l-2}) \xrightarrow{p_{m+1}(\rho_{l-2})} J_{m+1}(C^{l-1}) \xrightarrow{p_m(\rho_{l-1})} J_m(C^l)$$

is exact for $m \geq 0$. We set $D^l = \rho_l \circ j_1$.

The Spencer sequence is especially useful for the class of elliptic operators, which we now define. We can restrict our attention to fibers of all vector bundles over a fixed point of X; we shall use the same notation for a vector bundle and its fiber. If D is a differential operator from E to F of order k and $\xi \in T^*$, we denote by $\sigma_\xi(D): E \to F$ the map $e \mapsto (k!)^{-1}\sigma(D)(\xi^k \otimes e)$. We say that ξ is *noncharacteristic* for D if $\sigma_\xi(D)$ is injective.

THEOREM 7 (QUILLEN [5]). *Let D be a differential operator from E to F of order k, with the associated equation $R_k \subset J_k(E)$ formally integrable and g_{k+1} involutive. Then if $\xi \in T^*$, the following are equivalent:*

(i) *ξ is noncharacteristic for D;*

(ii) *the sequence*

$$0 \to C^0 \xrightarrow{\sigma_\xi(D^0)} C^1$$

is exact;

(iii) *the sequence*

$$0 \to C^0 \xrightarrow{\sigma_\xi(D^0)} C^1 \xrightarrow{\sigma_\xi(D^1)} C^2 \to \cdots \to C^n \to 0 \tag{4}$$

is exact.

We say that D is *elliptic* if there are no (real) nonzero characteristic vectors for D; a complex

$$0 \to \mathscr{C}^0 \xrightarrow{D^0} \mathscr{C}^1 \xrightarrow{D^1} \mathscr{C}^2 \to \cdots \to \mathscr{C}^n \to 0$$

is called an *elliptic complex* if the sequence (4) is exact for each nonzero $\xi \in T^*$. Thus if D is elliptic, its Spencer sequence is an elliptic complex. The main application of this result is the existence of C^∞ local solutions for analytic elliptic operators due to Spencer; in particular the Spencer sequence (3) is exact in this case (see [6], [3]).

REFERENCES

1. H. Goldschmidt, *Existence theorems for analytic linear partial differential equations,* Ann. of Math. **86**(2) (1967), 246–270.

2. ———, *Prolongations of linear partial differential equations*: I. *A conjecture of Elie Cartan*, Ann. Sci. École Norm. Sup. (4) **1** (1968), 417–444.

3. ———, *Prolongations of linear partial differential equations*: II. *Inhomogeneous equations,* Ann. Sci. École Norm. Sup. (4) **1** (1968), 617–625.

4. V. W. Guillemin and S. Sternberg, *An algebraic model of transitive differential geometry,* Bull. Amer. Math. Soc. **70** (1964), 16–47.

5. D. G. Quillen, *Formal properties of over-determined systems of linear partial differential equations,* Ph.D. Thesis, Harvard University, 1964.

6. D. C. Spencer, *Deformation of structures on manifolds defined by transitive, continuous pseudogroups*: I–II, Ann. of Math. (2) **76** (1962), 306–445.

STANFORD UNIVERSITY

ALGEBRAIC RESULTS ON CHARACTERISTICS

V. W. GUILLEMIN

In this lecture I want to discuss the noncharacteristic Cauchy problem for overdetermined systems of PDE's. Suppose we are given a smooth manifold, X, a submanifold, Y, vector bundles, E and F, on X and a differential operator $D: E \to F$. Let g be a cross-section of F satisfying appropriate compatibility conditions. We want to solve the equation

$$Df = g \tag{*}$$

and, in addition, prescribe f and certain derivatives of f along the submanifold, Y. We will only consider this problem for Y *noncharacteristic*. This term has roughly the same meaning for overdetermined systems as for a single equation, namely that the Taylor series of f at any point of Y can be completely determined by the equation (*) providing we know f and a finite number of its derivatives along Y.

I will try to make the definition of "noncharacteristic" a little more precise: If D is of order k and T^* is the cotangent bundle of X we have the symbol mapping $\sigma(D): E \otimes S^k(T^*) \to F$. Let N be the normal bundle of the submanifold Y. Since N is a subbundle of $T^*|Y$, we get a commutative diagram

$$\begin{array}{ccc} E \otimes S^k(T^*) & \to & F \\ \uparrow & \nearrow_{\alpha} & \\ E \otimes S^k(N) & & \end{array} \tag{**}$$

We will say that Y is noncharacteristic if α is injective.

Let's see what this condition means in local coordinates. For simplicity we will assume D is a first order operator. Let $x_1, \ldots, x_n$ be a local coordinate system on X and suppose Y is locally characterized by the equations $x_1 = \ldots = x_k = 0$. Then $E \otimes N = E \otimes dx_1 \oplus E \otimes dx_2 \oplus \ldots \oplus E \otimes dx_k \cong \bigoplus_k E$. Let E' be the image of α in the diagram (**). By assumption E' is isomorphic to $E \otimes N$; so we can identify E' with $\bigoplus_k E$.

Now assume that F has a Hermitian structure and let E_0 be the orthogonal complement of E' in F; so we can write: $F = E_0 \oplus (\bigoplus^k E)$. We will introduce the projections, π_0, of F on E_0 and, π_i, of F on the ith E component of F. Also, we will set $D_0 = \pi_0 D$ and $D_i = \pi_i D$. By construction we get the following conditions on the symbols of these operators

$$\begin{cases} \sigma(D_i)\ (e \otimes dx_j) = \delta^i_j e, & i, j \le k, \\ \sigma(D_0)\ (e \otimes dx_j) = 0, & j \le k. \end{cases} \tag{***}$$

To see what these conditions imply on the operators themselves we need the following:

LEMMA. $\sigma(D)(e \otimes dx_i) = (Dx_i - x_iD)e$.

PROOF. LTR.

In terms of local coordinates we can write:

$$D_i = \sum_{\alpha=1}^{n} A_i^\alpha \frac{\partial}{\partial x_\alpha} + B_i \quad i = 0, \ldots, k$$

where the A's and B's are in Hom(E, F). If $j \leq k$ then, by the lemma:

$$D_i x_j - x_j D_i = A_i^j = \sigma(D_i)(e \times dx_j) = e \\ = 0$$

depending on whether $i = j > 0$ or not. Therefore, in local coordinates, these operators have the form:

$$D_0 = \sum_{\alpha>k} A_0^\alpha \frac{\partial}{\partial x_\alpha} + B_0$$

$$D_i = \frac{\partial}{\partial x^i} + \sum_{\alpha>k} A_i^\alpha \frac{\partial}{\partial x^i} + B_i.$$

Now the equation $Df = g$ is the same as the set of equations: $D_i f = \pi_i g$, $i = 0, \ldots, k$, which can also be written:

$$\text{(****)} \qquad \begin{aligned} \frac{\partial f}{\partial x_i} &= -\sum_{\alpha>k} A_i^\alpha \frac{\partial}{\partial x^\alpha} - B_i f + g \\ \sum_{\alpha>k} A_0^\alpha \frac{\partial f}{\partial x^\alpha} &= \pi_0 g . \end{aligned}$$

So we have shown:

Noncharacteristic $\Leftrightarrow$ *the equation* $Df = g$ *can be algebraically solved for the normal derivatives* $\partial f/\partial x_i$, $i \leq k$, *in terms of* f, g *and the tangential derivatives of* f.

Naively one could try to solve the equation $Df = g$ with initial data prescribed along Y as follows: Let $\bar{x}_0 = x_{k+1}, \ldots, x_n$ be the local coordinates system on Y corresponding to $x_1, \ldots, x_n$ and let $f_0 = f_0(\bar{x}_0)$ be the prescribed value of f on Y. (Because of the equation (****) we must require that $D_0 f_0 = \pi_0 g$.) First we will try to find a function $f_1 = f_1(\bar{x}_0, x_1)$ such that $f_1|_{x_1=0} = f_0$ and $D_1 f_1 = \pi_1 g$. Next we will try to find a function $f_2 = f_2(\bar{x}_0, x_1, x_2)$ such that $f_2|_{x_2=0} = f_1$ and $D_2 f_2 = \pi_2 g$; and so on, until we finally get a function $f = f_k$ defined on an open set of X. (Note that at each stage we are trying to solve a determined Cauchy problem. If, for example, the data are real analytic this is always possible.)

Of course f will not, in general, satisfy the equation $Df = g$ but only the equation $D_k f = \pi_k g$. To get $Df = g$ we will need some additional assumptions about D. Before discussing these assumptions it will be helpful to take a single example:

Let $\xi_1 = \partial/\partial x_1 + \mathrm{a}\, \partial/\partial x_3$ and $\xi_2 = \partial/\partial x_2 + \mathrm{b}\, \partial/\partial x_3$ be vector fields in R^3 and consider the Cauchy problem $\xi_1 f = \xi_2 f = 0$ with $f = f_0$ along $x_1 = x_2 = 0$. This will be solvable for arbitrary f_0 if and only if $[\xi_1, \xi_2] = 0$. Otherwise the only solutions of these equations are constants. We will show that a similar commutation condition comes into the general case.

We recall a few results discussed by Goldschmidt in his lectures: If the differential operator, $D: E \to F$, is "sufficiently regular" then there exists a differential operator $D': F \to F'$ such that

$$E \overset{D}{\to} F \overset{D'}{\to} F'$$

is formally exact. More generally there exists a formally exact resolution

$$E \overset{D}{\to} F \overset{D'}{\to} F' \overset{D''}{\to} F'' \to \text{etc.}$$

which terminates after n terms.

We will say that D is *in involution* if the operators D', D'', etc. can all be taken to be of first order. Goldschmidt has shown that a "sufficiently regular" operator is replaceable by an operator in involution with the same set of solutions; so this assumption is not too far-fetched.

THEOREM. *If D is in involution, and $D_0, D_1, D_2, \ldots, D_k$ are the operators introduced above then they satisfy the following commutation relations:*

(a) $D_0 D_i = 0 \bmod D_0$.

(b) $[D_i, D_j] = 0 \bmod D_0$.

NOTE. If A and B are differential operators $A = 0 \bmod B \Leftrightarrow \exists$ an operator C of order $=$ order $A -$ order B such that $A = CB$.

Rather than trying to prove the theorem which involves some rather technical facts from commutative algebra, I will show how one applies this theorem to the Cauchy problem. Henceforth I will assume all data are *real analytic*. I will also, for simplicity, just consider the homogeneous equation $Df = 0$.

Going back to the argument sketched above, the ith stage in that argument involved solving the determined Cauchy problem:

$$D_i f_i = 0, f_i|_{x_i = 0} = f_{i-1} \quad \text{where } f_i = f_i(\bar{x}_0, x_1, \ldots, x_i).$$

We know from (a) that $D_0 D_i = D_i' D_0 f_i = D_0 D_i f = 0$; and we can assume by induction that $D_0 f_i|_{x_i=0} = D_0 f_{i-1} = 0$. It is not hard to check that the surface $x_i = 0$ is noncharacteristic for D_i'; so by the uniqueness theorem for the classical noncharacteristic Cauchy problem with analytic data, $D_0 f_i = 0$. Now, applying (b) we get $D_i D_j f_i = D_j D_i f_i = 0$ for $j < i$; and, since we can assume by induction that $D_j f_i|_{x_i=0} = D_j f_{i-1} = 0$ for $j < i$ we get $D_j f = 0$ by a repetition of the previous argument. Finally, when $i = k$, we have $Df = 0$. Q.E.D.

I would like to mention one application of the theorem. Goldschmidt, in his lectures, has discussed various formal existence theorems for overdetermined systems. If one is working with real analytic data, these formal existence theorems can be replaced by genuine existence theorems. One way of proving these theorems is by induction on the dimension of the base manifold X. Let $D: E \to F$ be a "sufficiently regular" differential operator and suppose we want to solve the equation

$$(*) \qquad Df = g,$$

assuming this equation is formally solvable and the term on the right analytic. By Goldschmidt's theorems we can assume D is of first order and in involution. If Y is an analytic noncharacteristic surface of X then solving (*) is equivalent, by the argument above, to solving the equation

$$(**) \qquad D_0 f = \pi_0 g$$

on Y. One can assume by induction that if (**) admits formal solutions it admits convergent analytic solutions. Hence one can deduce the same property for (*).

MASSACHUSETTS INSTITUTE OF TECHNOLOGY

ON ESTIMATE $\|(A(D) + B(x))u(x)\| \geq c\|u\|$

MASATAKE KURANISHI

Let V and W be vector spaces with hermitian metric. Let A_j and B_j ($j = 1, \ldots, n$) be linear mappings over $\boldsymbol{C}$ of W into V. Denote by $C_0^\infty(\boldsymbol{R}^n, W)$ the vector space of W-valued functions on $\boldsymbol{R}^n$ of class C^∞ with compact support. We set $D_j = -i\partial/\partial x_j$. The problem we want to discuss is the following: Under what conditions on A_j and B_j we have, for a constant c, estimate

$$\left\| \sum_{j=1}^{n} (A_j D_j + B_j x_j) u(x) \right\| \geq c\|u\|$$

for all u in $C_0^\infty(\boldsymbol{R}^n, W)$, where $\| \quad \|$ denotes L^2-norm. The problem is related to the half-estimate in generalized Neumann problem, which Spencer discussed in his talk. Namely, Hörmander showed that the half-estimate is equivalent to our estimate (with uniform c) for a family of A_j and B_j constructed from a given elliptic complex, provided generalized complex characteristics of the elliptic complex are simple (cf. Hörmander, *Pseudo-differential operators and nonelliptic boundary problems*, Ann. of Math. (2) **83** (1966), pp. 129–209).

In order to simplify notations we set $P_j = A_j, P_{n+j} = B_j, T_j = D_j, T_{n+j} =$ multiplication operator x_j. We also set $T = (T_1, \ldots, T_{2n})$ and denote by P the map $P(\xi) = \sum_{j=1}^{2n} P_j \xi_j$ of $\boldsymbol{R}^{2n}$ into the vector space $L(W, V)$ of linear mappings over $\boldsymbol{C}$ of W into V. Thus we can write

$$\sum_{j=1}^{n} (A_j D_j + B_j x_j) u = \sum_{j=1}^{2n} P_j T_j u = P(T)u.$$

In the following, j and k will always denote indexes $1, \ldots, 2n$. T_j and T_k satisfy a relation:

$$T_j^* T_k - T_k^* T_j = i\theta_{jk} I$$

where $\theta_{jk} \in \boldsymbol{R}$, $\theta_{jk} + \theta_{kj} = 0$, and I denotes the identity mapping. We also introduce Laplacian of P:

$$\Delta_P(\xi) = P(\xi)^* P(\xi) \in L(W, W),$$

which is a $L(W, W)$-valued quadratic form on $\boldsymbol{R}^{2n}$. We employ the method used by Hörmander in the above quoted paper, by which our problem is reduced to analyze all $Q: R^{2n} \to L(W, V_1)$ such that $\Delta_P(\xi) = \Delta_Q(\xi)$. It goes as follows:

$$\begin{aligned} \|P(T)u\|^2 &= \sum_{j,k} \langle P_j T_j u, P_k T_k u \rangle = \sum_{j,k} \langle P_k^* P_j T_k^* T_j u, u \rangle \\ &= \sum_{j,k} (\tfrac{1}{2} \langle P_k^* P_j T_k^* T_j u, u \rangle + \tfrac{1}{2} \langle P_k^* P_j (T_j^* T_k - i\theta_{jk}) u, u \rangle) \\ &= \sum_{j,k} (\langle (P_k^* P_j + P_j^* P_k) T_k^* T_j u, u \rangle - \tfrac{1}{2} \langle i P_j \theta_{jk} u, P_k u \rangle). \end{aligned}$$

Note that $P_k^*P_j + P_j^*P_k = (\partial/\partial\xi_j\,\partial/\partial\xi_k)\Delta_P(\xi)$. Therefore, if we take any $Q:\mathbf{R}^{2n} \to L(W, V_1)$ such that $\Delta_P = \Delta_Q$ and if we make the same calculation as above for $\|Q(T)u\|^2$, we obtain the same first term in the final formula. Hence

$$\|P(T)u\|^2 - \|Q(T)u\|^2 = \sum_{j,k}\tfrac{1}{2}(-\langle iP_j\theta_{j,k}u, P_ku\rangle + \langle iQ_j\theta_{jk}u, Q_ku\rangle).$$

Thus we have

LEMMA. *If we can find a vector space V_1 and a linear map $Q:\mathbf{R}^{2n} \to L(W, V_1)$ such that*
(1) $\Delta_P = \Delta_Q$,
(2) $\sum_{j,k}(-\langle iP_j\theta_{jk}w, P_kw\rangle + \langle iQ_j\theta_{jk}w, Q_kw\rangle) \geq 2c^2|w|^2$
for all $w \in W$, then we have estimate

$$\|P(T)u\| \geq c\|u\| \qquad (u \in C_0^\infty(\mathbf{R}^{2n}, W)).$$

When dim $W = 1$, we can identify $L(W, V) = v$ by fixing a base of W. Hence orthogonal group of the underlying real vector space V acts on $L(W, V)$. If we set $Q(\xi) = \sigma P(\xi)$ where σ is an orthogonal transformation of v, then $\Delta_Q = \Delta_P$. In fact, we have enough such Q (or rather enough σ) so that we can obtain a necessary and sufficient condition for P to have our estimate (cf. Hörmander's paper quoted above). However, in the general case, it seems that we have to consider a wider class of Q. The problem of finding all Q such that $\Delta_P = \Delta_Q$ appears to be a complicated one. In the case $\Delta_P(\xi) = |\xi|^2I$, the problem in somewhat restricted form was discussed in Atiyah's lecture, where he noted that the problem is representation theory of Clifford Algebras. Here we will try to find deformations $P(t)$ of P so that $\Delta_{P(t)} = \Delta_P$.

To proceed further, it is more convenient to consider $p:\mathbf{R}^{2n} \otimes_R W \to V$ instead of P. P and p are related by

$$P(\xi)w = p(\xi \otimes w), \qquad P(T)u = p(\sum_j e_j \otimes T_ju)$$

where $\{e_1, \ldots, e_{2n}\}$ is the standard base of $\mathbf{R}^{2n}$. On $\tilde{W} = \mathbf{R}^{2n} \otimes W$ we consider hermitian metric induced by that of W and by the standard euclidean metric of $\mathbf{R}^{2n}$. Set $\tilde{p} = (p^*p)^{1/2}$. Then it is easy to see that $\|\tilde{P}(T)u\| = \|P(T)u\|$. Hence we can assume without loss of generality that $V = \tilde{W}$ and p is hermitian. By the same reason we can take $V_1 = \tilde{W}$. So we consider deformation $p(t) \in L(\tilde{W}, \tilde{W})$ of p such that $\Delta_{P(t)} = \Delta_P$. In order to write down the condition, we define $B(q, r) \in L(\tilde{W}, \tilde{W})$ for $q, r \in L(\tilde{W}, \tilde{W})$ by the formula:

$$\begin{aligned} 4\langle B(q, r)(\xi \otimes w), \eta \otimes v\rangle &= \langle q(\xi \otimes w), r(\eta \otimes v)\rangle \\ &+ \langle r(\xi \otimes w), q(\eta \otimes v)\rangle + \langle q(\eta \otimes w), r(\xi \otimes v)\rangle \\ &+ \langle r(\eta \otimes w), q(\xi \otimes v)\rangle. \end{aligned}$$

$B(q, r)$ is bilinear over $\mathbf{R}$ in q and r, and $B(q, r) \in SA(\tilde{W})$, the linear space of self-adjoint transformation of $\tilde{W}$. Moreover, $B(q, r)$ satisfies the relation:

$$\langle B(q, r)(\xi \otimes w), \eta \otimes v\rangle = \langle B(q, r)(\eta \otimes w), \xi \otimes v\rangle.$$

So we denote by F the vector space of all $h \in SA(\tilde{W})$ such that

$$\langle h(\xi \otimes w), \eta \otimes v\rangle = \langle h(\eta \otimes w), \xi \otimes v\rangle.$$

We regard $B(q, r)$ as $\boldsymbol{F}$ valued. We see easily that the condition $\Delta_Q = \Delta_R$ is equivalent to the condition $B(q, q) = B(r, r)$. Hence if we write

$$p(t) = p + tp_1 + t^2 p_2 + \dots,$$

our condition is expressed in a sequence of equations:

$$\begin{aligned} &B(p, p_1) = 0, \\ &B(p, p_2) + \tfrac{1}{2}B(p_1, p_1) = 0, \\ &\quad\vdots \\ &B(p, p_l) + \tfrac{1}{2}\sum_{h=1}^{l-1} B(P_h, P_{l-h}) = 0, \\ &\quad\vdots \end{aligned}$$

The first equation is not bad. But we have to choose solutions p_1 so that the second equation is solvable in p_2. So we are naturally interested in knowing the image of the map $q \mapsto B(p, q)$, or its orthogonal complement in $\boldsymbol{F}$. (Note that $SA(\tilde{W}) \supseteq \boldsymbol{F}$ has metric $\langle k, h\rangle = \operatorname{tr} k^*h$.) Here, we use the formula:

$$\langle B(q, r), h\rangle = \mathscr{R} \operatorname{tr} q^*rh = \mathscr{R} \operatorname{tr} r^*qh \quad (h \in \boldsymbol{F}, q, r \in L(\tilde{W}, \tilde{W})).$$

Therefore $h \in (\text{Image } B(p, \cdot))^{\perp} \rightleftarrows \langle B(p, q), h\rangle = 0$ for all $q \in L(\tilde{W}, \tilde{W}) \rightleftarrows \mathscr{R} \operatorname{tr} q^*ph = 0$ for all $q \in L(\tilde{W}, \tilde{W}) \rightleftarrows ph = 0$. Thus we find that p_1 must satisfy, beside $B(p, p_1) = 0$, the condition

$$\operatorname{tr} p_1^* p_1 h = 0 \quad (\text{for all } h \in \boldsymbol{F} \text{ such that } ph = 0).$$

This suggests that we consider

$$\begin{aligned} G_p = \{&g \in SA(\tilde{W});\ g \text{ positive semidefinite}, \\ &\text{and } \mathscr{R} \operatorname{tr} gh = 0 \text{ for all } h \in \boldsymbol{F} \text{ such that } ph = 0\}. \end{aligned}$$

G_p is a convex closed subset of $SA(\tilde{W})$, and our condition can be rewritten as $p_1^* p_1 \in G_p$. To proceed further, we use the following

LEMMA. *Let G be a convex closed subset of the linear space of selfadjoint operators of a vector space over C. Assume that any g in G is positive semidefinite. Take g_0 in G with the maximum rank. Then* image $g_0 \supseteq$ Image g *for all g in G.*

Image g_0 is uniquely determined by G, which we denote by Image G. Since p is in G_p, Image $p \subseteq$ Image G_p, i.e.

$$\ker p \supseteq (\text{Image } G_p)^{\perp}.$$

(Note that p is hermitian.) Then condition $p_1^* p_1 \in G_p$ implies that Image $p_1^* p_1 \subseteq$ Image G_p, i.e.

$$\ker p_1 \supseteq (\text{Image } G_p)^{\perp}.$$

Conversely, if $B(p, p_1) = 0$ and $\ker p_1 \supseteq (\text{Image } G_p)^{\perp}$, then we can modify p_1 so that we can find p_2 satisfying the equation. This more or less gives us a way of avoiding the first obstruction. However, we have further obstructions to contend with, and these can fall back to p_1 to impose further conditions. This situation forces us to consider all obstructions at once. Namely, we consider a decreasing infinite sequence of subspaces $\ker p = U_0 \supseteq U_1 \supseteq \dots$ and assume that, for any deformation $p(t)$ of p such that $\Delta_{p(t)} = \Delta_p$, $p_r|U_r = 0$ for all r. Then we employ a similar argument as above to find a new sequence $\ker p = U_0^{(1)} \supseteq U_1^{(1)} \supseteq \dots$ such that $U_r^{(1)} \supseteq U_r$ and $p_r|U_r^{(1)} = 0$. In order to describe how to define such $U_r^{(1)}$, we introduce some notations. For a linear subspace K of $\tilde{W}$, we set

$$Y(K) = \{h \in \boldsymbol{F}; ph|K^{\perp} = 0\}.$$

For linear subspaces $H \subseteq \boldsymbol{F}$ K_1 and $K_2 \subseteq \tilde{W}$,

$$\begin{aligned} &Y(H, K_1, K_2) = \{h \in H; h(K_2^{\perp}) \subseteq K_1\}, \\ &G(H, K) = \{g \in SA(K) \subseteq SA(\tilde{W});\ g \text{ positive semidefinite and } \operatorname{tr} gh = 0 \text{ for all } h \in H\}, \\ &Z(H, K) = (\text{Image } G(H, K))^{\perp} \supseteq K^{\perp}. \end{aligned}$$

Thus $G_p = G(Y(0), \tilde{W})$. Set

$$\begin{aligned} &L_1 = H_1 = Y(U_2),\ U_1^{(1)} = Z(H_1, U_2^{\perp}) + U_1, \\ &L_2 = Y(U_4),\ H_2 = Y(L_2, U_1^{(1)}, U_2),\ U_2^{(1)} = Z(H_2, U_4^{\perp}) + U_2, \\ &\qquad \vdots \\ &L_r = Y(U_{2r}),\ H_r = \bigcap_{s=1}^{r-1} Y(L_r, U_s^{(1)}, U_{2r-s}), \\ &\qquad U_r^{(1)} = Z(H_r, U_{2r}^{\perp}) + U_r, \\ &\qquad \vdots \end{aligned}$$

Then $\ker p \supseteq U_1^{(1)} \supseteq U_2^{(1)} \supseteq \dots$ and $U_r^{(1)} \supseteq U_r$. We can show that $p_r U_r^{(1)} = 0$ for all r. Starting from the case $U_1 = U_2 = \dots = \{0\}$, we define $U_r^{(1)}$, $U_r^{(2)} = U_r^{(1)(1)}$, $\dots$, $U_r^{\infty} = \bigcup_{s=1}^{\infty} U_r^{(s)}$. We have

$$\ker p = U_0^{\infty} \supseteq U_1^{\infty} \supseteq \dots.$$

By the construction, $p_r|U_r^{\infty} = 0$ for all $p(t) = p + tp_1 + \dots + t^r p_r + \dots$ such that $\Delta_{p(t)} = \Delta_p$. We can also construct almost all such $p(t)$.

After we finished analyzing deformations $p(t)$, we have to determine which $p(t)$ will give us the estimate. Here we can proceed by a similar argument as above. The final result of this analysis is fairly complicated and, therefore, may be useless. For the sake of simplicity we will state *conditions by which we obtain the estimate already in* $p + tp_1 + t^2 p_2$ (other terms are of order t^3 and hence will not affect the estimate). For this we have to introduce again a number of notations. For $K \subseteq \tilde{W}$, set

$$J(K) = \{(c, b) \in SA(W) \oplus \boldsymbol{F};\ p(b - i\theta \otimes c)|K^{\perp} = 0\}$$

where θ is the linear transformation $\theta e_j = \theta_{jk} e_k$ (where $\{e_1, \dots, e_{2n}\}$ is the standard base of $\boldsymbol{R}^{2n}$ and θ_{jk} is the constant appearing in the relation of T_j), and

$$Y'(K) = \text{the projection of } J(K) \text{ to the factor } SA(W).$$

For a linear subspace $H' \subseteq SA(W)$, set

$$G'(H') = \{g \in SA(W);\ g \text{ positive semidefinite and } \operatorname{tr} gc = 0 \text{ for all } c \text{ in } H'\}.$$

For $q, r \in L(\tilde{W}, \tilde{W})$, $C(p, q)$ denotes the element in $SA(W)$ such that

$$\langle w, C(p, q)w\rangle = \sum_{j,k} \operatorname{Im}\langle p(\theta_{kj} f_j \otimes w), q(f_k \otimes w)\rangle$$

where Im indicates the imaginary part.

Proposition. *If* Image $G'(Y'(U_1^\infty)) = W$, *we have the estimate. If* Image $G'(Y'(U_1^\infty)) \subsetneq W$, *the following condition is sufficient: We can find* $a_1 \in L(Y_1, Y_1)$ (*where* $Y_1 = U_1^\infty \cap (U_2^\infty)^\perp$), $q_1 \in L((U_1^\infty)^\perp$, Image $p)$, *and* $C_2 \in SA(W)$ *such that*
(1) $C(p, q_1) \in G'(Y'(U_1^\infty))$,
(2) $\langle C_2, c\rangle = \mathscr{R} \operatorname{tr}(q_1^* q_1 + a_1^* a_1)(b - i\theta \otimes c)$ *for all* $(b, c) \in J(U_2^\infty)$,
(3) $C_2 > 0$ *on* (Image $C(p, q_1))^\perp$.

The condition is actually a little better than it looks: By the construction we can always find a_1, q_1, C_2 which satisfy (1) and (2), and the condition is to say that among them there is one with condition (3).

In the case $\dim W = 1$, $W = \boldsymbol{R}^{2n} \otimes \boldsymbol{C} = \boldsymbol{C}^m$. Then $U_1^\infty = U_2^\infty = \ldots$ is equal to the complex vector subspace generated by $\ker p \cap (\boldsymbol{R}^{2n} \otimes 1)$. In this case, the above condition is equivalent to a necessary and sufficient condition Hörmander obtained in the paper quoted before. His condition is as follows: (Since A_j, $B_j \in L(\boldsymbol{C}, V) = V$, we will consider $A_j, B_j \in V$.)
(1) $\sum_{j=1}^n \langle A_j, v\rangle\langle B_j, v'\rangle$ is not symmetric in $v, v' \in V$,
(2) $\mathscr{R} i \sum_{j=1}^\infty \langle B_j, v\rangle\langle v, A_j\rangle$ has positive eigenvalues.
The fact that Hörmander's condition for $\dim W = 1$ is so beautifully stated may suggest that there is a way of concisely formulating our condition for the general case, which is so messy at this stage that I did not have the heart to write it down.

Addendum. The condition on $r = q - p$ for $\Delta_p = \Delta_q$ is

$$B(p, r) + \tfrac{1}{2}B(r, r) = 0,$$

which is the integrability of almost complex structures when we replace r by (1, 0)-tangent vector valued differential forms of type (0, 1), $B(p, r)$ by $\bar{\partial} r$, and $B(r, r)$ by bracket $[r, r]$. Therefore it seems that we can apply the same argument used in constructing the universal family of deformations of a complex manifold, provided we replace the diffeomorphism group by the unitary group operating on the image space $\tilde{W}$. This seems to give hope to improve presentation of our condition.

Columbia University

QUANTIZATION OF THE DE RHAM COMPLEX

I. E. SEGAL

Introduction. The notion of "quantized differential form" on an affine space was introduced in [**1**]. It seems to form a natural extension of the conventional notion of differential form, especially from the standpoints of operator and differential equation theory, and arises in connection with quantum field theory. The word "quantization" does not refer to any specific mathematical procedure, but is essentially merely descriptive of a theory in which the role of functions has been taken over by operators on these functions. I shall give a C^∞ treatment, but a little reflection will show that the subject could be undertaken equally well from a real-analytic, complex-analytic, or algebraic viewpoint, etc., but not from a C^0-standpoint.

The de Rham complex is isomorphic to a subcomplex of the quantized complex, and there is a corresponding natural homomorphism of the classical cohomology algebra of a given manifold into the quantized cohomology algebra, which is an injection at least in dimension one. On the other hand, the quantized forms exist nontrivially up to dimension $2n$, n being the dimension of the manifold in question. In addition, in the case of a Riemannian manifold, they combine naturally with a different type of affine complex which is based on symmetric forms and Clifford algebra rather than antisymmetric forms and exterior algebra to form a locally simple doubly-quantized complex, properly containing the singly-quantized complex. It is not yet known whether the new C^∞ connectivity invariants which emerge are independent of the classical cohomological invariants; one difficulty in their treatment arises from the circumstance that the quantized forms transform canonically only under nonsingular C^∞ mappings. The quantized cohomology depends additionally on the choice of an algebra of operators on the manifold in question—differential, pseudodifferential, integral, etc.— and its closer study should involve in particular some illumination of the relations between these various types of operators.

Elementary motivation. For brevity, I shall refer to conventional differential forms as C-forms and to the new kind as Q-forms. A simple way to introduce C-forms is in connection with the differential equation $\partial y/\partial x_i = f_i$ $(i = 1, \dots, n)$ in R^n. This equation has a solution if and only if the C-form $\omega = \sum_i f_i dx_i$ is closed, and in this event, $dy = \omega$; one says, "On R^n, a form is exact if and only if it is closed". Observing now the elementary facts: (a) the mapping $h \to M_h$ which carries a function h into the operation, $g \to hg$, of multiplication by h, is an algebraic isomorphism; (b) if X is any vector field, the commutator $[X, M_g]$ is

M_{X_g}; the cited differential equation may be formulated as the problem of finding a multiplication operator Y such that

$$(*) \qquad [\partial/\partial x_i, Y] = M_{f_i} \qquad (i = 1, \ldots, n).$$

It then becomes visible that equation (*) may be considered a special case of the following equation for an unknown operator Y:

$$(**) \qquad [\partial/\partial x_i, Y] = A_i, \quad [M_{x_i}, Y] = B_i \qquad (i = 1, \ldots, n),$$

where the A_i and B_i are given operators. Indeed, equation (*) is precisely equivalent to the special case in which all B_i vanish; for the equations $[M_{x_i}, Y] = 0$, i.e. that Y commutes with all multiplications by coordinates, imply under the mildest regularity conditions, that Y is necessarily multiplication by a function, say y. The remaining equations then imply that $M_{(\partial y/\partial x_i)} = A_i$, showing that there can be a solution only if A_i is multiplication by a function f_i, and equation (*) is recovered.

Now the equation (**) has an extremely simple theory as regards its formal aspects. In an algebraic setting (see [**1**]) it can be shown that a solution exists if and only if

$$[A_i, M_{x_j}] = [B_i, \partial/\partial x_j]; \quad [A_i, A_j] = [B_i, B_j] = 0 \qquad (i, j = 1, \ldots, n);$$

and the solution Y is then unique within an additive scalar operator. Having made this and some related observations, it is then natural to introduce a generalized type of differential form which will permit the foregoing result to be summarized in the statement "On R^n, a generalized 1-form is exact if and only if it is closed"; to consider the extension of the theory to forms of higher degree, etc.

Q-forms on an affine space. The theory proceeds most compactly and canonically without the disjunction of the first-order operators in question into vector fields $\partial/\partial x_i$ on the one hand and multiplication operators M_{x_i} on the other; this disjunction is required only to exhibit explicitly the subsumption of the C-forms under the Q-forms. Let us therefore begin with a given real linear vector space $\mathscr{L}$, together with a given real nondegenerate antisymmetric bilinear form A on $\mathscr{L}$. ($\mathscr{L}$ may be thought of as the direct sum $\mathscr{M} \oplus \mathscr{M}^*$ of an underlying linear space $\mathscr{M}$ with its dual $\mathscr{M}^*$, and A as the corresponding induced form: $A(x + f, x' + f') = f(x') - f'(x)$; $x, x' \in \mathscr{M}$; $f, f' \in \mathscr{M}^*$.) The enveloping algebra $\mathscr{E}$ of $\mathscr{L}$ together with a unit e, modulo the relations $[x, y] = A(x, y)e$ is then central simple, and isomorphic to the algebra of all linear differential operators with polynomial coefficients on a suitable linear space (such as the cited space $\mathscr{M}$).

DEFINITIONS. An *r-Q-form* over the system $(\mathscr{L}, A)$ is an antisymmetric r-linear mapping from $\mathscr{L}$ into $\mathscr{E}$, r being a positive integer; a 0-Q-form is defined as an element of $\mathscr{E}$. The *derivative* of a given such form K is the $(r + 1)$-linear map on $\mathscr{L}$ to $\mathscr{E}$ given by the equation

$$dK(z_1, \ldots, z_{r+1}) = (r + 1)A_{r+1}([z_{r+1}, K(z_1, \ldots, z_r)]),$$

where A_{r+1} indicates the operation of antisymmetrization with respect to the indices $1, 2, \ldots, r + 1$. If H is an s-Q-form, the product $H \wedge K$ is defined by the equation

$$(H \wedge K)(z_1, \ldots, z_{r+s}) = A_{r+s}(H(z_1, \ldots, z_s)K(z_{s+1}, \ldots, z_{s+r})).$$

A Q-form is an element of the (algebraic) direct sum $\mathscr{K}(\mathscr{L}, A)$ of the spaces $\mathscr{K}_r(\mathscr{L}, A)$ of all r-Q-forms, over nonnegative integral values of r; and d and $\wedge$ are extended (uniquely) by linearity to all of $\mathscr{K}(\mathscr{L}, A)$. A Q-form K is called closed if $dK = 0$, and exact if $K = dH$ for some $H \in \mathscr{K}(\mathscr{L}, A)$.

THEOREM 1. *$\mathscr{K}(\mathscr{L}, A)$ is an associative algebra; it is invariant under d; $d^2 = 0$; and the usual algebraic rules involving d and $\wedge$ are valid.*

THEOREM 2. (GENERALIZED POINCARÉ LEMMA). *A Q-form is closed if and only if it is exact.*

As already indicated in a special case, the C-forms with polynomial coefficients on any maximal singular subspace $\mathscr{M}$ of $\mathscr{L}$ (i.e. one on which A vanishes identically) map naturally into $\mathscr{K}(\mathscr{L}, A)$, in such a way that d is carried over into conventional differentiation of exterior forms, and $\wedge$ into the conventional wedge product. One could now proceed to discuss analogs of conventional problems concerning partial differential equations, with unknown functions replaced by unknown operators, and Cauchy data or boundary values by given operators rather than given functions; a number of results expressed in terms of commutators may be expected to generalize. The problem of treating more general operators than the differential operators with polynomial coefficients, of which $\mathscr{E}$ essentially consists, is closely related to the problem of adapting Q-forms to nonlinear manifolds, to which I now turn.

Q-forms on manifolds. A fairly natural generalization of the space $\mathscr{L}$ occurring in the treatment of the affine case is the direct sum of the space $\mathscr{X}$ of all sufficiently regular vector fields (not merely vector displacements) with the space $\mathscr{M}$ of all multiplication operators (not merely multiplications by linear functionals). Formally speaking, one has little choice in the matter, in the absence of any given special structure on the given manifold M (it is of course another matter if this is given as a homogeneous space, or as complex analytic; in these cases, $\mathscr{X}$ and $\mathscr{M}$ may be correspondingly restricted). The passage from the space of infinitesimal translations to all vector fields, etc., represents a drastic change in the formal structure of the theory, and it is not at all obvious that the notion of Q-form can be effectively extended. Indeed, the extension depends on the introduction of a suitable kind of localization restriction for Q-forms, which restriction must in the first instance satisfy the basic desiderata of: (i) being satisfied by C-forms, (ii) being invariant under differentiation, d, (iii) being satisfied by arbitrary operators considered as 0-Q-forms.

For logical clarity and useful generality, it is helpful to begin in an abstract algebraic way. Let $\mathscr{M}$ be a given commutative topological algebra over the complex field; let $\mathscr{X}$ denote a given (Lie-) algebra of continuous derivations on $\mathscr{M}$, which is invariant under left multiplication by the multiplication operators M_a for $a \in \mathscr{M}$, where $M_a x = ax (x \in \mathscr{M})$. Let $\mathscr{E}$ denote a given algebra of continuous linear transformations on $\mathscr{M}$, which is invariant under multiplication on the right and left by elements of both $\mathscr{M}'$ and $\mathscr{X}$, where $\mathscr{M}' = [M_a : a \in \mathscr{M}]$. Associated with a given such system $(\mathscr{M}, \mathscr{X}, \mathscr{E})$ is a complex $\mathscr{K}$ obtained as follows. An r-Q-form is defined as an r-linear antisymmetric function on the space $\mathscr{L} = \mathscr{X} \oplus \mathscr{M}$, having its values in $\mathscr{E}$ and satisfying the condition that as a function of any one variable (the others being held fixed), it is a derivation, relative to multiplication as a partially defined operation in $\mathscr{L}$. In other terms, it is required that if $z_0, z_1, \ldots, z_r$ are in $\mathscr{L}$, and if $z_0 z_1 \in \mathscr{L}$ (as will be the case if $z_0 \in \mathscr{M}$ but not in general otherwise), then

$$(\asymp)\ K(z_0 z_1, z_2, \ldots, z_r) = z_0 K(z_1, z_2, \ldots, z_r) + K(z_0, z_2, \ldots, z_r) z_1 .$$

The algebraic direct sum $\mathscr{K}$ of the spaces $\mathscr{K}_r$ of all r-Q-forms ($r = 0, 1, \ldots$; a 0-Q-form is defined as an element of $\mathscr{E}$) is defined as the set of all Q-forms. If K is any r-Q-form, its derivative dK is defined as the function on $\mathscr{L} \times \ldots \times \mathscr{L}$ ($(r + 1)$ factors) to $\mathscr{E}$ given by the equation

$$dK(z_1, \ldots, z_{r+1}) = \textstyle\sum_{i=1}^{r+1} (-1)^i [K(z_1, \ldots, \hat{z}_i, \ldots, z_{r+1}), z_i]$$
$$-\textstyle\sum_{i<j} (-1)^{i+j} K([z_j, z_i], z_1, \ldots, \hat{z}_i, \ldots, \hat{z}_j, \ldots, z_{r+1}).$$

If L is an s-Q-form, the product $K \wedge L$ is the function on the $(r + s)$-fold product $\mathscr{L} \times \ldots \times \mathscr{L}$ given by the equation

$$(K \wedge L)(z_1, \ldots, z_{r+s}) = A_{r+s}(K(z_1, \ldots, z_r) L(z_{r+1}, \ldots, z_{r+s})).$$

As earlier, the operations d and $\wedge$ extend uniquely by linearity to all of $\mathscr{K}$, and one has the

THEOREM 3. *$\mathscr{K}$ is an associative with partially defined multiplication; it is invariant under d; $d^2 = 0$; and the usual algebraic rules involving d and $\wedge$ are valid to the extent that $\wedge$ is defined with K.*

The complex $\mathscr{K}$ may be called the *quantized* de Rham complex over $(\mathscr{M}, \mathscr{X}, \mathscr{E})$, in distinction to its subcomplex $\mathscr{H}$ consisting of "C-forms", where an r-Q-form K is defined as a C-form in case $K(z_1, \ldots, z_r)$ commutes with every element of $\mathscr{M}'$ for all values of $z_1, \ldots, z_r$, and vanishes if any argument z_i is in $\mathscr{M}'$. The terminology derives of course from the important special case in which $\mathscr{M}$ is the algebra of all C^∞ functions on the given C^∞ manifold M, $\mathscr{X}$ is the Lie algebra of all C^∞ vector fields on M, and $\mathscr{E}$ is the enveloping algebra of $\mathscr{X}$ and $\mathscr{M}'$. In this case, $\mathscr{H}$ is isomorphic with the complex of all C^∞ differential forms on M, via the mapping which carries a given such form Ω into the Q-form K given by the equations

$$K(X_1, \ldots, X_r) = M_{\Omega(X_1, \ldots, X_r)} \quad (X_1, \ldots, X_r \in \mathscr{X})$$
$$K(z_1, \ldots, z_r) = 0 \text{ if any } z_i \in \mathscr{M}';$$

this mapping intertwines the present operations d and $\wedge$ with the conventional ones.

On the other hand, this algebra $\mathscr{E}$ is less convenient as a topological algebra than the normed algebra of all C^∞ integral operators of compact support, which otherwise enjoys similar properties. A variety of other canonical algebras having the required properties could be cited, but the cited normed algebra seems well adapted also for analytical considerations in the direction of the quantization of the Hodge theory, and is the only one I shall consider here.

In extending the Poincaré lemma to Q-forms on R^n, the same difficulty as in the case of C-forms arises: in general, the antiderivative of a closed form of compact support (or square-integrable) can not be chosen to have compact support (or to be square-integrable). To contain this difficulty, let a form K be called "relatively exact" in case for any given compact set C, there exists a Q-form H such that $dH(z_1, \ldots, z_n) = K(z_1, \ldots, z_n)$ provided all the z_i are supported by C. Then with the indicated choices for $\mathscr{M}$, $\mathscr{X}$, and $\mathscr{E}$ as the indicated normed algebra, one has

THEOREM 4. *A closed Q-form on R^n is relatively exact.*

The proof is attained through the use of the Weyl transform, which is a unitary map from the space of all Hilbert-Schmidt operators T on $L_2(G)$ onto the space $L_2(G \oplus G^*)$, G being an n-dimensional vector group with dual G^* (see [2]). This transform is applicable by virtue of the observation that bracketing of T with infinitesimal translations or multiplications by coordinates correspond to multiplication of the Weyl transform by coordinates (cf. [3]). Antiderivation is thereby reduced to division, consistency being assured by the relations expressing closure; and the only problem is the singularity at the origin of the coordinates in $G \oplus G^*$. This singularity can be eliminated by modifying the kernel of T outside a compact set, and employing the Weyl transform in a fashion analogous to the possible employment of the Fourier transform in connection with the corresponding theorem for C-forms.

Symmetric quantization in a Riemannian space. In addition to the antisymmetric quantization involved in the foregoing, there is another species which is entirely analogous except for the one important difference of involving symmetric, rather than antisymmetric, forms. If $\mathscr{L}$ is a given linear space with a nondegenerate real symmetric form S, rather than antisymmetric form A, a corresponding (symmetric) r-Q-form may be defined as a symmetric multilinear mapping K from the r-fold product $\mathscr{L} \times \ldots \times \mathscr{L}$ to the Clifford algebra $\mathscr{E}$ over $(\mathscr{L}, S)$. With definitions for d and $\wedge$ which are entirely analogous to the earlier ones, except that antisymmetrization is replaced by symmetrization, and the commutator by a generalized commutator (which is the ordinary commutator for elements of even parity with elements of $\mathscr{L}$, and the anticommutator for elements of odd parity), there again results an acyclic complex $K^+(\mathscr{L}, S)$.

Now if $(\mathscr{L}, A)$ and $(\mathscr{L}', S)$ are two pairs consisting of given real linear vector spaces with given nondegenerate bilinear forms A and S (respectively antisymmetric and symmetric), the direct product $K^{-}(\mathscr{L}, A) \times K^{+}(\mathscr{L}', S)$ may be similarly treated, and indeed a unified treatment may be given for the complex $K(\mathscr{L}'', B)$ associated with a given real linear vector space $\mathscr{L}''$ and nondegenerate form B, provided B is the direct sum of an antisymmetric with a symmetric form. The commutator or generalized commutator need only be replaced by the operation $[u'', z'']_*$ defined for $u'' \in \mathscr{E}''$ and $z'' \in \mathscr{L}''$ by linearity together with the equations

$$[u \times u', z + z']_* = [u, z] \times u' + u \times [u', z']_+,$$

where $[u', z']_+$ denotes the anticommutator $u'z' + z'u'$.

If now M is a given C^∞ Riemannian manifold, one may first form a complex analogous to the purely symmetric one defined earlier for the pair $(\mathscr{L}, S)$, consisting of the sections of the Clifford bundle over M (the set of all pairs consisting of an element of M together with an element of the Clifford algebra over the tangent space at the point relative to the symmetric form defined by the metric). With the obvious extension of the notion of derivative, this complex is again locally acyclic. This complex may in turn be joined in a locally trivial fashion with the quantized de Rham complex earlier defined to form a doubly quantized de Rham complex in which both complexes are naturally imbedded.

This complex is in a certain sense locally acyclic, and its cohomology algebra therefore appears as a true connectivity invariant. The precise relations between this doubly quantized cohomology algebra, the two singly quantized cohomology algebras, and the classical de Rham cohomology, as well as their dependence on the precise algebra $\mathscr{E}$ of operators involved, are presently uncertain. It is however plausible to conjecture that (1) the natural homomorphism of the de Rham cohomology algebra into the quantized (either singly-antisymmetrically or doubly) cohomology algebra is an injection; (2) in the case of a compact manifold, the quantized cohomology is finite-dimensional. The first of these is easily seen in dimension 1, at least. The second may well follow from an extension of the Hodge theory to Q-forms with Hilbert-Schmidt operator coefficients, with a corresponding integration theory in which the trace of the operator plays the same role as the integral of a function in the classical theory.

References

1. I. Segal, *Quantized differential forms,* Topology **7** (1968), 147–172.

2. ———, *Transforms for operators over a locally compact abelian group*, Math. Scand. **13** (1963), 31–43.

3. R. Lavine, *The Weyl transform—Fourier analysis of operators in L_2 spaces*, Ph. D. Thesis, Massachusetts Institute of Technology, June 1965.

Brandeis University and
Massachusetts Institute of Technology

UNCOUPLED OVERDETERMINED SYSTEMS

D. C. SPENCER

1. **Introduction.** One of the main problems concerning overdetermined operators is the existence of local solutions. To make this problem clear, we require some remarks.

Let X be a differentiable manifold of dimension n, where "differentiable" will always mean "differentiable of class C^∞". If E is a vector bundle over X we shall, for simplicity, also write E for the sheaf of germs of differentiable sections of E. The interpretation of E as a sheaf of germs will only occur in contexts where differentiation is involved and such occurrences will always be clear. Thus, if E and F are vector bundles over X, a differential operator $D:E \to F$ is a map (linear over the constants) of the sheaf of sections of E into the sheaf of sections of F.

We recall, from preceding lectures, that we have the following theorem:

THEOREM 1.1 *Let $D:E \to F$ be a formally integrable (linear) differential operator. Then there exist a vector bundle G and differential operator $D':F \to G$ such that $D' \cdot D = 0$ and the complex*

$$E \overset{D}{\to} F \overset{D'}{\to} G \tag{1.1}$$

is formally exact and exact if and only if the following complex attached to the differential operator is exact at degree 1, namely the sequence

$$0 \to \Theta \to C^0 \overset{D^0}{\to} C^1 \overset{D^1}{\to} C^2 \overset{D^2}{\to} \cdots \overset{D^{n-1}}{\to} C^n \to 0, \tag{1.2}$$

where Θ is the kernel of $D:E \to F$ (sheaf of solutions of the homogeneous system), $C^0 = C^0_m = R_m$ (equation of the given operator) and, for $i \geq 1$,

$$C^i = C^i_m = (\Lambda^i T^* \otimes R_m)/\delta(\Lambda^{i-1} T^* \otimes g_{m+1}).$$

Here m is chosen sufficiently large to insure that the δ-cohomology vanishes (stable range).

The existence of local solutions is the exactness of (1.1).

The purpose of this lecture is to describe a condition which, hopefully, is a reasonable generalization, to overdetermined systems, of the notion of uncoupled operators in the determined case. This condition was first suggested by I. M. Singer. Most of the results presented here are contained in the Stanford thesis of B. MacKichan (see [**4**]) whose work completes systematically and extends some observations of V. W. Guillemin based on Singer's idea.

The condition, which takes the form of a bound for the operator in the δ-cohomology (δ-estimate), and various estimates derived from it by MacKichan, are described in §2. In §3 it is pointed out how the δ-estimate implies commutation of the symbols with their adjoints in Guillemin's local decomposition of the operator D^0 in (1.2). Finally, in §4, it is indicated how the δ-estimate implies the solvability of the Neumann problem which, in turn, shows that the cohomology of (1.1) (obstruction to the existence of local solutions) is a finite dimensional space. Recently W. J. Sweeney [7] has shown that, if the δ-estimate holds for an elliptic operator, then on a sufficiently small disc defined with respect to a fixed coordinate system, the harmonic space for the Neumann problem vanishes in positive degrees. Hence ellipticity and the δ-estimate imply the existence of local solutions (exactness of (1.1)).

2. **The δ-estimate.** We shall adhere strictly to the notation and definitions introduced in the previous sessions of this seminar, and shall not redefine here concepts previously introduced. We suppose throughout that $D:E \to F$ is a formally integrable differential operator of order k. We have the symbol maps of the sequence (1.2), namely

$$\sigma_i(D^l): S^{i+1}T^* \otimes C^l \to S^i T^* \otimes C^{l+1},$$

and we denote the kernel of $\sigma_i(D^l)$ by g^l_{i+1}. We remark that g^0_{i+1} is isomorphic to g_{m+i+1}.

We assume that riemannian metrics have been chosen on X, hence along the fibers of the cotangent bundle T^* of X, and along the fibers of the other vector bundles involved.

DEFINITION 2.1. We say that the operator $D:E \to F$ satisfies the δ-estimate if and only if, on the sequence

$$0 \to g^0_2 \xrightarrow{\delta} T^* \otimes g^0_1 \xrightarrow{\delta} \Lambda^2 T^* \otimes C^0,$$

we have the inequality

$$\|\delta\chi\|^2 \geq \tfrac{1}{2}\|\chi\|^2 \tag{2.1}$$

for all $\chi \in (T^* \otimes g^0_1) \cap \ker \delta^*$, where δ^* is the adjoint of δ (defined in terms of the metric).

We have

PROPOSITION 2.1. *The operator $D:E \to F$ satisfies the δ-estimate if and only if*

$$\|\delta\chi\|^2 \geq \tfrac{1}{2}(m+1)^2\|\chi\|^2$$

for all $\chi \in (T^* \otimes g_{m+1}) \cap \ker \delta^*$. Moreover, if D satisfies the δ-estimate then, for $i, l \geq 1$, we have

$$\|\delta\chi\|^2 \geq (l^2/i)\|\chi\|^2$$

for $\chi \in (\Lambda^{i-1}T^* \otimes g^0_l) \cap \ker \delta^*$.

Here and elsewhere δ^* denotes the adjoint of the operator indicated by the assertion.

As an immediate corollary of Proposition 2.1, we see that the sequences

$$0 \to g_l^0 \xrightarrow{\delta} T^* \otimes g_{l-1}^0 \xrightarrow{\delta} \dots \xrightarrow{\delta} \Lambda^{l-1}T^* \otimes g_1^0 \to \Lambda^l T^* \otimes C^0$$

are exact, and hence g_1^0 is involutive or equivalently g_{m+1} is involutive.

The following result should also be noted:

PROPOSITION 2.2. *If D satisfies the δ-estimate, then*

$$\|\delta\chi\|^2 \geq \tfrac{1}{2}\|\chi\|^2$$

for all $\chi \in (T^* \otimes g_1^l) \cap \ker \delta^*$.

The next proposition shows that the constant 1/2 occurring in the δ-estimate is distinguished.

PROPOSITION 2.3. (i) *If* $\|\delta\chi\|^2 \geq c(1, 1)\|\chi\|^2$ for $\chi \in (T^* \otimes g_1^0) \cap \ker \delta^*$, then $\|\delta\chi\|^2 \geq c(1, l)\|\chi\|^2$ for $\chi \in (T^* \otimes g_l^0) \cap \ker \delta^*$, where

$$c(1, l) = (1 - (l - 1)^2/4c(1, l - 1))l^2.$$

In particular, if $c(1, 1) = 1/2$, then $c(1, l) = l^2/2$.

(ii) If $\|\delta\chi\|^2 \geq c(1, l)\|\chi\|^2$ for $\chi \in (T^* \otimes g_l^0) \cap \ker \delta^*$, then $\|\delta\chi\|^2 \geq c(i, l)\|\chi\|^2$ for $\chi \in (\Lambda^i T^* \otimes g_l^0) \cap \ker \delta^*$, where

$$c(i, l) = \frac{i^2 c(i - 1, l) - l^2}{(i + 1)(i - 1)}.$$

In particular, if $c(1, l) = l^2/2$, then $c(i, l) = l^2/(i + 1)$.

From (i) of Proposition 2.3 we see that the constant 1/2 in the δ-estimate is the least constant which insures that $c(1, l) > 0$ for all l, and from (ii) that 1/2 is the least value for $c(1, 1)$ such that $c(i, l) > 0$ for all i and l.

3. **Commutation of symbols with adjoints.** We begin by recalling Guillemin's local decomposition of the operator D^0 in (1.2) (see [1]). For simplicity we drop the superscript and write simply D for the operator $D^0: C^0 \to C^1$, and we write D' for $D^1: C^1 \to C^2$. Since the given differential operator from E to F will not occur explicitly, no confusion will arise by writing D for D^0.

Given a point x_0 of X, let U_{x_0} be a maximal noncharacteristic subspace of $T_{x_0}^*$. We can choose a local coordinate $(x^1, x^2, \dots, x^n)$ on a neighborhood of x_0 such that $dx^1, dx^2, \dots, dx^k$ form a basis for U_{x_0} at the point x_0. Then $dx^1, dx^2, \dots, dx^k$ span a maximal noncharacteristic subspace U_{x_0} of T_x^2, and $dx^{k+1}, dx^{k+2}, \dots, dx^n$ span a complementary subspace W_x of T_x^*, at each point x of some neighborhood of x_0. Let U denote the sub-bundle of T^* (restricted to the neighborhood) spanned by $dx^1, dx^2, \dots, dx^k$. Since dx^i is noncharacteristic for $1 \leq i \leq k$, the map $\sigma_{dx^i}(D)$: $C^0 \to C^1$ is injective and we write

$$E_i = \sigma_{dx^i}(D)(C^0), \quad i = 1, 2, \dots, k.$$

We have the direct-sum decomposition $C^1 = E_0 + E_1 + \dots + E_k$ where E_0 is a complement of $E_1 + \dots + E_k$, and we let

$$\pi_i : C^1 \to E_i, \quad i = 0, 1, 2, \dots, k,$$

be the projections. Define

$$D_0 = \pi_0 \circ D, \qquad D_i = \sigma_{dx^i}(D)^{-1} \circ \pi_i \circ D \quad (1 \le i \le k),$$

where

$$D_0 : C^0 \to E_0, \qquad D_i : C^0 \to C^0 \quad (1 \le i \le k).$$

Then we have the Guillemin (direct-sum) decomposition

$$D = D_0 + \sum_{i=1}^{k} \sigma_{dx^i}(D) D_i.$$

We recall the following proposition:

PROPOSITION 3.1. *There exist differential operators*

$$D'_{ij} : C^1 \to C^0, \qquad D'_i : C^1 \to C^1, \quad 1 \le i, j \le k,$$

such that

$$(3.1) \qquad \begin{aligned} [D_i, D_j] &= D'_{ij} D_0 \\ D_0 D_i &= D'_i D_0 \end{aligned} \qquad 1 \le i, j \le k$$

where $[D_i, D_j] = D_i D_j - D_j D_i$. *In particular,* D_i *maps* $\ker D_0$ *into* $\ker D_0$ *and* $D_i D_j = D_j D_i$ *on* $\ker D_0$.

The commutation relations (3.1) yield corresponding relations for the symbols. Let η be an arbitrary covector. Then

$$(3.2) \qquad [\sigma_\eta(D_i), \sigma_\eta(D_j)] = \sigma_\eta(D'_{ij})\sigma_\eta(D_0), \qquad \sigma_\eta(D_0)\sigma_\eta(D_i) = \sigma_\eta(D'_i)\sigma_\eta(D_0).$$

Hence $\sigma_\eta(D_i)$ maps $\ker \sigma_\eta(D_0)$ into $\ker \sigma_\eta(D_0)$, and

$$(3.3) \qquad \sigma_\eta(D_i)\sigma_\eta(D_j) = \sigma_\eta(D_j)\sigma_\eta(D_i) \text{ on } \ker \sigma_\eta(D_0).$$

We remark that these relations are trivial if η belongs to the maximal noncharacteristic subspace since $\sigma_\eta(D_0)$ then vanishes.

After these preliminaries, we can state the theorem.

THEOREM 3.1 (MACKICHAN [4]). *Suppose that the given operator from E to F satisfies the* δ*-estimate. Let U be a maximal noncharacteristic subspace of* T^*, *and let* D_i, $0 \le i \le k$, *be the local differential operators constructed in terms of U. Then, if* η *is a covector not belonging to U we have, on* $\ker \sigma_\eta(D_0)$, *the commutation relations*

$$\sigma_\eta(D_i)\sigma_\eta(D_j^*) = \sigma_\eta(D_j^*)\sigma_\eta(D_i), \quad 1 \le i, j \le k,$$

where $\sigma_\eta(D_j^*) = \sigma_\eta(D_j)^*$ *denotes the adjoint of* $\sigma_\eta(D_j)$, D_j^* *the adjoint of* D_j, *defined in terms of the metrics.*

We shall indicate briefly how the δ-estimate enters into the proof of Theorem 3.1. We suppose that η is a covector not belonging to the maximal noncharacteristic subspace U in terms of which $D_i (0 \leq i \leq k)$ are defined and we write, for simplicity,

$$A_i = \sigma_\eta(D_i), \quad i = 1, 2, \ldots, k.$$

Moreover, let A_0 be the identity map of C^0 and set $B = \sum_{i=0}^k A_i^* A_i$. It is obvious that B is selfadjoint and positive-definite (since A_0 is the identity), and B is an automorphism of $\ker \sigma_\eta(D_0)$. A calculation shows that, if

$$\chi = \sum_{i,j=0}^{k} \xi^j \otimes \xi^i \otimes A_i c_j$$

where $\xi^0 = \eta$, $\xi^i = dx^i$ for $i = 1, 2, \ldots, k$ (basis of U), and $c_j \in \ker \sigma_\eta(D_0)$, then $\delta^* \chi = 0$ if and only if

$$\sum_{j=0}^{k} A_j^* B c_j = 0. \tag{3.4}$$

Moreover,

$$\|\delta\chi\|^2 - \tfrac{1}{2}\|\chi\|^2 = - \sum_{i,j=0}^{k} \langle A_j c_i, A_i c_j \rangle.$$

The δ-estimate is therefore equivalent to the assertion that (3.4) implies

$$\sum_{i,j=0}^{k} \langle A_j c_i, A_i c_j \rangle \leq 0. \tag{3.5}$$

From (3.5) the desired commutation relations of Theorem 3.1, namely $A_i A_j^* = A_j^* A_i$ on $\ker \sigma_\eta(D_0)$, are obtained by a calculation which we omit.

The point is that the δ-estimate seems to imply inequalities of the type (3.5). An inequality of similar form will be derived from the δ-estimate in the next section.

4. **The Neumann problem.** Let X be a smooth riemannian manifold of dimension, n, and let $\Omega \subset X$ be a compact, n-dimensional manifold whose boundary is imbedded differentiably in X as a compact $(n-1)$-dimensional submanifold. The metric on X induces metrics on Ω and on its boundary $\omega = \partial\Omega$. For simplicity we drop the superscripts on the differential operators of the sequence (1.2) and write simply D for D^i. No confusion will arise between these operators and the given operator from E to F, since the latter will never occur explicitly. Next, we choose an inner product along the fibers of the bundle $J_m(E) \to X$, where m is a fixed integer which is sufficiently large to insure that the δ-cohomology vanishes (stable range). This metric, together with the riemannian metric, induces, for each i, inner products along the fibers of the bundles $\Lambda^i T^* \otimes J_m(E)$ and hence along the fibers of each of the sub-bundles $\Lambda^i T^* \otimes R_m$, $\Lambda^i T^* \otimes g_m$ and $\delta(\Lambda^i T^* \otimes g_{m+1})$. Finally, by choosing a suitable splitting (see [**6**]), we have the isomorphism

$$C^i = C^i_{m+1} \cong (\Lambda^i T^* \otimes R_m) + \delta(\Lambda^i T^* \otimes g_{m+1}).$$

(Here m has been replaced by $m + 1$ for notational convenience.) Hence we obtain, for each i, an inner product along the fibers of C^i.

We denote by $\Gamma(\Omega, C^i)$ the space of all differentiable sections of C^i over Ω; these sections are smooth up to and including the boundary ω of Ω and can be extended smoothly across ω. If $u, v \in \Gamma(\Omega, C^i)$ we have the inner product

$$\langle u, v\rangle = \textstyle\int_\Omega \langle u, v\rangle_x dv$$

where dv is the riemannian volume element and $\langle \ldots \rangle_x$ is the inner product along the fiber of C^i over the point $x \in X$. We denote by $L_2(\Omega, C^i)$ the Hilbert space of all square-integrable sections over Ω of C^i.

Let D^* be the formal adjoint of the operator D. We define the Neumann space to be the graded space $\boldsymbol{N} = \oplus_i \boldsymbol{N}^i$ where $\boldsymbol{N}^i$ is the subspace of $\Gamma(\Omega, C^i)$ composed of all sections u of $\Gamma(\Omega, C^i)$ satisfying the pair of boundary conditions

$$\langle Dv, u\rangle = \langle v, D^*u\rangle \quad \text{for each } v \in \Gamma(\Omega, C^{i-1}), \tag{4.1}$$

$$\langle Dv, Du\rangle = \langle v, D^*Du\rangle \quad \text{for each } v \in \Gamma(\Omega, C^i). \tag{4.2}$$

Define $\boldsymbol{H} = \oplus_i \boldsymbol{H}^i$ to be the graded subspace of $\boldsymbol{N}$ which is annihilated by the laplacian $DD^* + D^*D$. Since, for $u \in \boldsymbol{N}^i$, we have by (4.1) and (4.2)

$$\langle (DD^* + D^*D)u, u\rangle = \|Du\|^2 + \|D^*u\|^2,$$

we see that $\boldsymbol{H}^i = \{u \in \boldsymbol{N}^i | Du = D^*u = 0\}$.

Definition 4.1. We say that the Neumann problem is solvable on Ω (for the given operator from E to F) if, for each $i \geq 1$, $\boldsymbol{H}^i$ is closed in the space $L_2(\Omega, C^i)$, and there exist bounded operators $N: L(\Omega, C^i) \to L(\Omega, C^i)$ mapping $\Gamma(\Omega, C^i)$ into $\boldsymbol{N}^i$ such that:

(i) $NH = HN = 0$, where $H: L_2(\Omega, C^i) \to \boldsymbol{H}^i$ is the orthogonal projection;

(ii) each element $u \in \Gamma(\Omega, C^i)$ can be written in the form

$$u = DD^*Nu + D^*DNu + Hu \tag{4.3}$$

where, in view of (4.1) and (4.2), the terms are mutually orthogonal;

(iii) the commutation relation $DN = ND$ holds.

We remark that (iii) is a consequence of (i) and (ii).

By (iii) the decomposition (4.3) yields a cochain homotopy $I - H = D(D^*N) + (D^*N)D$, and the cohomology of the sequence

$$0 \to \Gamma(\Omega, \Theta) \to \Gamma(\Omega, C^0) \xrightarrow{D} \Gamma(\Omega, C^1) \xrightarrow{D} \ldots \xrightarrow{D} \Gamma(\Omega, C^n) \to 0$$

is therefore isomorphic to $\boldsymbol{H}^+ = \oplus_{i>0} \boldsymbol{H}^i$. If the Neumann problem is solvable on Ω, then the cohomology of the sequence (see (1.1)) $\Gamma(\Omega, E) \xrightarrow{D} \Gamma(\Omega, F) \xrightarrow{D'} \Gamma(\Omega, G)$ is isomorphic to H^1, i.e. $\ker D'/\mathrm{im} D = \boldsymbol{H}^1$. It is to be expected that, in the case where Ω is a sufficiently small disc, $\boldsymbol{H}^1$ vanishes.

We have the following important theorem of J. J. Kohn and L. Nirenberg [**3**, §2, Theorem 5].

THEOREM 4.1. *Suppose that the operator from E to F is formally integrable and elliptic, and that there is a constant* $c > 0$ *such that, for any element u of* $\Gamma(\Omega, C^i)$, $(i \geq 1)$ *satisfying the boundary condition* (4.1), *we have*

$$\|Du\|^2 + \|D^*u\|^2 + \|u\|^2 \geq c\textstyle\int_\omega \|u\|_x^2 ds, \tag{4.4}$$

where ds denotes the volume element in the boundary ω *of* Ω. *Then the Neumann problem is solvable on* Ω *for the given operator.*

We shall now describe briefly how the δ-estimate implies (4.4) for domains Ω satisfying certain convexity conditions. For simplicity we shall assume that Ω is covered by a single coordinate $(x^1, x^2, \ldots, x^n)$, although this assumption is not necessary in any way.

Let r be a differentiable function on X such that:

(i) $r(x) = 0$ if and only if $x \in \omega$ (boundary of Ω);

(ii) $r(x) \geq 0$ for $x \in \Omega$;

(iii) $|dr| = 1$ on ω.

Then an element u of $\Gamma(\Omega, C^i)$ satisfies the boundary condition (4.1) if and only if

$$\sigma_{dr}(D)^*u = 0 \quad \text{on } \omega. \tag{4.5}$$

Let u be an element of $\Gamma(\Omega, C^i)$, where $i \geq 1$, which satisfies (4.5). We integrate by parts the term $\|D^*u\|^2$ in a way which has now become familiar (compare Kohn [2]). In order to formulate the result of this integration by parts, we introduce the map

$$(1 \otimes \sigma(D))^* : T^* \otimes C^i \to T^* \otimes T^* \otimes C^{i-1}.$$

Let v be an element of $T^* \otimes C^i$. Then, in terms of the coordinate $(x^1, x^2, \ldots, x^n)$,

$$v = \sum_{l=1}^{n} dx^l \otimes v_l$$

where $v_l \in C^i$, and

$$(1 \otimes \sigma(D))^*v = \sum_{l,m=1}^{n} dx^l \otimes dx^m \otimes A_m^* v_l$$

where

$$\sigma(D)^*v_l = \sum_{m=1}^{n} dx^m \otimes A_m^* v_l.$$

We let

$$S : T^* \otimes T^* \otimes C^{i-1} \to T^* \otimes T^* \otimes C^{i-1}$$

be the switching operator where $S(dx^l \otimes dx^m \otimes c) = dx^m \otimes dx^l \otimes c$. For simplicity let $Q(u, u) = \|Du\|^2 + \|D^*u\|^2 + \|u\|^2$. Then the integration by parts yields an inequality of the form

$$cQ(u, u) \geq \langle S(1 \otimes \sigma(D))^*du, (1 \otimes \sigma(D))^*\text{du}\rangle + L(u) \tag{4.6}$$

where $c > 0$ and $L(u)$ is a "Levi form". Namely, denote scalar products over the boundary by attaching a subscript ω to $\langle \ldots \rangle$, i.e., $\langle \ldots \rangle_\omega$ is the scalar product in the boundary and $\|\ldots\|_\omega$ is the corresponding norm. Then, in the case where the operator has constant coefficients (with respect to the coordinate), $L(u)$ has the form

$$L(u) = \sum_{l,m=1}^{n} \langle \frac{\partial^2 r}{\partial x^l \partial x^m} A_l^* u, A_m^* u \rangle_\omega .$$

The convexity requirement for Ω is that there exist a constant $c' > 0$ such that

$$L(u) \geq c' \|u\|_\omega^2. \tag{4.7}$$

We suppose that (4.7) is satisfied; then (4.6) can be written in the form

$$\|u\|_\omega^2 \leq -c\langle S(1 \otimes \sigma(D))^* du, (1 \otimes \sigma(D))^* du\rangle + c' Q(u, u)$$

where c and c' are positive constants. Therefore (4.4) holds if there is a positive constant c such that

$$-\langle S(1 \otimes \sigma(D))^* du, (1 \otimes \sigma(D))^* du\rangle \leq cQ(u, u). \tag{4.8}$$

Next, the inequality (4.8) will hold *a fortiori* if it holds in the pointwise sense, namely if, for every $v \in T^* \otimes C^i$, we have the pointwise inequality

$$-\langle S(1 \otimes \sigma(D))^* v, (1 \otimes \sigma(D))^* v\rangle_x \leq c\{\|\sigma(D)v\|_x^2 + \|\sigma(D^*)v\|_x^2\}. \tag{4.9}$$

In fact, suppose that (4.9) holds, take $v = du$ where u is an element of $\Gamma(\Omega, C^i)$ and integrate (4.9) over Ω. Then we obtain (4.8) since $\|\sigma(D)du\|^2 + \|\sigma(D^*)du\|^2$ is dominated by a constant multiple of $Q(u, u)$.

Finally, it is easily seen that (4.9) holds if

$$\langle S(1 \otimes \sigma(D))^* v, (1 \otimes \sigma(D))^* v\rangle_x \geq 0 \tag{4.10}$$

for all $v \in T^* \otimes C^i$ satisfying $\sigma(D)v = 0$ and $\sigma(D^*)v = 0$. We drop the condition $\sigma(D^*)v = 0$. Then we have the main result:

THEOREM 4.2 (MACKICHAN [**4**]). *The inequality* (4.10) *for all* $v \in g_1^i = \ker \sigma(D^i)$ *is equivalent to the estimate of Proposition* 2.2, *namely* $\|\delta\chi\|^2 \geq \frac{1}{2}\|\chi\|^2$ *for all* $\chi \in (T^* \otimes g_1^i) \cap \ker \sigma^*$. *Hence, if the given operator from E to F satisfies the* δ*-estimate, then* (4.4) *holds on* Ω *for all* $u \in \Gamma(\Omega, C^i)$, $i \geq 1$, *satisfying* (4.5), *provided that* Ω *satisfies* (4.7).

Let $v \in T^* \otimes C^i$; then, in terms of the coordinate, $v = \sum dx^1 \otimes v_1$ where $v_1 \in C^i$, and we have

$$\langle S(1 \otimes \sigma(D))^* v, (1 \otimes \sigma(D))^* v\rangle = \sum_{l,m=1}^{n} \langle A_l^* v_m, A_m^* v_l\rangle. \tag{4.11}$$

Thus the δ-estimate implies the inequality

$$\sum_{l,m=1}^{n} \langle A_l^* v_m, A_m^* v_l\rangle \geq 0 \tag{4.12}$$

for all $v = \sum dx^1 \otimes v_1 \in g_1^i = \ker \sigma(D^i)$. The form of this inequality is similar to that of (3.5).

THEOREM 4.3 (SWEENEY [7]). *Suppose that the given operator from E to F is elliptic and satisfies the δ-estimate. Then, on a sufficiently small coordinate disc Ω (defined in terms of a fixed coordinate system), the harmonic space of the Neumann problem vanishes in degrees. In particular, the sequence*

$$\Gamma(\Omega, E) \xrightarrow{D} \Gamma(\Omega, F) \xrightarrow{D'} \Gamma(\Omega, G)$$

is exact (since its cohomology is isomorphic to $\boldsymbol{H}^1$), and hence (1.1) *is exact (local solutions exist).*

The δ-estimate holds in the cases where the Neumann problem is known to be solvable, for example in the case of the $\bar{\partial}$-operator (Cauchy-Riemann operator in several complex variables). The solvability of the $\bar{\partial}$-Neumann problem on strongly pseudoconvex domains was first proved by Kohn [2(a), (b)] on the basis of an inequality of Morrey [5(a)].

Problems. (These problems came out of the seminar on overdetermined systems.)

1. (Guillemin) Given $\varepsilon > 0$, does every involutive operator $D:E \to F$ satisfy, for suitable choice of metric the δ-estimate (2.1) of Spencer's talk with $\frac{1}{2}$ replaced by $\frac{1}{2} - \varepsilon$ for any $\varepsilon > 0$? (The converse is true.)

2. (Goldschmidt-Kuranishi) Given a differential operator $D:E \to F$, under what conditions can one find a differential operator $D':G \to E$ such that the sequence

$$G \xrightarrow{D'} E \xrightarrow{D} F$$

is formally exact (see Theorem 3 of Goldschmidt's talk)? The operator D must be formally exact. Find conditions on D so that this sequence is an elliptic complex.

3. (Guillemin) In the analytic category, generalize the Cartan-Kähler theory with initial data prescribed on characteristic surfaces, that is, study the analogue of the Goursat problem for overdetermined systems.

4. (Guillemin) Are generic characteristics belonging to lower dimensional components of the characteristic variety of a formally integrable operator integrable?

5. Are formally exact elliptic complexes locally exact?

6. (Spencer) Are formally exact elliptic complexes exact over
   (a) a sufficiently small open ball;
   (b) a sufficiently small closed ball;

over what domains with appropriate "convexity" conditions are these complexes exact?

7. (Ehrenpreis) Is the "generic" formally exact complex of real operators locally exact?

8. (Sweeney) Are formally exact elliptic complexes

$$E \xrightarrow{D} F \xrightarrow{D'} G,$$

where D, D' are of arbitrary order and E is a line bundle, locally exact? If D is of first order, the answer is yes in most cases (Nirenberg).

9. When is the D-Neumann problem (see Spencer's talk) for an elliptic complex a coercive (or elliptic) boundary value problem? (This problem can be reduced to an algebraic question by well known methods.)

10. (Calabi) Consider the nilpotent Lie group

$$\begin{pmatrix} 1 & x & t \\ 0 & 1 & y \\ 0 & 0 & 1 \end{pmatrix}$$

with parameters t, x, y. The partial differential operator

$$L[u] = \partial_t u - \partial_x^2 u - \partial_y^2 u + 2x\partial_y\partial_t u - x^2\partial_t^2 u$$

where $u = u(t, x, y)$, is invariant under left translations of this group; at each point it is equivalent to the parabolic heat operator, but, even locally, the characteristic 1-form is $dt + 2dy$ which is not integrable. Is any fundamental solution real analytic away from the singularity as in the case of elliptic operators?

REFERENCES

1. V. W. Guillemin, *Some algebraic results concerning the characteristics of overdetermined partial differential equations*, Amer. J. Math. **90** (1968), 270–284.

2. J. J. Kohn, *Harmonic integrals on strongly pseudo-convex manifolds*, I, Ann. of Math. **78** (1963), 112–148.

3. ———, *Harmonic integrals on strongly pseudo-convex manifolds*, II, Ann. of Math. **79** (1964), 450–472.

4. ———, *Boundaries of complex manifolds*, Proc. Conference on Complex Analysis, Univ. of Minnesota Press, 1964.

5. J. J. Kohn and L. Nirenberg, *Non-coercive boundary value problems*, Comm. Pure App. Math. **18** (1965), 443–492.

6. B. B. MacKichan, *A generalization to overdetermined systems of the notion of diagonal operators*, Thesis, Stanford University, 1968.

7. C. B. Morrey, *The analytic embedding of abstract real-analytic manifolds*, Ann. of Math. **68** (1958), 159–201.

8. ———, *The $\bar{\partial}$-Neumann problem on strongly pseudo-convex manifolds*, Outlines of Joint Soviet–American Symposium on Partial Differential Equations, Novosibirsk, 1963, pp. 171–178.

9. D. C. Spencer, *Deformation of structures on manifolds defined by transitive, continuous pseudogroups*, III, Ann. of Math. **81** (1965), 389–450.

10. W. J. Sweeney, *A uniqueness theorem for the Neumann problem*, Ann. of Math. (to appear).

PRINCETON UNIVERSITY

TRIANGULAR OPERATORS

W. J. SWEENEY

Our topic here is the Neumann problem for those overdetermined elliptic operators whose symbol matrices are triangular. We shall show that the usual *a priori* estimates may fail to solve the Neumann problem for such operators, even in cases where C^∞ existence theorems are easily obtained by other means.

We begin by describing the Neumann problem. Let A, B, C be Hermitian vector bundles over a Riemannian manifold X, and let

$$0 \to A \overset{\mathscr{D}}{\to} B \overset{\mathscr{E}}{\to} C \tag{1}$$

be an elliptic complex consisting of first order differential operators $\mathscr{D}$ and $\mathscr{E}$. Let Ω be a compact submanifold of X with C^∞ boundary, Ω having the same dimension as X. Let $S: L_2(\Omega, A) \to L_2(\Omega, B)$ and $T: L_2(\Omega, B) \to L_2(\Omega, C)$ denote, respectively, the L_2 closures of $\mathscr{D}: C^\infty(\Omega, A) \to C^\infty(\Omega, B)$ and $\mathscr{E}: C^\infty(\Omega, B) \to (\Omega, C)$. We now state

Neumann problem. To find a bounded operator N on $L_2(\Omega, B)$, mapping $C^\infty(\Omega, B)$ into itself, such that

$$(SS^* + T^*T)N = I - H \tag{2}$$

on $L_2(\Omega, B)$, where H is orthogonal projection onto $\ker(SS^* + T^*T)$.

We remark that (2) is to be understood in the operator theoretic sense. Thus if N satisfies (2), then, in particular, Nu must belong to the domain of $SS^* + T^*T$ for each $u \in L_2(\Omega, B)$. For $u \in C^\infty(\Omega, B)$ this means that $v = Nu$ satisfies a certain first order boundary condition, say $b(v) = 0$ on $\partial\Omega$, which one can compute explicitly from Stokes' formula. If $u \in C^\infty(\Omega, B)$ is orthogonal to $\ker(SS^* + T^*T)$, then $v = Nu$ is a C^∞ solution to the boundary value problem

$$(\mathscr{D}\mathscr{D}^* + \mathscr{E}^*\mathscr{E})v = u \quad \text{on } \Omega,\ b(v) = 0 \quad \text{on } \partial\Omega. \tag{3}$$

If we impose the normalizing condition

$$HN = NH = 0 \tag{4}$$

on N, then $v = Nu$ is the unique solution which is orthogonal to $\ker(SS^* + T^*T)$. When $A = 0$, $B =$ trivial line bundle, $C =$ cotangent bundle, $\mathscr{D} =$ exterior derivative, then (3) is the classical Neumann problem.

One function of the Neumann problem is to investigate the homology of the sequence

$$C^\infty(\Omega, A) \overset{\mathscr{D}}{\to} C^\infty(\Omega, B) \overset{\mathscr{E}}{\to} C^\infty(\Omega, C). \tag{5}$$

In fact, the Neumann boundary conditions on Nu are precisely the right conditions to make

$$\begin{aligned} u &= (SS^* + T^*T)Nu + Hu \\ &= \mathscr{D}\mathscr{D}^*Nu + \mathscr{E}^*\mathscr{E}Nu + Hu \end{aligned} \tag{6}$$

a decomposition of $u \in C^\infty(\Omega, B)$ into mutually orthogonal terms. Using this orthogonality one shows that $\mathscr{E}^*\mathscr{E}Nu = 0$ when $\mathscr{E}u = 0$ and that the homology of (5) is isomorphic to $\ker(SS^* + T^*T)$. One of the aims in the theory is to solve the Neumann problem when Ω is a sufficiently small ball in Euclidean space and to show, in addition, that $\ker(SS^* + T^*T) = 0$. By solving this "local Neumann problem" one establishes the exactness of (5) for small Ω and hence the local solvability of the equation $\mathscr{D}f = g$ if g satisfies $\mathscr{E}g = 0$.

The following theorem, a special case of theorems due to J. J. Kohn and L. Nirenberg, provides a method of solving the Neumann problem:

THEOREM 1. *If the Dirichlet norm* $Q(u) = (\|\mathscr{D}^*u\|^2 + \|\mathscr{E}u\|^2 + \|u\|^2)^{1/2}$ *is completely continuous with respect to the* L_2 *norm on the space* $C^\infty(\Omega, B) \cap \operatorname{dom} S^*$, *then the Neumann problem is solvable.*

The hypothesis of complete continuity means that if a sequence $\{u_j\} \subset C^\infty(\Omega, B) \cap \operatorname{dom} S^*$ is bounded in the norm Q, then a subsequence converges in the L_2 norm. This will be true, for example, if the estimate

$$\|u\|_{1/2} \le cQ(u) \tag{7}$$

holds for all $u \in C^\infty(\Omega, B) \cap \operatorname{dom} S^*$, where $\|\cdot\|_{1/2}$ is the Sobolev norm of order 1/2 on Ω.

If $\dim X = 1$ or 2, then (7) is not difficult to verify. If fiber $\dim A = 1$, then the pointwise methods discussed above by Spencer apply, and (7) holds for Ω satisfying the appropriate convexity conditions. The local solvability of the equation $\mathscr{D}f = g$, in this case, is also a consequence of the Frobenius-Nirenberg theorem.

We now turn to an example of an elliptic complex (1) where $X = R^3$ and A, B, C are product bundles with complex fiber dimension 2, 4, 2 respectively. Namely, we define $\mathscr{D}: A \to B$ and $\mathscr{E}: B \to C$ by

$$\begin{aligned} \mathscr{D}(u, v) &= \left(\frac{\partial v}{\partial \bar z}, \frac{\partial v}{\partial t}, \frac{\partial u}{\partial \bar z}, \frac{\partial u}{\partial t} - \frac{\partial v}{\partial y}\right), \\ \mathscr{E}(w_1, w_2, w_3, w_4) &= \left(\frac{\partial w_1}{\partial t} - \frac{\partial w_2}{\partial \bar z}, \frac{\partial w_3}{\partial t} - \frac{\partial w_1}{\partial y} - \frac{\partial w_4}{\partial \bar z}\right), \end{aligned} \tag{8}$$

where (x, y, t) is the coordinate on R^3, $Z = x + iy$, and $\partial/\partial \bar z = \frac{1}{2}((\partial/\partial x) + i(\partial/\partial y))$. It is easy to check that (8) defines an elliptic complex (1).

The equation $\mathscr{D}(u, v) = (w_1, w_2, w_3, w_4)$ can be written in triangular form:

$$\begin{aligned} \mathscr{D}'v &= (w_1, w_2) \\ \mathscr{D}'u - (0, \partial v/\partial y) &= (w_3, w_4), \end{aligned} \tag{9}$$

where $\mathscr{D}'v = (\partial v/\partial \bar{z}, \partial v/\partial t)$. If A', B', C' are product bundles over R^3 with fiber dimension 1, 2, 1, respectively, and if $\mathscr{E}' : B' \to C'$ is defined by $\mathscr{E}'(w_1, w_2) = \partial w_1/\partial t - \partial w_2/\partial \bar{z}$, then

$$A' \xrightarrow{D'} B' \xrightarrow{\mathscr{E}'} C' \tag{10}$$

is an elliptic complex. If $\Omega \subset R^3$ is the unit ball, then the estimate (7) is easily established for (10) and, in fact, the sequence

$$C^\infty(\Omega, A') \xrightarrow{D'} C^\infty(\Omega, B') \xrightarrow{\mathscr{E}'} C^\infty(\Omega, C') \tag{11}$$

is exact.

THEOREM 2. *With $\mathscr{D}$ and $\mathscr{E}$ defined by* (8) *the sequence*

$$C^\infty(\Omega, A) \xrightarrow{D} C^\infty(\Omega, B) \xrightarrow{\mathscr{E}} C^\infty(\Omega, C)$$

is exact.

PROOF. Assuming that $(w_1, w_2, w_3, w_4) \in C^\infty(\Omega, B)$ satisfies $\mathscr{E}(w_1, w_2, w_3, w_4) = 0$, we must find $(u, v) \in C^\infty(\Omega, A)$ satisfying (9). To do this note that $\mathscr{E}'(w_1, w_2) = 0$ so that by the exactness of (11) there exists $v \in C^\infty(\Omega, A')$ satisfying $\mathscr{D}'v = (w_1, w_2)$. Next note that $(w_3, w_4) + (0, \partial v/\partial y)$ is also in the null space of $\mathscr{E}'$ so that again we can find $u \in C^\infty(\Omega, A')$ with $\mathscr{D}'u = (w_3, w_4) + (0, \partial v/\partial y)$. The desired (u, v) has now been found.

We now ask whether the $\frac{1}{2}$-estimate (7) holds for $\mathscr{D}$ and $\mathscr{E}$ over Ω. The following theorem will be useful.

THEOREM 3. *Assume that* (7) *holds for* $A \xrightarrow{\mathscr{D}} B \xrightarrow{\mathscr{E}} C$. *If $f \in$ dom S is orthogonal to* ker S, *then f belongs to the Sobolev space $H_{1/2}(\Omega, B)$*

According to Theorem 3, the estimate (7) implies that the equation $\mathscr{D}f = g$ can always be solved, if $\mathscr{E}g = 0$, with a gain of one-half a derivative. Note that the proof of Theorem 2 cannot be extended to show that our operator $\mathscr{D}$ has this property. If $(w_1, w_2, w_3, w_4) \in L_2(\Omega, B)$ is in the kernel of $\mathscr{E}$, then we may solve $\mathscr{D}'v = (w_1, w_2)$ for $v \in H_{1/2}(\Omega, A')$; but then $(w_3, w_4) + (0, \partial v/\partial y)$ may belong only to $H_{1/2}(\Omega, B)$ and perhaps the solution u to $\mathscr{D}'u = (w_3, w_4) + (0, \partial v/\partial y)$ will belong only to $L_2(\Omega, A')$. In fact, we have

THEOREM 4. *Let $g = (0, (1 - z)^{-\alpha}, 0, 0)$, where $\alpha = 5/4 - \varepsilon$, $\varepsilon > 0$ small. Then $g \in$ range S, but if $f \in$ dom S, f orthogonal to* ker S, *and $Sf = g$, then $f \notin H_{1/2}(\Omega, A)$.*

The proof of Theorem 4 is computational and somewhat involved. The following pair of theorems are easier to prove and are actually stronger.

THEOREM 3′. *Assume that the Dirichlet norm Q is completely continuous. Let $\{f_m\}$ be a sequence in* dom S *satisfying $\|Sf_m\| \le 1$, f_m orthogonal to* ker S. *Then a subsequence of $\{f_m\}$ converges in L_2 norm.*

THEOREM 4′. *Assume that Q is completely continuous for the complex defined by* (8). *Let a_m be the $L_2(\Omega)$ norm of $z^m = (x + iy)^m$ and let $g_m = (0, z^m/a^m, 0, 0)$. Then g_m is in the range of S, but if $Sf_m = g_m$ and if f_m is orthogonal to* ker S, *then no subsequence of $\{f_m\}$ converges in L_2 norm.*

Theorems 4′ and 5′ show that the Dirichlet norm is our example is not completely continuous so that Theorem 1 does not apply. On the other hand, we have already seen (Theorem 2) that local exactness holds.

If we consider the complex $A \overset{\mathscr{D}}{\to} B \overset{\mathscr{E}}{\to} C$ with $\mathscr{D}$ defined by

$$\mathscr{D}(u, v, w) = \left(\frac{\partial v}{\partial \bar{z}}, \frac{\partial v}{\partial t}, \frac{\partial u}{\partial \bar{z}}, \frac{\partial u}{\partial t} - \frac{\partial v}{\partial y}, \frac{\partial w}{\partial \bar{z}} - \frac{\partial u}{\partial y}\right)$$

then one can show that for some $g \in L_2(\Omega, B)$ satisfying $\mathscr{E}g = 0$ the equation $\mathscr{D}f = g$ can be solved for f belonging only to $H_{1/2}(\Omega, A)$. Hence the sequence

$$L_2(\Omega, A) \overset{S}{\to} L_2(\Omega, B) \overset{T}{\to} L_2(\Omega, C)$$

is not exact. Since $\mathscr{D}$ is of triangular form, however, the argument of Theorem 2 shows that

$$C^\infty(\Omega, A) \overset{\mathscr{D}}{\to} C^\infty(\Omega, B) \overset{\mathscr{E}}{\to} C^\infty(\Omega, C)$$

is exact.

Proofs of Theorems 3′ and 4′ are contained in our note *A non-compact Dirichlet norm*, Proc. Nat. Acad. Sci. **58** (1967), 2193–2195. The results of Kohn-Nirenberg cited above are contained in their paper *Non-coercive boundary value problems*, Comm. Pure Appl. Math. **18** (1965), 443–492.

PRINCETON UNIVERSITY

REGULARITY THEOREMS FOR SOLUTIONS OF ELLIPTIC POLYNOMIAL EQUATIONS

K. UHLENBECK

Our goal is to prove regularity for solutions of nonlinear systems in classes which are invariant under coordinate transformations in the image space as well as the domain space of the functions. We will then be able to use our results to get smoothness theorems for critical points of certain integrals which satisfy condition C on some Sobolev manifolds of maps from a compact manifold M to a manifold N without boundary. Some of these results are well known, but I do not know of any general treatment. Eliasson has proved similar results for $p = 2$.

M will be a compact n-manifold, possibly with boundary, with a measure $\partial\mu$. L_k^p denotes the Sobolev space of functions whose first k-derivatives are p-integrable. If $k > n/p$ and N is a manifold without boundary, $L_k^p(M, N) \subseteq C^0(M, N)$ is a Banach manifold with a submanifold $L_{k_{\partial\lambda}}^p(M, N)$ of maps with Dirichlet boundary values of $\lambda \in C^\infty(M, N)$. The reference for the geometry of these manifolds is Palais [**4**]. In general, $x \in M$, s denotes a vector-valued function, and coefficients are linear maps on suitable spaces and are C^∞ functions of the domain and image. Local notation in coordinate patches is implied by expressions in $D_x^\alpha s = (\partial/\partial x_1)^{\alpha_1} \dots (\partial/\partial x_n)^{\alpha_n} s$ where $|\alpha| = \sum_{i=1}^n |\alpha_i|$. A term of weight w and order k can be written as sums $C(x, s(x))(D_x^{\alpha_1} s, \dots, D_x^{\alpha_n} s)$ for $|\alpha_i| \leq k$ and $\sum |\alpha_i| \leq w$.

We first give examples of our systems which arise from an embedding $N \subseteq R^N$ and p-linear integrals on $C^\infty(M, R^N)$ restricted to $C^\infty(M, N)$.

(I–A)
$$(p = 2) \quad J(s) = \int_M B(s, s)_x d\mu.$$

$$B(s, s)_x = \sum_{|\beta| \leq k; |\alpha| \leq k} a^{\alpha,\beta}(x)(D_x^\beta s, D_x^\alpha s)$$

is a strongly elliptic form.

(I–B)
$$(p > 2) \quad J(s) = \int_M |A(s)|^p d\mu.$$

A is a linear elliptic (overdetermined) system of order k.

If $k > n/p$, J will extend to a smooth function on $L_{k_{\partial\lambda}}^p(M, N)$ which satisfies condition C, at least if N is compact, so we get critical points in L_k^p. By checking the form in the local vector bundle neighborhood structure of $M \times N$, we find that these functions are special cases of smooth integrals which do not depend on any linear fibre structure and have the form in any vector bundle neighborhood of $M \times N$:

$$J(s) = \int_M B^s(s, s)_x + C(s)_x d\mu.$$

(II–A)

$$B^s(v, w)_x = \sum_{|\alpha| = |\beta| = k} a^{\alpha,\beta}(x, s(x))(D_x^\alpha v, D_x^\beta w)$$

is strongly elliptic, $a^{\alpha,\beta}$ is a C^∞ matrix function, and $C(s)_x$ is a sum of $C(x, s(x))$ $(D_x^{\alpha_1}s, \ldots, D_x^{\alpha_n}s)$ for $\sum|\alpha_i| \leq 2k$ and $|\alpha_i| < k$ for $i > 1$.

$$J(s) = \int_M |A^s s + C(s)|^p d\mu$$

(II–B)

$$A^s(v)_x = \sum_{|\alpha| \leq k} a^\alpha(x, s(x)) D_x^\alpha v$$

is an elliptic system and $C(s)$ is polynomial of order $k - 1$ and weight k. In addition to the imbedding problems (I) we can define integrals of this type depending on the Riemannian structures for M and N which satisfy condition C on $L^p_{k_{\partial\lambda}}(M, N)$ ($k > n/p$ if $\partial M \neq \emptyset$ and N is complete, or N is compact [**5**]).

EXAMPLE 1. $\mathscr{E}_k^p(s) = \int_M |E_k(s)|^p_{s(x)} d\mu$.

$E_k(s)$ corresponds to $\{D_x^\alpha s\}_{1 \leq |\alpha| \leq k}$, but differentiating covariantly. $\mathscr{E}_1^p(s) = \int_M |ds|^p_{s(x)} d\mu$ is the extension of the usual energy whose critical points for $M = [0, 1]$ are geodesics.

EXAMPLE 2. We let $\mathscr{E}_1^2(s) = \int_M |ds|^2_{s(x)} d\mu$ and let $-\int_M (\Delta(s)_x, u_x)_{s(x)} = d\mathscr{E}^2_{1s}(u)$ so $\Delta(s) \in \Gamma(s^*T(N))$. $\int |\Delta(s)_x|^p_{p(x)} + |ds|^p_{x,s(x)} d\mu$ satisfies condition C if $p > n$.

If we write down the variation for the integrals (II) in any vector bundle neighborhood $\eta \subseteq M \times N$, the critical points of J satisfy the following equations:

(III–A)
$$\int_M B^s(s, u)_x + \tilde{C}^s(u)_x d\mu = 0 \quad \text{for } u \in C_0^\infty(\eta)$$

B^s is as in II–A and $\tilde{C}^s(u)_x =$ sums of the form

$$\tilde{C}(x, s(x))(D_x^{\alpha_1} s, \ldots, D_x^{\alpha_n} s, D_x^\beta u)$$

for $\sum|\alpha_i| + |\beta| \leq 2k$ and $|\alpha_i|$ or $|\beta| < k$.

(III–B)
$$\int_M |A^s s + C(s)|^{p-2}(As + C(s), A^s s + \tilde{C}^s(u)) = 0$$

for all $u \in C_0^\infty(\eta)$. A^s and $C(s)$ have the same form as in (II–B) and $\tilde{C}^s(u)_x$ is the sum of terms $C(x, s(x))(D_x^{\alpha_1}s, \ldots, D_x^{\alpha_n}s, D_x^\beta u)$ for $0 \leq |\beta| < k$, $1 \leq |\alpha_i| \leq k$ and $|\beta| + \Sigma|\alpha_i| \leq k$.

The regularity theorems we can prove easily with linear techniques correspond to those for our nonrestricted examples in I. The critical points of I–A are C^∞, but to get any results for I–B, for which the variation is $A^*|As|^{p-2}As$, we must use the ellipticity of A^*, and then we shall get only interior regularity. We choose a vector bundle so the Dirichlet Boundary values are zero.

THEOREM A. *If $s \in L^q_{k_0}(\eta)$ for $k > n/q$ and $q \geq 2$ satisfies an equation of type* III–A (*we do not assume $k > n/2$*) *then $s \in C^\infty(\eta)$.*

THEOREM B. *If $s \in L^q_k(\eta)$ for $k > n/p$ and $q \geq p$ satisfies an equation of type* III–B (*we do not assume $k > n/p$*) *and A^s and $(A^s)^*$ are both elliptic, then $s \in C^{k,\alpha}$ on the* *interior of M. Any error terms $C(x, s(x))(D_x^{\alpha_1}s, \ldots, D_x^{\alpha_n}s, D_x^{\beta}u)$ with $|\alpha_i| \leq k$, $|\beta| \leq k$, $|\alpha_i| + |\beta| \leq pk$ and $|\alpha_i|$ or $|\beta| < k$ can be added, and the theorem is still true.*

PROOF OF THEOREM A. We use the usual ladder argument, but we shall have to jack up q until $k - 1 > n/q$. Our equation is:

$$\int (B^s(s, u) + \tilde{C}^{(s)}(u)) = 0$$

for $u \in C_0^\infty(\eta)$ which can be rewritten

$$B^s(s) + (\tilde{\tilde{C}}(s) = 0$$

after integrating by parts. Because $s \in L_k^q$, the cofficients of the elliptic bilinear form B^s are $L_k^q \subseteq C^\alpha$, and $\tilde{\tilde{C}}(s)$ is better than L_{-k}^q, say $L_{-k}^{q+\varepsilon}$. We can use a linear regularity theorem to get $s \in L_k^{q+\varepsilon}$. Once $k - 1 > n/q$, we find that $s \in L_{k+R}^q$ (for $R \geq 0$) implies the coefficients of B^s are C^{R+1} and the error term is in L_{-k+R+1}^q, so we conclude $s \in L_{k+R+1}^q$. These results are good up to the boundary.

The proof of Theorem B needs more steps. We assume A^s is an elliptic (not overdetermined) operator from $\eta \to \eta'$, but $C(s)$ may go to a larger bundle $\xi \supseteq \eta'$ in order to apply the results to the linear imbedding case. P denotes a projection $\xi \to \eta'$. We find that integrating by parts gives

$$(A^s)^* |A^s(s) + C(s)|^{p-2}(A^s(s) + PC(s)) + \tilde{\tilde{C}}(s) = 0$$

(note $(A^s)^*$ acts only on η'). Again $\tilde{\tilde{C}}(s)$ is left in divergence form and is really of lower order, better than $L_{-k}^{q/(p-1)}$, and the coefficients of A^s and $(A^s)^*$ [which we leave as $(A^s)^*v = D_x^\alpha a^\alpha(\ , s)v$] are L_k^q if $s \in L_k^q$ and $k > n/q$. We use a theorem for linear equations to conclude $V(s) \in L_0^{q/(p-1)+\varepsilon}$. A pointwise inequality shows

$$\|A^s s + PC(s)\|_{L_0^{q+\varepsilon(p-1)}}^{p-1} \leq \|V(s)\|_{L_0^{q/(p-1)+\varepsilon}}$$

and since $C(s)$ is of lower order, $A^s s \in L_0^{q+\varepsilon(p-1)}$. The coefficients of A^s are smooth enough to imply $s \in L_k^{q+\varepsilon(p-1)}$. By this method, we can get $s \in L_k^Q$ for any $Q < \infty$. We find then that $\tilde{\tilde{C}}(s) \in L_{-k+1}^Q$, which implies $V(s) \in L_1^Q \subseteq C^\alpha$ for $\alpha < 1$. $PC(s) \in L_1^Q \subseteq C^\alpha$ also.

Now we give a pointwise argument to show

$$|A^s(s)_x - A^s(s)_y|^{p-1} \leq C|V(s)_x - V(s)_y| + k|C(s)_x - C(s)_y|$$

or

$$\|A^s(s)\|_{C^{\alpha/(p-1)}}^{p-1} \leq C\|V(s)\|_{C^\alpha} + k\|C(s)\|_{C^\alpha}.$$

Finally $s \in C^{k,(p-1)/d}$, because A^s will have smooth enough coefficients. These results are not good up to the boundary of M.

To show the above pointwise inequality, we observe the following inequality for any vectors X and Y in R^N with $P: R^N \to R^M \subseteq R^N$ the orthogonal projection:

$$|X - Y|^p \leq c(\||X|^{p-2}PX - |Y|^{p-2}PY| + (p-1)(|X|^{p-2} + |Y|^{p-2})|(I - P)X - Y|)\|X - Y\|.$$

To get the desired inequality, we let

$$X = A^s(s)x + C(s)x, \qquad Y = A^s(s)y + C(s)y.$$

Notice k depends on max $|A^s(s)x + C(s)x|$, which we can find a bound for.

Applications. The critical points we can find using condition C for the case $p = 2$ (see I–A and II–A) are C^∞. If we consider the problem II–B with A^s elliptic, or I–B with A an elliptic linear operator (not overdetermined) having scalar symbol, the critical points which lie in L_k^p are $C^{k,\alpha}$ in the interior. The critical points of Example 2 are $C^{2,\alpha}$ in the interior.

REMARKS. (1) The proofs of Theorem A and B depend on showing that the error terms are in the spaces claimed, and on proving regularity theorems for elliptic operators with non-C^∞ but at least C^α coefficients. These details are discussed in the appendix. If we assume $s \in L_k^q \cap C^\alpha$, we can dispense with the assumption $k > n/q$.

(2) The same techniques will certainly give results on variants of these integrals, including nonpolynomial ones. We may also extend them to classes of integrands of order k and weight$_t$ $p(k-t)$ for $t < k$. A term of order k and weight$_t$ w is of the form

$$a(x, d_x^t s)(D_x^{\alpha_1} s, \ldots, D_x^{\alpha_r} s) \text{ for } \sum(|\alpha_i| - t) \leq w \text{ and } t \leq |\alpha_i| \leq k.$$

One must only require the same sort of ellipticity for all $(x, d_x^t s)$ and that the original solution is in L_k^q for $k - t > n/q$.

(3) The proofs also give us uniform estimates of norms. If we start with a set of solutions bounded in L_k^q, for $k > n/q$, the solutions will be bounded in all L_{k+r}^q for Theorem A and bounded in $C^{k,\alpha}$ of the interior domains for Theorem B.

(4) The problem for $p > 2$ and A^s overdetermined cannot be done by this linear method. Of particular interest is the problem $\mathscr{E}_1^p(s) = \int_M |ds|^p du$ for $p > n$. I have only been able to show, using Harnack inequalities, that a critical point is Lipschitz in the interior. Morrey has also gotten this result for $\int_M (1 + |ds|^2)^{p/2} du$.

Appendix. Estimates for lower order polynomial terms and regularity theorems for linear equation with L_k^p coefficients.

We will state our theorems for domains in R^n for simplicity of notation, although they are true for n-manifolds. L_k^p and C^α will represent $L_k^p(G, R^m)$ and $C^\alpha(G, R^m)$ respectively, where G is a compact domain in R^n and m is the dimension of the system involved.

Our first lemmas are aimed at showing that the lower order terms in equation III-A and III-B are effectively of lower order [**4**].

MULTIPLICATION LEMMA. *The multilinear maps $M: \bigoplus_{i=1}^r C^\infty \to C^\infty$ defined by $M(f_1, \ldots, f_r)_x = f_1 \cdot \ldots \cdot f_r(x)$ extends to a continuous multilinear map*

$$M: \bigoplus_{i=1}^r L_{k_i}^{p_i} \to L_l^q \quad \text{if } \sum_{i \in \mathscr{D}} (k_i/n - 1/p_i) \geq l/n - 1/q$$

for $\mathscr{D}$ any subset of $\{1, \ldots, r\}$. We assume $k_i/n - 1/p_i \neq 0$ and $k_i \geq 0$, although l may be negative.

COROLLARY. *Let $(k-\alpha)/n > 1/q$, $0 < \alpha \leq 1$, $R \geq k$ and $q \geq p$. Then M extends to a multilinear map $M: \bigoplus_{i=1}^{r} L^q_{R-a_i} \to L^{q/(p-1)}_{R-b-\alpha}$ whenever $\sum a_i + b \leq pk$, $a_i \leq k$, $b \leq k$ and either a_i or b is less than k.*

This can be directly applied to the lower order terms in Theorems A and B.

The original estimates of Agmon, Douglis and Nirenberg [2] for elliptic systems with coercive boundary conditions assume $C^{k+\alpha}$ coefficients, whereas the coefficients for our systems naturally lie in some Sobolev space $L^p_k \subseteq C^\alpha$. In the application of the regularity theorem, we only need Dirichlet boundary conditions; however this method of proof works for coercive boundary conditions. Let

$$Lu = \sum_{|\beta| \leq j; |\alpha| \leq k} D^\beta a^{\alpha,\beta} D^\alpha u$$

be an elliptic system on a compact domain G in R^n, $a^{\alpha,\beta}$ will always be at least Hölder continuous, and

$$\lambda = \max_{x \in G;\, \gamma \in R^n;\, |\gamma|=1} \left|\det \sum_{|\alpha|=k, |\beta|=j} \gamma^\beta \bar{a}^{\alpha,\beta}(x) \gamma^\alpha \right|^{-1} < \infty.$$

L_{x_0} denotes the constant coefficient operator

$$L_{x_0} u = \sum_{|\beta| \leq j; |\alpha| \leq k} a^{\alpha,\beta}(x_0) D^{\beta+\alpha} u.$$

Lemma 1 is the refined estimate which allows us to extend the regularity theorems. Lemma 2 is proved in [3].

LEMMA 1. *Let L and L_{x_0} denote the two systems above, $a^{\alpha,\beta} \in L^Q_s$ for $s/n - 1/Q > 0$, $l \leq s$ and $q < Q$. If support $u \subset B_\delta(x_0) = \{x : |x - x_0| < \delta\}$, then*

$$\|(L - L_{x_0})u\|_{L^q_{-j+l}} \leq c\, \delta^\varepsilon \|u\|_{L^q_{k+l}}.$$

c depends on $\|a^{\alpha,\beta}\|_{L^Q_s}$ but not on δ.

PROOF. $\|(L - L_{x_0})u\|_{L^q_{-j+l}} \leq \sum_{\alpha,\beta} \sum_{|\gamma| \leq l} \|D^\beta (a^{\alpha,\beta} - a^{\alpha,\beta}(x_0)) D^\alpha u\|_{L^q_0}.$

We use the product rule for derivatives and a Hölder inequality to estimate each term, using the fact that support $u \subset B_\delta(x_0)$

$$\|D^\beta (a - a(x_0)) D^\alpha u\|_{L^q_0} \leq \sum_{j=0}^{|\gamma|} \|a - a(x_0)\|_{L^{q(j)}(B_\delta(x_0))} \|D^\alpha u\|_{L^{p(j)}_{|\gamma|-j}}.$$

We must have $1/p(j) + 1/q(j) = 1/q$. We can in fact choose $j/n - 1/q + 1/p(j) > 0$ and $(s-j)/n - 1/Q + 1/q(j) > 0$ precisely because $s/n - 1/Q > 0$. Then

$$\|D^\alpha u\|_{L^{p(j)}_{|\gamma|-j}} \leq c\, \|u\|_{L^q_{k+l}}$$

and

$$\begin{aligned} \|a - a(x_0)\|_{L^{q(j)}_j(B_\delta(x_0))} &< c\delta^{\varepsilon'/q(j)} \|a - a(x_0)\|_{L^{q(j)+\varepsilon'}_j} \\ &\leq c'' \delta^\varepsilon \|a\|_{L^Q_s} \end{aligned}$$

when ε is small enough and $j \neq 0$. But when $j = 0$, $q(j) = \infty$ and we find that

$$\max_{x \in B_\delta(x_0)} |a(x) - a(x_0)| < \delta^\varepsilon \|a\|_{C^\varepsilon}$$
$$\leq c\delta^\varepsilon \|a\|_{L^Q_s} \quad \text{if } \varepsilon \text{ is small.}$$

LEMMA. 2. *Let* $\| \quad \|_{L^q_{k,\delta}} = \sum_{r=0}^k \delta^{-r} \| \quad \|_{L^q_{k-r}}$ *be a parameterized family of norms for* L^q_k. ϕ_δ *denotes the operation of multiplication by the function* $\phi(x/\delta)$, *where* ϕ *is a* C^∞ *function with compact support. Then* $\|\phi_\delta \cdot u\|_{L^q_{k,\delta}} \leq C\|u\|_{L^q_{k,\delta}}$ *and* C *does not depend on* δ.

THEOREM (REGULARITY FOR ELLIPTIC SYSTEMS WITH COEFFICIENTS ON A SOBOLEV SPACE). *Let* G *be a compact domain in* R^n, $Lu = \sum_{|\beta| \leq j; |\alpha| \leq k} D^\beta a^{\alpha,\beta} D^\alpha u$ *be an elliptic system on* G. *From the conditions*

(i) $a^{\alpha,\beta} \in L^Q_s(G)$ *for* $s/n - 1/Q > 0$,
(ii) $u \in L^p_k(G)$,
(iii) $Lu \in L^q_{-j+l}(G)$,
(iv) $q < Q$, $l \leq s$,

it follows that $u \in L^q_{k+l}(G')$ *for any interior domain* G' *of* G *and*

$$\|u\|_{L^q_{k+l}(G')} \leq c(\|a\|_{L^Q_s(G)}, \lambda)(\|u\|_{L^p_k(G)} + \|Lu\|_{L^q_{l-j}(G)}).$$

(REGULARITY UP TO THE BOUNDARY). *If, in addition* G *is smooth,* $u \in L^p_{(k+j)/2,0}(G)$ *and* L *is compatible with the Dirichlet boundary conditions (in particular if* L *is strongly elliptic), then we can take* $G' = G$.

The proof of the theorem follows Morrey [**3**]. From a partition of unity argument, we need only prove the theorem for u having support in an arbitrarily small ball $B_\delta(x_0)$ about x_0 in G. Furthermore, the additional difficulty encountered at the boundary is in constructing kernel operators on a half-space as inverses to L_{x_0} to the proper boundary value spaces. Once we assume the existence of these, the proof on the boundary follows that in the interior, so we assume x_0 is an interior point.

Let P_{x_0} be a pseudo-differential operator of order $-k - j$ which is an inverse for L_{x_0} up to an operator of order $-\infty$. We also assume support $P_{x_0}V \subseteq$ some compact set and the support $u \subseteq B_\delta(x_0)$.

$$L_{x_0}u + (L - L_{x_0})u = Lu.$$

Apply P_{x_0} to get

$$u + P_{x_0}(L - L_{x_0})u = P_{x_0}Lu + u - P_{x_0}L_{x_0}u$$

which is the same as

$$u + P_{x_0}(L - L_{x_0})\Phi_\delta u = P_{x_0}Lu + (I - P_{x_0}L_{x_0})u.$$

Here Φ_δ denotes multiplication by $\Phi((x - x_0)/\delta)$ for $\Phi(x) \equiv 1$, $|x| \leq 1$ and $\Phi(x) \equiv 0$ for $|x| \geq 2$. The right side of our equation is in L^q_{k+l}, because P_{x_0} is of order $-k - j$, $Lu \in L^q_{-j+l}$ and $I - P_{x_0}L_{x_0}$ is of order $-\infty$ on functions with small supports. Lemmas 1 and 2 are used to show that

$$\|P_{x_0}(L - L_{x_0})\Phi_\delta v\|_{L^q_{k+1,\delta}} \leq \tfrac{1}{2}\|v\|_{L^q_{k+1,\delta}},$$
$$\|P_{x_0}(L - L_{x_0})\Phi_\delta v\|_{L^p_{k,\delta}} \leq \tfrac{1}{2}\|v\|_{L^p_{k,\delta}}$$

when we choose δ small enough. It is then true that $I + P_{x_0}(L - L_{x_0})\Phi_\delta$ is an isomorphism on both $L^p_k(R^n)$ and $L^q_{k+1}(R^n)$, so $u \in L^p_k$ must also be in L^q_{k+1}. To get the inequality of norms it is only necessary to keep track of all the inequalities.

REFERENCES

1. S. Agmon, *The L_p approach to the Dirichlet problem,* Ann. Scuola Norm. Sup. Pisa **13** (1959), 405–448.

2. S. Agmon, A. Douglis and L. Nirenberg, *Estimates near the boundary for the solutions of elliptic differential equations satisfying general boundary values* II, Comm. Pure Appl. Math. **1** (1964), 35–92.

3. C. B. Morrey, *Multiple integrals in the calculus of variations,* Springer, New York, 1966.

4. R. S. Palais, *Foundations of global non-linear analysis,* Benjamin, New York, 1968.

5. K. Uhlenbeck, *The calculus of variations and global analysis,* thesis, Brandeis, 1968.

MASSACHUSETTS INSTITUTE OF TECHNOLOGY

PROBLEMS IN NONLINEAR HYPERBOLIC EQUATIONS

J. GLIMM AND P. LAX

The equations we consider have the form

$$u_t + f(u)_x = 0. \tag{*}$$

(*) is a nonlinear system of equations in one space variable; the equations are in conservation form. In general they govern nonlinear motions of a one dimensional nondissipative medium (gas dynamics, nonlinear elasticity, nonlinear optics, plasmas). The equations (*) state that mass and momentum, etc. are conserved. In general $u = u_1 \ldots u_N$ is an N-tuple. For gas dynamics,

$$u = \begin{cases} v \text{ momentum} \\ \rho \text{ density} \\ \text{(constant entropy)} \end{cases} \qquad \text{or } u = \begin{cases} v \text{ momentum} \\ \rho \text{ density} \\ E \text{ total energy} \end{cases}$$

and so $N = 2$ or $N = 3$. The nonlinearity can be described simply by saying that the characteristic speed c depends on u. Of course for gas dynamics one knows that the speed of sound, c, depends on pressure (and temperature).

For single equations ($N = 1$) the theory is in good shape, due primarily to the work of Hopf, Lax, and Oleinik. The main features of this work follow. One imposes a convexity condition on f and a supplementary condition on the solutions (entropy condition). There is then an existence and uniqueness theorem for solutions of the Cauchy problem. The solutions contain jump discontinuities (shock waves) even when the data are smooth. The shock waves propagate along curves (in the x, t plane). They may increase or decrease in strength or combine. They can never split into two or more shock waves and they can never disappear entirely. Because of the entropy condition, the solutions are not reversible in time, and for large time a solution will approach a very simple asymptotic state. The mapping from data at time $t = 0$ to data at time $t = T > 0$ is compact (using the sup norm and solutions constant off a bounded set). If one omits the entropy condition, then uniqueness fails (at least for certain data containing jump discontinuities). In summary, we emphasize that the main phenomena which one observes do not occur for linear hyperbolic equations and can be explained by very simple calculations, starting from the simple equation

$$u_t + \tfrac{1}{2}(u^2)_x = 0.$$

These phenomena can all be traced back to the fact that c, the sound speed, depends on u, and they are essentially nonlinear in their nature.

For systems ($N > 1$) there is work of Glimm and of Glimm and Lax. Existence is proved for data near a constant state (near in the sup norm or sup + total var. norm), and the decay of the solution to its simple asymptotic state (u = const.) is proved. There are quite a number of problems here since one expects the main features of the case $N = 1$ to go over. In particular we mention:

The convexity condition. For small changes in u, the correct convexity condition on f has been formulated by Lax as the nonvanishing of certain first partial derivatives of f with respect to u. These conditions are probably not sufficient globally—for large changes in u. A simple but exacting test can be proposed for candidates for a convexity condition. We define a Riemann problem as a Cauchy problem in which the data consists of two states ($u = u_r$ and $u = u_l$) separated by a jump discontinuity at the origin. The convexity condition on f should then imply that any Riemann problem has a solution in terms of centered waves, and that such a solution is unique if one excludes rarefaction shocks. So far as we know, interesting examples of (*) meet this Riemann problem test. It may be necessary to restrict f in other ways, for example in its rate of growth as u tends to infinity, its smoothness. (See Smoller, Michigan Math J. (to appear).)

Uniqueness of solutions. Solutions to (*) are not unique, so an entropy condition is needed. We do not know of a general formulation of an entropy condition for the system (*), but there are several possible approaches and the main ideas seem clear. One can use a differential inequality on the solution (Oleinik, Uspehi Mat. Nauk pp. 167–176) or an integral inequality (Godunov, J. Comp. Math and Math. Phys. 1961, pp. 623–637 (Russian)) or restrictions on the geometry of the characteristics (Lax, Comm. Pure Appl. Math. 1957, pp. 537–566). The distinction between these three should be made at first on the basis of technical convenience; it would also be nice to know whether they each define equivalent restrictions on the solutions of (*). Having settled on an entropy condition, one can try to prove uniqueness. On the basis of Oleinik's work, cited above, a proof may have something to do with an existence theorem for linear hyperbolic systems with bad coefficients.

The problem of invariant functionals. We consider initial data with compact support. The corresponding solution $u(x, t)$ has then the same property as function of x for all t. Because of the conservation form of the equations, the quantities

$$\int_{-\infty}^{\infty} u_j(x, t)dx, \quad j = 1, \ldots, n$$

are *invariant functionals*, i.e., they do not depend on t.

It is natural to ask if there are other invariant functionals. The answer is yes, and these can be constructed from the asymptotic form of solutions for large t. This asymptotic form is in the shape of n so-called N-waves, one corresponding to each characteristic speed, and consisting of a j-refraction wave bracketed by two j-shocks, $j = 1, \ldots, n$. The strength of the j-shocks is $a_j/\sqrt{t}$ and $b_j/\sqrt{t}$, where a_j and b_j are numbers determined by the initial data. Clearly the a_j, b_j are invariant functionals of the initial data; this shows that there are at least $2n$ invariant functionals.

For a single conservation law these functions can be determined explicitly; their sum is the conserved quantity $\int u dx$. It would be interesting to study the nature of these functionals for systems, and to find others.

Generic properties of solutions. We ask what properties of solutions are valid for almost all solutions (in the sense of category). If we consider solutions with L_∞, bounded variation or continuous data and use the corresponding norms on the data to give a topology on the space of solutions, then it seems that almost every solution is discontinuous at a dense set of points in the x, t plane. For C^k data and the C^k topology ($1 \leqq k \leqq \infty$) this would not be true, but it may be that almost every solution for discontinuities dense in a half-space $t \geqq t_0$, t_0 depending on the solution. In the converse direction, it may be that solutions with piecewise analytic data are themselves piecewise analytic provided the equations are analytic.

Zero viscosity limit. Consider the equation

$$u_t + f(u) = \nu D^u{}_{xx} \tag{**}$$

where D is a positive semidefinite matrix. One choice of D will correspond to the presence of viscous friction in our medium. Other D's (for example $D = I$, the unit matrix) may correspond to an "artificial viscosity" introduced in order to stabilize a numerical computation scheme. In either case one would like to know that solutions of (**) converge to solutions of (*) as ν tends to zero. So far as we know, it is not known whether solutions will necessarily exist for all $t > 0$ for (**), and this question would certainly have to be settled first. Since (**) is parabolic or degenerate parabolic, we would expect its existence theory to be better behaved than that of (*); perhaps one could prove existence without restriction to nearly constant data. Presumably the solutions of (**) are smooth. There are examples of nonlinear parabolic equations whose solutions blow up after a finite interval, so some restrictions on the growth of f for large u may be required. In case of doubt, one can take an example of (**) arising in physics, for example gas dynamics; even in these special cases, we do not know of an existence theorem for (**).

Solutions for (*) with data not near a constant. To proceed in general, one would have to find the right global conditions on f. However, it is not necessary to wait. There are several explicit examples of (*) with reasonable f's and for none of these cases is it known whether there are solutions for general data and all $t > 0$. There is a decay of solutions which may be helpful here. The decay is for small time and yields

$$\text{const. } t^{-1}$$

as a bound for the total variation in the solution in a fixed bounded interval at time t, even if the data at time zero is merely L_∞. For single equations this decay was proved by Oleinik and for 2×2 systems with data near a constant by Glimm and Lax. There is another decay phenomena for small times which has not been rigorously demonstrated. After a short time interval, most of the compression waves (especially the large ones) will have broken and will have formed shock waves, while all rarefaction type discontinuities will have spread to form rarefaction waves. As a result, at most points the solution will consist of a rarefaction wave (whose strength is approximately proportional to t^{-1}). Now suppose that a rarefaction and a shock wave of the same family meet and cancel (it is this cancellation which yields the t^{-1} decay above). This interaction will also produce a "back wave", that is, a wave of the opposite family, and in several explicit examples of (*) the back wave must be a compression wave. We now take as hypothesis that the back wave to be a compression wave. By our comments above it is created in a milieu consisting (at most points) of rarefaction waves, and must be cancelled immediately. Thus once the shocks have broken, there should be very little exchange of variation between left waves and right waves. The reason decay (for small time) seems important for existence theorems is that when two waves interact, in general each will get larger and some principle is needed to counteract or bound the growth resulting from these interactions.

Two or three space dimensions with circular or spherical symmetry. Because of the singularity at the origin, this would present new problems and new phenomena. Nearly constant data would no longer lead to nearly constant solutions in the sup norm, but it might in some L_p norm. For which p? This problem is important because it serves as a natural testing ground for conjectures about 2 and 3 dimensional problems in general.

Viscous problems in 2 or 3 dimensions. Because parabolic problems may be more tractable than hyperbolic ones, (**) in several space variables might be attempted. One need not posit great generality in the form of the equation in order to find a challenging problem; it should be sufficient to consider the compressible Navier Stokes equations.

MASSACHUSETTS INSTITUTE OF TECHNOLOGY AND
COURANT INSTITUTE OF MATHEMATICAL SCIENCES

HAMILTONIAN MECHANICS ON LIE GROUPS AND HYDRODYNAMICS

J. MARSDEN AND R. ABRAHAM

Introduction. Some of the most classical and important examples in mechanics are systems whose configuration space is a Lie group. The particular examples we have in mind are the rigid body (on the Lie group SO(3)) and the perfect fluid (on the Lie group of volume preserving diffeomorphisms).

Most of what we have to say is classical and well known. What we do is to put it in the language of global analysis with perhaps some simplification. Our sources are mainly the papers of Arnold and Blancheton [**3**], [**4**].

The paper is divided into two parts. In the first we present the general theory. In the second we describe the case of hydrodynamics. Some connections will be made with the calculus of variations in the future. In addition, a more complete exposition of the present work will appear in lecture note forms shortly [**11**].[1]

1. **Abstract theory.** Let G be a Lie group.[2] By this we mean that G is a smooth manifold modelled on a locally convex topological vector space (locally convex since, amongst other reasons, the Hahn-Banach theorem is needed) and is also a group such that group multiplication and inversion are C^∞ mappings. The tangent bundle of G is denoted TG and the fiber over $x \in G$ is written T_xG.

Let g be a (weak) Riemannian metric[3] on G. This means that the tensor g is an inner product on each T_xG, but inducing a different topology in general. For each $x \in G$, $e_x \in T_xG$ and $f_x \in T_xG$ we write $\langle e_x, f_x\rangle = g(x)\cdot(e_x, f_x)$.

A *weak symplectic form* on a manifold M is a closed two form ω such that the mapping $\omega_b\colon TM \to T^*M$ defined by $\omega_b(e_x)\cdot f_x = \omega(x)\cdot(e_x, f_x)$ is injective on each fiber. (If ω_b is an isomorphism on each fiber, we call ω a *symplectic form.*)

For infinite dimensional mechanics (continuum mechanics and quantum mechanics for example), if one wishes to work with a symplectic form it is necessary to use domains for vectorfields in the same sense as occurs in semigroup theory; see [**8**]. If, however, one wishes to exploit differentiability directly and have the vectorfield defined and smooth in the usual sense, then it is necessary to use weak symplectic forms instead. The reason will become evident shortly. We shall use weak symplectic forms here.

Except in unusual and artificial circumstances (these can be obtained for the wave equation using the spaces in [**10**]), the manifolds one needs to get a smooth

[1] See also [**14**].

[2] The term "Lie Group" may be misleading since, in the infinite dimensional case, the usual Lie theorems do not hold. The term ILH Lie Group, or Fréchet Lie Group may be better.

[3] g is assumed smooth.

vectorfield are usually Fréchet and not Banach, the C^∞ functions for example. In that case the standard flow theorem for vectorfields is false. Instead one must use techniques of Browder, Kato and others. For the case of hydrodynamics with viscosity a good part of a classical book [6] is needed. Also, the flow is possibly only local, that is, cannot be extended for all time. For the nonviscous case we are concerned with, see Kato [5].[4]

Let M be a manifold and ω a weak symplectic form. A one form α *can be lifted* when there exists a vectorfield (unique) X on M such that $X^b = \alpha$, where $X^b(m) = 2\omega_b(m)\cdot X(m)$, $m \in M$. We write $X = \alpha^\#$.

For a smooth function $f\colon M \to \boldsymbol{R}$, such that df can be lifted, we write $X_f = (df)^\#$, and we call X_f a *Hamiltonian vectorfield.*

It will be necessary to recall a few theorems about Hamiltonian systems.

THEOREM 1. *Let (M, ω) be a weak symplectic manifold and X_H a Hamiltonian vectorfield with a local smooth flow F_t. Then*
(i) *$H \circ F_t = H$ (conservation of energy), and*
(ii) *$F_t^*\omega = \omega$, or F_t is symplectic (preserves the form ω).*

See [1] or [8] for the proof.

Recall that if M is a manifold then T^*M admits a natural weak symplectic structure given by (locally)

$$\omega(\alpha_x)\cdot((e_1, \alpha_1), (e_2, \alpha_2)) = [\alpha_2(e_1) - \alpha_1(e_2)]/2$$

for $e_i \in T_xM$ and $\alpha_i \in T_x^*M$. [ω is symplectic iff M is modelled on a semireflexive space.] See [8, Theorem 2.4].

THEOREM 2. *Let G be a Lie group and M a manifold with $\Phi\colon G \times M \to M$ a smooth action of G on M, which extends naturally to an action Φ^* of G on T^*M. Suppose X_H is a Hamiltonian vectorfield on T^*M and H is invariant under the action Φ^*. Then the following functions P_X are invariant under the flow of X_H. Let X be an infinitesimal generator of Φ, so X is a vectorfield on M and define $P_X\colon T^*M \to \boldsymbol{R}$ by $P_X(\alpha_m) = \alpha_m(X(m))$. (We call P_X the momentum of X.)*

For the proof see [8, Theorem 5.3]. This is the basic conservation law of mechanics.

In case g is a weak Riemannian metric on M inducing a map $g_b\colon TM \to T^*M$, then we deduce (using [8, Theorem 5.2]) that if X_H is a Hamiltonian vectorfield on TM (with respect to the form $\omega_g(e_x)\cdot((e_1, f_1), (e_2, f_2)) = [\langle f_2, e_1\rangle - \langle f_1, e_2\rangle]/2$) and H is invariant under the induced (adjoint) action on TM then the function $P_X(e_m) = \langle e_m, X(m)\rangle$ is invariant under the flow of X_H.

The latter situation is the one which arises naturally in the case of a Lagrangian system and will be of concern to us below.

Let us now return to the setting of Lie groups. We shall require our group and metric g to have certain regularity properties introduced as follows.

[4] The general nonviscous existence problem is settled in [**14**].

DEFINITION.[5] Let G be a Lie group and g a weak Riemannian metric on G. We say that g is *compatible* with G iff

(i) g is left invariant; that is, for each $x \in G$, $L_x^* g = g$ where L_x is the diffeomorphism defined by $L_x(y) = xy$;

(ii) the kinetic energy function $T: TG \to \boldsymbol{R}$ defined by $T(e_x) = \langle e_x, e_x \rangle / 2$ can be lifted to a (smooth) vectorfield X_T using the weak symplectic form

$$\omega_g((e_1, f_1), (e_2, f_2)) = [\langle f_2, e_1 \rangle - \langle f_1, e_2 \rangle]/2; \text{ and}$$

(iii) X_T possesses a local smooth flow (called the *geodesic flow* of g).

Also, we say that G is a *regular Lie group* iff every (smooth) left invariant vectorfield on G has a flow.

In the finite dimensional case these conditions are of course redundant. In the infinite dimensional case it is (iii) above which is difficult to verify.

The main conservation theorem is the following (the cases of hydrodynamics and a rigid body are due to Euler):

THEOREM 3. *Let G be a regular Lie group and X_H a Hamiltonian vectorfield on TG with H invariant by left translations ($H \circ TL_x = H$ for each $x \in G$, $TL_x: TG \to TG$ being the tangent map). Then for each $v \in T_eG$ (e = identity element of G) the function P_v is invariant under the flow of X_H, where*

$$P_v : TG \to \boldsymbol{R}$$

is defined by

$$P_v(u_x) = \langle T_e R_x \cdot v, u_x \rangle$$

where $u_x \in T_xG$, R_x is right translation by x and T_eR_x is the tangent of R_x evaluated at $e \in G$.

COROLLARY 4. *Let G be a regular Lie group and g a compatible weak Riemannian metric. Then the functions P_v are invariant under the geodesic flow of g. Further, this flow is a symplectic (local) diffeomorphism and conserves (kinetic) energy.*

PROOF. Let G act on itself by left translations. Each $v \in T_eG$ determines an exponential flow on G by assumption, and its derivative is the infinitesimal generator X. Let E_t be the exponential map of v. Then

$$X(y) = d(E_t(y))/dt = d(E_t(e) \cdot y)/dt \text{ at } t = 0.$$

Now $E_t(e) \cdot y = R_y \circ F_t(e)$ and so by the composite mapping theorem, $d(E_t(e) \cdot y)/dt = T_eR_y \cdot d(E_t(e))/dt = T_eR_y \cdot v$. The result is now an immediate consequence of the conservation theorem. ■

In practice it is usually most convenient to work in the Lie algebra T_eG by pulling back the flow to T_eG by left translation (in the so called "body coordinates"). The pull back of the vectorfield X_T to T_eG is the *Euler equations*. They are determined as follows:

---

[5] In the rest of section one, "left" and "right" can be interchanged, and this introduces a minus sign in Theorem 5 and Lemma 6.

THEOREM 5. *Let G be a (regular) Lie group, g a compatible metric and X_T the corresponding Hamiltonian vectorfield with flow $F_t: TG \to TG$ (which may be just local). Define H_t on T_eG by*

$$H_t(v) = T_{x(t)}L_{x(t)^{-1}} \cdot F_t v$$

(where defined), where $F_t v \in T_{x(t)}G$. Then H_t is a smooth flow on T_eG and has vectorfield Y uniquely determined by: $Y: T_eG \to T_eG$, $\langle Y(u), v\rangle = \langle [u, v], u\rangle$ where $[u, v]$ is the Lie bracket in T_eG.

PROOF. It is easily checked that H_t is a flow. To compute Y we must compute $dH_t(v)/dt$ at $t = 0$. For this we use the following:

LEMMA 6. *Let $x(t)$ be a smooth curve in G, $v \in T_eG$ and $v(t) = \mathrm{Ad}_{x(t)^{-1}} \cdot v$ where*

$$\mathrm{Ad}_x = T_x R_{x^{-1}} \circ T_e L_x = T_e(R_{x^{-1}} \circ L_x): T_eG \to T_eG.$$

Then we have

$$dv/dt = [v(t), T_{x(t)}L_{x(t)^{-1}} \cdot (dx/dt)].$$

We omit the proof, as it is more or less standard; see [11] or [13].

To prove the theorem, we start with

$$d\,\langle F_t u, T_e R_{x(t)} v\rangle/dt = 0$$

by the conservation law, for each $u, v \in T_eG$. By left invariance of g,

$$\begin{aligned}\langle F_t u, T_e R_{x(t)} \cdot v\rangle &= \langle T_{x(t)}L_{x(t)^{-1}} \cdot F_t u, \mathrm{Ad}_{x(t)^{-1}} v\rangle \\ &= \langle H_t u, \mathrm{Ad}_{x(t)^{-1}} v\rangle.\end{aligned}$$

Differentiating, using Leibnitz's rule and the lemma, gives at $t = 0$,

$$\langle Y(u), v\rangle + \langle u, [v, u]\rangle = 0$$

where we use the fact that $dx/dt = F_t u$ (which is because we have a second order equation). ■

If one finds the flow of Y on T_eG, the problem of finding F_t is solved using the equation

$$dx/dt = T_e L_{x(t)} \cdot H_t u = F_t u$$

and the fact that F_t is left invariant. For groups of diffeomorphisms and in particular for hydrodynamics, this is just a problem in ordinary differential equations.

COROLLARY 7. *If v_0 is a critical point of Y on T_eG then $x(t) = \exp(tv_0) \in G$ is the geodesic with initial value v_0.*

PROOF. We have $H_t v_0 = v_0$ and so

$$F_t v_0 = T_e L_{x(t)} \cdot v_0 = dx/dt.$$

This is the equation for $x(t) = \exp(tv_0)$. ■

By left invariance, the conservation law and conservation of energy pull back to the Lie algebra T_eG.

For many circumstances it is important to work in a Hilbert space. Therefore we could complete T_eG with respect to $\langle\, , \rangle$, to obtain a Hilbert space H. It is absolutely crucial to recognize that on H, the vectorfield Y is not smooth and is not everywhere defined. Also, the flow H_t is *not* defined on all of H. (This can be seen for the diffeomorphism group where T_eG corresponds to C^∞ vectorfields, so H would be L_2 vectorfields. For flows of nonsmooth vectorfields see [**9**].)

In the finite dimensional case, or if the energy function T had D^2T continuous on H (so Y would be smooth on H say), then definiteness of the quadratic form D^2T[6] at a stationary point v_0 would imply its stability (the quadratic form is computed in Arnold [**3**]). Unfortunately for hydrodynamics this is not the case, so it is unknown whether or not this criterion for stability is valid, as far as we know.

2. **Hydrodynamics of perfect fluids.** Let D be a compact orientable n-manifold with boundary (or without boundary). Let g_0 be a Riemannian metric on D and Ω a volume (orientation) on D (that is, a nonvanishing n-form) which is generally one derived from g_0. For tangent vectors on D we write $v \cdot u$ or $\langle v, u\rangle$ for the inner product with respect to g_0.

Let G be the group of volume preserving diffeomorphisms on D. Leslie [**6**], and Omori [**11**] show that the group of all diffeomorphisms has a structure modelled on a Fréchet space for which the group is a Lie group. The procedure has become a more or less standard one in manifolds of maps. See also [**2**]. A recent theorem of D. Ebin[7] tells us that G is also a Lie group (in fact a Lie subgroup). Then T_eG may be identified with C^∞ vectorfields X on D such that

(i) $\operatorname{div}_\Omega X = 0$, and

(ii) $i_* i_x \Omega = 0$ where $i\colon \operatorname{bd}(D) \to D$ is the inclusion map.

For the full diffeomorphism group, (i) is omitted.

Condition (ii) means that X is parallel to the boundary. Also, the Lie bracket in T_eG is the negative of the usual Lie bracket of vectorfields. This requires the observation that, under an action, the map taking the Lie algebra to the infinitesimal generator is an antihomomorphism, a standard result [**13**]. Letting G act naturally on D gives the stated result.

Define a weak Riemannian structure for G or all diffeomorphisms by setting

$$\langle X, Y\rangle = \int_D X \cdot Y \Omega$$

for X, Y vectorfields. Extend by right translation to all of G.[8]

CONJECTURE.[9] *On the regular Lie group G, $\langle\, , \rangle$ is compatible with G.*

[6] On the foliation described in [**3**].

[7] See [**14**] for the proof.

[8] For hydrodynamics one must use right invariant metrics, while it is customary to use left invariant ones for the rigid body.

[9] This conjecture is proven in [**14**].

That G is regular is simple. Namely if $X \in T_eG$, and X has flow F_t, then the map $f \mapsto f \circ F_t$ is the exponential map of X. This is easily seen. One cannot do this for general X which are just in L^2.

For compatibility of $g = \langle\, , \rangle$ one must show that we get a vectorfield X and that it has a local flow. From §1, it is enough to work with the Euler equations. We shall just show how to get the Euler equations Y (Theorem 5). Conjecture 8 is true for the full diffeomorphism group and suitable metrics. Details may be found in [**11**].

THEOREM 9. *For G described above, the vectorfield Y of Theorem 5 is given by: for each vectorfield X, there are C^∞ functions f and g (unique up to constants) such that*

$$\widetilde{Y(X)} = -i_X d(\tilde{X}) + df = -L_X\tilde{X} + dg$$

where $\tilde{X}$ denotes the one form obtained from X via g_0, i_X is the inner product, and L_X is the Lie derivative. Traditionally p, defined by $p = f - \langle X, X\rangle/2 = g + \langle X, X\rangle/2$ is called the pressure.[10]

PROOF. First, we note that for vectorfields X, X_0, Y that

$$\int_D \langle X, [X_0, Y]\rangle\Omega = \int_D i_{[X_0, Y]}\tilde{X}\Omega = \int_D (i_{[X_0, Y]} + L_Y i_{X_0})\tilde{X}\Omega$$

since $\int_D L_Y(i_{X_0}\tilde{X})\Omega = 0$ by Stoke's theorem and the boundary condition $i_Y\Omega = 0$ on bd(D). But $i_{[X_0, Y]} + L_Y i_{X_0} = i_{X_0} L_Y$ (see [**1**]), and so we finally obtain

$$\int_D \langle X, [X_0, X]\rangle\Omega = \int_D i_{X_0} i_Y d\tilde{X}\Omega = \int_D \langle Y, L_{X_0}\tilde{X}\rangle\Omega.$$

Of course in this the fact that the vectorfields are divergence free is essential. (On the full diffeomorphism group the equations are a little different.)[11] If this condition were omitted, the pressure term would be absent.

It is a classical theorem (Hodge theory)[12] that a vectorfield Z can be written uniquely

$$Z = Z_0 + \text{grad}(f)$$

where grad(f) = df, and Z_0 and grad(f) are orthogonal and $\text{div}_\Omega Z_0 = 0$, and Z_0 is parallel to the boundary. (The function f is obtained by solving Laplace's equation $\nabla^2 f = \text{div}\, Z$.) One easily sees the orthogonality directly, as follows:

For any X in T_eG, observe that $\langle X, \text{grad}(f)\rangle = \int X \cdot \text{grad}(f)\,\Omega = 0$ since

$$\int X \cdot \text{grad}(f)\,\Omega = \int L_X f\,\Omega = \int L_X(f\Omega) = \int di_X(f\Omega) = 0,$$

[10] The equations for Y are usually written in the equivalent form $dv/dt + \nabla_v v = dp$ where ∇ is the covariant derivative and v is an integral curve of Y.

[11] For the right invariant metric on the full diffeomorphism group, the equations are $dv/dt + v(\text{div} v) + L_v v = 0$. The existence problem is unknown for these equations.

[12] Cf. Morrey, *Multiple integrals in the calculus of variations*, Chapter 7 and also [**11**], [**14**].

using Stoke's theorem and $i_X\Omega = 0$ or bd(D). Now for the theorem we let f be such that $i_X d(\tilde{X}) - df$ is divergence free and then observe, by our remarks, that

$$\begin{aligned} -\langle Y(X), X_0\rangle = \langle [X, X_0], X\rangle &= \int_D i_{X_0} i_X d(\tilde{X})\Omega \\ &= \langle i_X d\tilde{X}, X_0\rangle \\ &= \langle i_X d\tilde{X} - \operatorname{grad}(f), X_0\rangle. \end{aligned}$$

Note that $i_X d(\tilde{X})$ might not be parallel to the boundary but $L_X(\tilde{X})$ is, and these differ by a gradient. By nondegeneracy we conclude the result.

The pressure thus constrains the motion to being divergence free.

Observe that the motion of a stationary point X_0 is just its own flow. See Corollary 7. This holds for harmonic vectorfields for example.

Finally we give a theorem classically known as *Kelvin's circulation theorem.* One should note that the theorem as proven in standard hydrodynamics books lacks rigor.

THEOREM 10. *Let $X(t)$ be the velocity vectorfield in T_eG above. Let l be a smooth closed loop in D (a compact 1-manifold) and $l_t = F_t(l)$, where F_t is the motion of the fluid (geodesic). Then $\int_{l_t} \tilde{X}(t)$ is independent of t.*

PROOF. Since F_t is a diffeomorphism,

$$\int_{l_t} \tilde{X}(t) = \int_l F_t^* \tilde{X}(t).$$

Now $d(F_t^* \tilde{X}(t))/dt = F_t^* L_X \tilde{X} + F_t^* d\tilde{X}/dt$, which by Theorem 9 equals $F_t^* dg = d(F_t^* g)$. Thus by Stoke's theorem,

$$\frac{d}{dt}\int_{l_t} \tilde{X}(t) = \int_l d(F_t^* g) = 0. \ \blacksquare$$

Curiously this is not true for the geodesic flow on the full diffeomorphism group.

This theorem is quite analogous to the circulation theorem for mechanics. It also holds for TG, so we give it

THEOREM 11. *Let M be a (weak) symplectic manifold with form $\omega = d\theta$. Let F_t be a smooth Hamiltonian flow (or local flow) on M. Let A be a compact two manifold in M with boundary $l =$ bd(A), and $l_t = F_t(l)$. Then $\int_{l_t} \theta$ is independent of t.*

PROOF. By Stoke's theorem,

$$\int_{l_t} \theta = \int_{A_t} \omega = \int_A F_t^* \omega = \int_A \omega,$$

since $F_t^* \omega = \omega$.

REFERENCES

1. R. Abraham and J. Marsden, *Foundations of mechanics,* Benjamin, New York, 1967.

2. R. Abraham, *Lectures on global analysis,* mimeographed, Princeton Univ. Press, Princeton, N.J., 1968.

3. V. Arnold, *Sur la geometric differentialle des groupes de Lie de dimension infinite et ses applications à l'hydrodynamique de fluides parfaits*, Ann. Inst. Fourier, (Grenoble), **16** (1966).

4. E. Blancheton, *Mechanique analytique des milieus continus*, Ann. Inst. H. Poincaré, Sect. A, 7 (1967), 189–213.

5. T. Kato, *On classical solutions of the two dimensional non-stationary Euler equation*, Arch. Rational Mech. Anal. **25** (1967), 188–200.

6. O. Ladyzhenskaya, *The mathematical theory of viscous incompressible flow*, Gordon and Breach, New York, 1963.

7. J. Leslie, *On a differential structure for the group of diffeomorphisms*, Topology, **6** (1967), 263–271.

8. J. Marsden, *Hamiltonian one parameter groups*, Arch. Rational Mech. Anal. **28** (1968), 362–396.

9. ———, *Generalized Hamiltonian mechanics*, Arch. Rational Mech. Anal. **28** (5), (1968), 323–361.

10. ———, *A Banach space of analytic functions for constant coefficient equations of evolution*, Canad. Math. Bull. **11** (1968).

11. ———, *Hamiltonian mechanics, infinite dimensional Lie groups, geodesic flows and hydrodynamics*, mimeographed lecture notes, University of California, Berkeley.

12. H. Omori, *On the group of diffeomorphisms on a compact manifold* (preprint).

13. P. Tondeur, *Introduction to Lie groups and transformation groups*, Springer Lecture Note Series, (1965).

14. D. Ebin and J. Marsden, *Groups of diffeomorphisms and the motion of an incompressible fluid* (to appear).

University of California, Berkeley and Santa Cruz

AUTHOR INDEX

Roman numbers refer to pages on which a reference is made to an author or a work of an author. *Italic numbers* refer to pages on which a complete reference to a work by the author is given. **Boldface numbers** indicate the first page of the articles in the book.

SUBJECT INDEX